HPLC

Practical and Industrial Applications

Edited by

Joel Swadesh, Ph.D.

CRC Press

Boca Raton   New York   London   Tokyo

Senior Editor:	Paul Petralia
Editorial Assistant:	Cindy Carelli
Project Editor:	Sarah Fortener
Assistant Managing Editor:	Gerry Jaffe
Marketing Manager:	Susie Carlisle
Cover design:	Denise Craig
PrePress:	Kevin Luong
Manufacturing:	Sheri Schwartz

Library of Congress Cataloging-in-Publication Data

HPLC : practical and industrial applications / edited by Joel Swadesh.
    p. cm.
    Includes bibliographical references and index.
    ISBN 0-8493-2682-6
    1. High performance liquid chromatography. 2. High performance
liquid chromatography--Industrial applications. II. Swadesh, Joel.
  QP519.9H53H694 1996
  543'.0894--dc20

                          96-9513
                            CIP

No claim to original U.S. Government works
International Standard Book Number 0-8493-2682-6
Library of Congress Card Number 96-9513
Printed in the United States of America 1 2 3 4 5 6 7 8 9 0
Printed on acid-free paper

Preface

Organization of the book

Books on chromatography are conventionally divided into theory, instrumentation, and practice, or into isocratic vs. gradient techniques, or by class of analyte. The organization of the present work is somewhat unconventional in that it is structured to facilitate problem-solving. The requirements of meeting product specifications and regulatory constraints within the boundaries of tight production schedules impose considerable discipline on industrial work. Industrial decisions move so quickly that sessions in the library and extended research in the laboratory are often not options. In the present work, information is clustered around certain topics in a manner to aid rapid problem-solving.

With the increasing emphasis on research productivity, academic scientists may also find value in a text oriented to problem-solving. Increasingly, students in chemistry, biochemistry, engineering, and pharmaceutics help to fund their education with short-term industrial positions. Some academic laboratories now perform contract work for industry to augment basic research funds. Students who choose to enter industry find they must now be very independent, since mentors are a rare commodity in the workplace. Some companies are turning to temporary employees, requiring extremely rapid learning on the part of those coming in for a limited period of time. These changes in the university and in industry argue for a modification — but not a "dumbing down" of the traditional educational approach. Beginners, students, temporary workers, and experienced scientists confronted with a new area need to get up to speed quickly, comfortably, and with a genuine sense of mastery.

At one stage in my career, I operated an applications laboratory for a company that produced chromatographic standards, columns, and instruments. Each day, urgent calls would come in from companies of all kinds. Many of the calls were simple technical questions. A significant number of calls, however, came from scientists who needed to become overnight experts in an area of chromatography. Expertise, of course, requires more than an understanding of the theory. Having a full bibliography describing precedents, listing the suppliers of columns and instruments, and developing a feel for the strengths and limitations of a particular kind of chromatography are all necessary aspects of expertise.

Accordingly, this book is organized to facilitate rapid absorption of a particular area of chromatography. The first chapter is a general chapter on instrumentation, theory, and laboratory operations, designed for the non-specialist unexpectedly drafted into analytical chemistry. A brief survey of absorbance, fluorescence, and refractive index detectors is presented. Pumps and columns are also described. Detailed information on specialty detectors, such as electrochemical, viscosimetric, and light-scattering detectors, is presented in later chapters in association with those chromatographic modes with which they are commonly used. The second chapter is designed for the traditional analytical chemist who is transferred into the manufacturing environment. It covers process sampling and analysis. The third chapter describes process chromatography.

The remaining four chapters are on specialties within separations technology, i.e., reversed phase chromatography, ion exchange chromatography, gel permeation chromatography, and capillary electrophoresis. Each of these chapters includes an introductory section to outline the key features of the technique, a thorough bibliography and list of precedents, and detailed examples of one or more applications, presented from the viewpoint of an industrial scientist. Specialty detectors are also described in these chapters. The chapter on ion exchange discusses electrochemical detectors, and the chapter on gel permeation chromatography describes light-scattering and viscosimetric detectors.

Inspiration for this book

> "...[D]rug development cannot be managed in the traditional sense. The 'managers' must rather be strong leaders, accomplished and respected scientists themselves, who must exhibit broad vision, long-term perspective, trust in other professionals, and the ability to inspire others. ... The public and the ethical industry are best served by decisions based on good science, adherence to high standards, and independent, expert review. ... If the industry starts with high quality science, effective analyses, and honest, responsive presentations, its regulatory problems will be few."[1]

It was with this quotation from Cuatresecas that I closed my previous work on industrial HPLC,[2] and it was the inspiration to write this book. Cuatresecas rightly refutes the misconception that dedication, quality, vision, trust, and honesty are inimical to profit. By historical accident, the American drug industry was driven by regulation to develop quality standards at a period in time when other segments of industry were degrading their scientific organizations. Experience makes it plain that, over the long run, profit flows to organizations that insist on the highest standards in every aspect of business.

Industrial scientists are partners in the production of goods and can positively influence the process from the early stages of research to technical support of a finished product. The decision to bring a production process from benchtop to large scale is a momentous one, requiring the commitment of huge amounts of capital and human resources. Analytical chemistry is critically important in the development process and beyond, serving to referee the production process. Changes in production feedstocks or processing conditions, planned or not, lead to changes in the ultimate product. Some of the changes may be beneficial and others deleterious. It is to a company's great advantage to define the chemical and physical limits within which a product's properties are desirable and beyond which they are not. When such limits are well defined, failure of the product in the field is far less likely.

The purpose of this book is to examine analytical HPLC as it is actually used in industry. Rather than focus on the technical issues alone, the book acknowledges that technical issues are inseparably intertwined with nontechnical issues. Managerial and regulatory knowledge, project planning, purchasing, reasoning and presentation of data, teaching skills, legal knowledge, and ethical issues are all integral parts of the day-to-day lives of ordinary scientists. Learning such skills is both essential to working effectively in industry and difficult. For the student, the academic bias toward technical excellence sometimes conflicts with the need for excellence in organizational, teamwork, and leadership skills. There are some excellent works on general and scientific management, but much of that writing is wanting in integrating the theory of management with the realities of the workplace. The present work addresses some of these nontechnical subjects.

Also, there is the creative side to science that ultimately decides the morale and energy of a scientific organization. This is as true for the development side of the organization as for the research side. Sir James Black, one of the great industrial scientists of our time, put it this way: "There is just no shortage on the shop floor … of ideas, exciting ideas," but … "some kind of aphasia … develops as you go up the company. … They are expressing lack of trust in the scientific process."[3] He adds that "it is very hard for big corporations not to contaminate [small drug discovery units] with too much control."

While it is sometimes recognized that intelligence and creativity are useful in basic research, they are less frequently rewarded in areas such as production and quality control. The mechanics of production are much more complex than generally credited and can be disturbed by changes in feedstock, instrumentation, or personnel. It is precisely in such "routine" areas as quality control that an alert analyst can detect a failing production process promptly and diagnose the means by which failure occurred. Given the amount of documentation required to monitor a production process, bright, capable scientists can contribute substantially to the bottom line by devising reliable and meaningful assays, writing clear procedures that can be transmitted easily, and arranging convenient archives for data retrieval.

The technical aspects of analytical HPLC are the principal focus of the present work. The goal is to impart the generalist's breadth with the specialist's depth. One would think that it would be easy for an analyst in one industry to transfer his skills to an unrelated industry. Often, it is not so easy. Although there are common threads in the issues involved in industrial processes, there is essential, highly specialized knowledge associated with each manufacturing process. While no book can hope to do justice to all of the aspects of analytical HPLC, it is my hope that this book will be of special service to students just entering industry, to those displaced from positions in one industry seeking to retrain in another, and to those, like myself, who simply enjoy understanding the big picture of how things are made.

References

1. Cuatresecas, P., Preface, in *Drug Development,* 2nd ed., Hammer, C. E., Ed., CRC Press, Boca Raton, FL, 1990.
2. Benedek, K. and Swadesh, J., HPLC of proteins and peptides in the pharmaceutical industry, in *HPLC in the Pharmaceutical Industry,* Fong, G. W. and Lam, S. K., Eds., Marcel Dekker, New York, 1991.
3. Schuber, S., An interview with Sir James Black, *Pharm. Technol.,* March, 48, 1989.

The Editor

Joel K. Swadesh, Ph.D., is Adjunct Assistant Professor at the Department of Veterinary Sciences, University of Massachusetts at Amherst. He graduated from the University of New Mexico in Albuquerque in 1977, and went on to obtain a Ph.D. in physical chemistry from Duke University in 1981. While at Duke, he attended classes in the Graduate School of Business. He served as a Postdoctoral Fellow in the laboratory of Harold A. Scheraga at the Department of Chemistry of Cornell University, receiving a Fellowship from the National Institutes of Health from 1982–1984 for the study of protein refolding. In 1984–1985, he held a postdoctoral position with Mortimer M. Labes at Temple University Department of Chemistry in Philadelphia. From 1985–1988, he was Associate Senior Investigator in the Department of Analytical, Physical, and Structural Chemistry at SmithKline & French Laboratories in King of Prussia, PA, where he participated in the testing of seven biopharmaceutical and peptide products in the areas of thrombolysis, gastric disorders, and vaccines. From 1988–1990, he managed a technical applications laboratory dealing with chromatography and detectors at Polymer Laboratories in Amherst, MA. From 1991–1993, he was the Group Leader of the Analytical Biochemistry group at Alpha-Beta Technology, which brought the polymeric carbohydrate immunomodulator Betafectin™ into the clinic. Since 1990, he has been Assistant Adjunct Professor at the University of Massachusetts at Amherst, assisting in the training of graduate and postdoctoral students.

Dr. Swadesh is a member of the American Chemical Society, the New York Academy of Sciences, and the American Association for the Advancement of Science. He is the recipient of fellowships from the Tennessee Eastman Company and the National Institutes of Health. He has served as an invited speaker at Northeastern University, University of Massachusetts (Amherst), Smith College, Kyoto University, and Nagoya City University. He is the author of 21 publications and four posters in the areas of drug development, high performance liquid chromatography, liquid crystals, and statistical mechanics. He prepared and is currently applying for a patent for a drug delivery device intended for treatment of autoimmune and chronic inflammatory disease.

Contributors

Kálmán Benedek, Ph.D.
Chief Scientific Officer
HTS Associates, Inc.
Thousand Oaks, California

Rajesh G. Beri, Ph.D.
Process Development Associate
Cell Culture Department
Lonza Biologics, Inc.
Portsmouth, New Hampshire

András Guttman, Ph.D.
Vice President, Research and
    Development
Genetic BioSystems, Inc.
San Diego, California

Laurel S. Hacche, Ph.D.
Quality Assurance Manager
Allergan Pharmaceuticals
Irvine, California

Jeffrey R. Larson
Research Associate
The Dow Chemical Company
Midland, Michigan

Carl F. Martin
Senior Professional
Allergan Pharmaceuticals
Irvine, California

Cynthia A. Maryanoff, Ph.D.
Chemical Development Department
The R. W. Johnson Pharmaceutical
    Research Institute
Spring House, Pennsylvania

Patricia Puma, Ph.D.
Associate Director
Hybridon, Inc.
Worcester, Massachusetts

Rekha D. Shah
Scientist
Chemical Development Department
The R. W. Johnson Pharmaceutical
    Research Institute
Spring House, Pennsylvania

Joel K. Swadesh, Ph.D.
Assistant Adjunct Professor
Department of Veterinary and
    Animal Science
University of Massachusetts
Amherst, Massachusetts

James E. Tingstad, Ph.D.
Green Valley, Arizona

Contents

Dedication

This book owes much to the scientific comments of Prof. Ira Krull of Northeastern University and Prof. Peter Uden of the University of Massachusetts, Amherst. I likewise thank Prof. Charles Lochmüller of Duke University for giving me a solid introduction to separation science. I am grateful to Prof. Roy Lewicki at the business school of Duke University, who gave me valued formal training in human organization, and to my parents, Prof. Morris Swadesh and Dr. Frances L. Quintana, who gave me a foundation to understand truly what the theory meant.

At SmithKline & French Research Laboratories, at Polymer Laboratories, and at Alpha-Beta Technology, I worked on industrial projects ranging from human therapeutics to recycling plastics and gained greatly from the experience. In particular, I thank Dr. Andrew Blow of Polymer Laboratories for helping to facilitate the production of this book and Dr. Cynthia Randall of SmithKline Beecham for helping me bring it to completion.

chapter one

Introduction

Jeffrey R. Larson, James E. Tingstad, and Joel K. Swadesh

0-8493-2682-6/97/$0.00+$.50
© 1997 by CRC Press, Inc.

1.1   Overview

This chapter briefly introduces chromatographic practice, with more detailed information presented in subsequent chapters of the book. General information regarding analytical pumps, columns, packing materials, and commonly used detectors, such as those based on ultraviolet (UV) and visible absorbance, refractive index, and fluorescence are given in this chapter. Specialty detectors and trends in automated sample processing are described in Chapter 2. Preparative chromatography is described in Chapter 3. Electrochemical detection is described in Chapter 5, in association with ion exchange chromatography. Viscosimetric, light-scattering, and evaporative light-scattering detectors are described in Chapter 6, the chapter on gel permeation chromatography. The present chapter describes general chromatographic theory, while more detailed discussion of the theory of strong adsorption is reserved for the chapter on ion exchange, where moment theory on band shape is presented in association with reversed phase chromatography. Theory applicable to electrophoretic separations is presented in the final chapter. The present chapter also describes working in and managing an analytical laboratory.

1.2   Pumps

The performance characteristics of the chromatographic pump and gradient maker fundamentally define and limit the kind of separations that can be performed on a liquid chromatographic system. Preparative chromatographic apparatus is briefly described in Chapter 3. The most critical performance characteristics of analytical pumps are flow rate reproducibility, flow rate range, and pressure stability. Most of the commercially available analytical pumps are reciprocating piston devices,[1] which use one or two pistons to pressurize the mobile phase. Usually, the pistons are rods formed of an abrasion-resistant material, such as sapphire. The Hewlett-Packard® Model 1090 liquid chromatograph, using a diaphragm piston, represents a significantly different pump design. Low-pressure metering pumps introduce solvent into the diaphragm, which then delivers solvent with relatively little pulsation.

Piston rods in a conventional high performance instrument typically are driven by a cam. The cam smooths the delivery of the mobile phase. Dual pistons allow one cylinder to recharge while compression of the other maintains the operating pressure. More recently, electronic flow sensing has been used to continuously adjust and maintain control of the flow rate. Check valves ensure that flow is unidirectional, as well preventing a drop in pressure during the recharge cycle.

A second type of pump is the syringe pump, such as the Model 100-DM produced by ISCO. Gradient formation at the high-pressure side of the check valves is accomplished by mixing the flow from two or more pumps in a chamber that promotes turbulent flow. Gradient formation at the low pressure side is usually accomplished by means of a solenoid switching valve, which intersperses pulses of one solvent with pulses of another. Eldex® Laboratories (Napa, CA) has added the MicroPro™ pump to its product line.[2] As column manufacturers have demonstrated the ability to obtain high performance separations in short columns at reduced pressure, low-pressure syringe pumps such as the FPLC® (Pharmacia Biotech AB; Uppsala, Sweden) system have become increasingly adaptable to high performance liquid chromatography (HPLC) separations.

In the last decade, some systems, such as the Dionex DX-500, have been manufactured with a flow path using corrosion-resistant materials such as polyetheretherketone (PEEK®, ICI Americas; Wilmington, DE), rather than the traditional stainless steel. Since stainless steel is prone to corrosion by salts, particularly halides, the introduction of titanium, ceramic, and PEEK® was welcomed by those performing chromatography in aqueous systems, particularly in biological applications. PEEK®, however, is not useful in applications requiring pressures greater than about 4000 psi.

The cost of a chromatographic system is usually determined by the number of pumps and the number of pistons per pump. A single-piston pump will exhibit pulsation and flowrate variation. Since most detectors are sensitive to pressure and flow rate fluctuations, single-piston pumps normally are used in the least demanding applications. Dual piston reciprocating pumps are a relatively low-cost means of reducing pressure fluctuations, and this design is widely used. Further pulse dampening can be accomplished by several means. The simplest is to insert several meters of tubing ahead of the injector. A diaphragm device to reduce pulsation is available from SSI. The lowest fluctuation in pressure and flow rate may be found in syringe pump systems. Even in isocratic separations, dual syringe systems are perhaps preferable, since otherwise the flow rate falls while the syringe is recharging, causing the baseline to shift. For gradient separations, dual pistons are, of course, required.

Flowrate stability is an important characteristic in isocratic systems, especially for molecular weight determination by gel permeation (size exclusion) chromatography. As has been described, pressure pulsations lead to fluctuations in flow rate. However, over the course of a day, flow rate can vary due to the completeness of solvent degassing. The mobile phase is usually degassed by sonication, by imposing a vacuum, or by displacing atmospheric gases such as nitrogen and oxygen with helium, which has a very low solubility in most mobile phases. On-line degassing through gas-permeable/liquid-impermeable membranes has become increasingly popular.

Over the course of a day, a degassed mobile phase can reabsorb gas from the atmosphere. Since the dissolved gas is far more compressible than the mobile phase, the piston will deliver progressively less mobile phase per stroke, leading to a drop in flow rate. For this reason, it is preferable to bubble helium through the mobile phase continuously, to maintain a pressure of helium over the mobile phase, or to continuously degas using gas-permeable membranes. The disadvantage of continuous helium degassing is that the composition of a multicomponent mobile phase may change due to evaporation of the more volatile component. Also, high purity helium is expensive, and lower purity grades can contaminate the mobile phase. Systems that continuously degas the mobile phase by passing it through a gas-permeable membrane that is under vacuum include the Alltech (Deerfield, IL) Model NO-OX® membrane degassing system, which has a 15-ml dead volume and claims to reduce dissolved oxygen to about 1 ppm, a feature that may be especially useful for electrochemical detection.

Temperature variation may also be a relevant factor in flowrate stability. Since the viscosity of the solvent is temperature dependent, wide swings in the ambient temperature can directly affect pump performance. The direct effects of temperature on pump performance usually are far smaller, however, than the effects on retention and selectivity; therefore, control of column temperature is generally sufficient to obtain high reproducibility.

Most analytical pumps are designed to operate best at about 0.1 to 10 ml/minute, the lower value being useful for overnight equilibration, and the upper value for purging the system lines. A flow rate of 1 ml/minute is usually ideal for columns about 2 to 10 mm in diameter. In recent years, narrow-bore (1 to 2 mm) and microbore (<1 mm) columns have come into more general use. For these systems, flow rates from 1 to 100 μl/minute are often required. Few pumps function well at the lower end of their rated specification; therefore, many laboratories use flow splitters to adapt conventional pumps to microbore capability. In isocratic systems, solvent recycling may make this a cost-effective approach, but in gradient separations, solvent waste quickly makes the purchase of a low-flow pump cost effective. Syringe pumps such as the ISCO Model 100-D perform well at low flow rates (1 to 100 μl/minute). In industrial processing, described in Chapter 3, column diameters are much greater, and much higher flow rates are required. High flow rates are generally accomplished at low pressure, so peristaltic pumps or low-pressure syringe pumps of the type used in the Pharmacia BioPilot® system are adequate.

Gradient linearity and repeatability are essential for many demanding applications, such as peptide mapping of proteins. As has been described, mixing by means of a solenoid valve introduces pulses of the solvents used to form the gradient. Unless mixing is complete, the solvent composition will fluctuate slightly about the specified value, leading to erratic elution profiles. Excessive dead volume between the locus of gradient formation and the head of the column leads to gradient rounding. This dead volume also lengthens the time that it takes a gradient that has been formed to propagate to the column. Although the consequent increase in run time may be insignificant for ordinary chromatography, the delay can be substantial for microbore work. Finally, run-to-run variability in gradient formation can cause comparison of different runs to be difficult. Run-to-run variability is sometimes traceable to the performance of the microprocessor that controls gradient making, but it may also be due to chromatographic variables, such as pre-equilibration time.

1.3 Columns

High performance columns are classified as being open tubular or packed bed. Although it has been shown that extremely high efficiencies can be attained on open tubular columns of narrow diameter,[3-7] packed-bed columns have found more general use.[8,9] Packed-bed microbore columns have found some utility, particularly for cases in which high efficiency, extreme sensitivity, or low solvent flow are required.[10-12] Packed-bed columns 1 to 10 mm in diameter are most widely used, since the performance is extremely rugged and pump performance requirements are minimally demanding.

Packed-bed analytical columns are filled with particles about 3 to 10 μ in diameter. Larger particles, typically 20 to 50 μ in diameter, are used in preparative applications. The particles typically are formed of incompressible materials such as silica, alumina, graphitic carbon, or rigid polymers.[13,14] Particles smaller than 3 μ can be used in analytical functions and are commercially available from Micra. A number of column materials are homogeneous, i.e., there is no phase bonded to the base material. Unmodified silica is a homogeneous phase used as in normal phase chromatography, while poly(styrene-divinyl benzene) is a homogeneous phase when used as a gel permeation material in organic solvents or as a reversed phase material in mixed aqueous-organic solvents. Cellulose and modified cellulose, which are commonly used in low-pressure applications, have found some application in chiral separations. Cyclodextrins are also a common homogeneous chiral phase. A most unusual homogeneous phase, used for high performance ultrafiltration, is formed from plant cell clusters.[15]

The material most widely used as the base material, or *chromatographic support*, for bonded phases is silica, and the materials most widely used as

bonded phases are alkyl silanes. Polymethylmethacrylate (PMMA), methacrylate, poly(styrene-divinyl benzene) (PS-DVB), hydroxyethylmethacrylate (HEMA), alumina, carbon, and other polymeric and inorganic materials
have been used as base materials. The particles may be regular or irregular
in shape. They may be permeable or impermeable to flow, and the surface
may be smooth or irregular; an irregular surface has a larger effective surface
area than a smooth one. Particles that are permeable to flow are said to have
an *internal surface*. Gel-type polymers, comprised of a linear polymeric chain,
are called *microporous* and can be conceptualized as a loosely wound ball of
yarn. Polymers formed from suspension polymerization form aggregates of
microparticles and are called *macroporous*. Microparticles may also be
agglomerated onto the surface of a larger, rigid particle to form a pellicular
resin.

The surface area of a particle that is accessible to a given analyte depends
on the analyte as well as the particle. Macromolecules may be unable to
penetrate to the internal surfaces or even to the more constricted surface
irregularities of a particle. For this reason, chromatographic packings are
often categorized according to *pore size*, which may be conceptualized as the
average size of solvent-accessible pockets in the particle. One simple method
of measuring the pore size is to heat a solvent-moistened particle on an
analytical balance and measure the loss of weight of evaporated solvent, a
technique known as thermogravimetry.[16] The bulk liquid readily evaporates
near the boiling point, while liquid trapped inside pores evaporates less
quickly. Pore size measurements are useful as rough guides, but not as
absolute measures, of the appropriateness of a column material for a given
separation. Clearly, if the molecular radius of the analyte is larger than the
pore, it will be unable to access it. However, molecular shape is rarely
perfectly spherical, and electrostatic or other interactions between the analyte
and the particle may influence the entry of an analyte into the pore. For these
reasons, a molecule nominally of a given average radius may not be able to
penetrate a pore of equivalent size.

It is possible to coat, to graft onto, or even to encapsulate the chromatographic support with another material, called the *stationary phase* or, since
bonding is such a common procedure, the *bonded phase*. Typical stationary
phases are aliphatic and phenyl moieties for reversed phase chromatography; amines and diols are used for normal phase chromatography; unmodified or alkylated amines are used for anion exchange chromatography;
sulfonates or carboxylates are used for cation exchange chromatography;
and affinity ligands, such as protein A and heparin, are used for affinity
chromatography. Other ligands, including bovine serum albumin, are used
in chiral chromatography. The phases, both homogeneous or bonded, that
have been used for HPLC are summarized in Table 1.[2] Some examples of
commercially available columns are given in Table 2.

Table 1 Composition of Some Common Chromatographic Phases

Material	Homogeneous (H) or bonded (B)	Type of chromatography
Silica	H	Normal phase, gel permeation
Alumina	H	Normal phase
Graphite	H	Reversed phase
Poly(styrene-divinyl benzene)	H	Reversed phase, gel permeation
Alkyl silane	B	Reversed phase
Phenyl	B	Reversed phase
Cyano	B	Reversed phase
Amino	B	Normal phase, ion exchange
Alkylated amino	B	Ion exchange
Sulfonate	B	Ion exchange
Alkyl sulfonate	B	Ion exchange
Diol	B	Normal phase
Poly(methylmethacrylate)	H	Gel permeation

1.4 Chromatographic modes

1.4.1 Overview

The chromatographic column is often conceptualized as a stationary bed immersed in a rapidly flowing mobile phase, with stagnant pools of the mobile phase situated in the pores of the packing material. A film of strongly bound solvent is clustered around the stationary phase. This is an oversimplification, as flow can affect the volume of the bed or the accessible pore volume. In a very simple formulation, then, liquid chromatography is the movement of the analyte along a path, with its rate of movement being determined by a competition between residence in the flowing mobile phase or in the immobile, stagnant pools, perhaps even bound to the stationary phase. The differential rate of migration of two analytes, therefore, is determined by their relative tendencies to interact with the stationary phase and surrounding stagnant or bound solvent. In the absence of any interaction between the stationary phase and the analyte, there are *hydrodynamic*, or flow, effects. Principal among hydrodynamic effects are those consequent from the difference in effective column void volumes due to exclusion from pores that two analytes may experience.

From the viewpoint of molecular interactions, the number of fundamentally distinct chromatographic stationary phases is very limited.[17] One mechanism for adsorption to the stationary phase is solvophobic, or mobile-stationary phase transfer free energy effects, in which the adsorption of an analyte to the stationary phase liberates bound solvent. There is often an accompanying enthalpic component to such binding through dispersion interactions. Another mechanism for adsorption is that of specific interactions,

Table 2 Examples of Commercially Available HPLC Columns

Mode	Manufacturer	Name	Packing	Particle size	Note
Normal phase	YMC, Inc.	YMC-Pak	Silica	5 μ, 120 Å	
	Perkin-Elmer (Norwalk, CT)	Silica, Spheri-5®	Silica	5 μ, 80 Å	
	Waters (Milford, MA)	μBondapak™-NH₂	Amine-bonded silica	10 μ, 125 Å	3.5% C not endcapped
	Macherey-Nagel (Düren, Germany)	Nucleosil™ 100 Diol	Diol on silica	7 μ, 100 Å	
	BTR Separations (Wilmington, DE)	Zorbax® Pro 10/60 CN	Alkyl cyano phase on silica	10 μ, 60 Å	
Reversed phase	TosoHaas (Tokyo, Japan)	SuperODS	C-18 on silica	2 μ	Fast separations
	Micra		C-18 on silica	1.5 μ nonporous	
	DyChrom (Santa Clara, CA)	Shiseido RP-Capcell™	Polymeric C8 or C18 on silica	3–5 μ, 120–300 Å	
	Polymer Labs (Church-Stretton, U.K.)	PLRP-S™ 100	Unmodified PS-DVB	5 μ, 100 Å	
	LC Packings (San Francisco, CA)	Fusica™ II	C-4, -8, or -18 on silica	3–5 μ	Microbore
Ion exchange	Dionex (Sunnyvale, CA)	CarboPak™ PA-1	Polystyrene		
	Shodex™ (Tokyo, Japan)	SC1011	Sulfonated PS	7 μ	Ca⁺² form, exclusion 10³

Type	Manufacturer	Product	Material	Particle size	Notes
	TosoHaas (Tokyo, Japan)	DEAE-5PW™	Methacrylate-DEAE	10 μ, 1000 Å	
	BioRad (Hercules, CA)	Aminex® HPX-87H	Sulfonated PS-DVB	9 μ	H+ form, exclusion 10³
Gel permeation	Shodex™ (Tokyo, Japan)	KB80M	Cross-linked methacrylate	(not listed)	Aqueous, exclusion 10⁷
	Polymer Labs (Church-Stretton, U.K.)	PL-Gel™	PS-DVB	5 μ, 50Å	
	TosoHaas (Tokyo, Japan)	G1000-HXL	Porous PS-DVB	5 μ, 40 Å	Organic
	TosoHaas (Tokyo, Japan)	G3000-SWXL	Silica	5 μ, 250 Å	Aqueous
	Waters™ (Milford, CA)	Styragel® HR-3	PS-DVB	5 μ	Organic, exclusion 30 × 10³
Chiral	Regis (Morton Grove, IL)	Whelk-O1®	4-(3,5-dinitrobenzamido)-tetrahydrophenanthrene	5 μ, 100 Å	
	DaiCel (Tokyo, Japan)	Chiralcel® OD	Cellulose on silica	10 μ	
	Phenomenex™ (Torrance, CA)	Chirex™ 3019	S-tert-Leu and S-1α-naphthylethylamine on aminopropyl silica		Pirkle type
	J. T. Baker® (Phillipsburg, NJ)	DNBPG	(R)-N-(3,5)-dinitro benzoyl phenylglycine on silica	5 μ	Pirkle type

Note: PS = polystyrene, DVB = divinyl benzene, DEAE = diethylaminoethyl.

which are typically primarily enthalpy-driven attractions between definable groups on the stationary phase and on the analyte. These notably include hydrogen-bonding, dipolar, and electrostatic interactions. Another mechanism of interaction between stationary phase and analyte is that of a reversible chemical transformation, in which a chemical reaction, such as disulfide interchange, is involved in the binding of the analyte to the stationary phase. The field of chromatographic science, however, has developed an extensive nomenclature to further differentiate chromatographic phases. These are presented in swift panorama to give the reader a sense for the range of chromatographic types.

1.4.2   Gel permeation/size exclusion

The stationary phase in gel permeation (also called size exclusion) chromatography contains cavities of a defined size distribution, called pores. Analytes larger than the pores are excluded from the pores and pass through the column more rapidly than smaller analytes. There may be secondary effects due to hydrophobic adsorption, ionic interaction, or other interactions between the stationary phase and analyte. Gel permeation and non-ideal interactions in gel permeation are described more fully in Chapter 6.

1.4.3   Normal phase

In normal phase chromatography, the analyte interacts with the stationary phase, typically through hydrogen bonding or polar interactions. Silica and alumina have long been used in normal phase chromatography.[18] These phases are compatible with nonaqueous solvent systems and are suitable for the separation of many organic compounds. Normal phase still is routinely used for the separation of simple organic compounds. The separation of 4-nitrobenzo-2-oxa-1,3-diazole derivatives of a number of glycero- and sphigolipids,[19] arachidonic acid metabolites,[20] and diglycerides and ceramide[21] are typical separations performed by normal phase chromatography. Coupled with a size exclusion column, normal phase may also be useful for copolymer analysis.[22]

Silica has often been modified with silver for *argentation chromatography* because of the additional selectivity conferred by the interactions between silver and π-bonds of unsaturated hydrocarbons. In a recent example, methyl linoleate was separated from methyl linolenate on silver-modified silica in a dioxane-hexane mixture.[23] Bonded phases using amino or cyano groups have proved to be of great utility. In a recent application on a 250 × 1-mm Deltabond® (Keystone Scientific; Bellefonte, PA) Cyano cyanopropyl column, carbon dioxide was dissolved under pressure into the hexane mobile phase, serving to reduce the viscosity from 6.2 to 1 MPa and improve efficiency and peak symmetry.[24] It was proposed that the carbon dioxide served to suppress the effect of residual surface silanols on retention.

1.4.4 Reversed phase and hydrophobic interaction chromatography

In reversed phase chromatography (RPLC), the analyte adsorbs to the stationary phase through the hydrophobic effect. Reversed phase chromatography is described in much greater detail in Chapter 4. Fluorocarbons are finding application as durable reversed phase materials, with the branched polyfluorocarbon Neos (Shiga, Japan) Fluofix columns exhibiting slightly less retention than ODS in the separation of phenols, halophenols, and polyphenolic flavonoids such as hesperidin, naringin, quercitrin, hesperitin, quercitin, naringenin, and kaempferol.[25] Argentation chromatography, mentioned above, is also used in reversed phase separations of linoleic and oleic acids by the simple addition of silver ion to the mobile phase.[26]

Hydrophobic interaction chromatography (HIC) can be considered to be a variant of reversed phase chromatography, in which the polarity of the mobile phase is modulated by adjusting the concentration of a salt such as ammonium sulfate. The analyte, which is initially adsorbed to a hydrophobic phase, desorbs as the ionic strength is decreased. One application demonstrating extraordinary selectivity was the separation of isoforms of a monoclonal antibody differing only in the inclusion of a particular aspartic acid residue in the normal, cyclic, or iso forms.[27] The uses and limitations of hydrophobic interaction chromatography in process-scale purifications are discussed in Chapter 3.

1.4.5 Ion exchange/electrostatic interaction chromatography

In conventional ion exchange chromatography (IEC), electrostatic interactions between an analyte and stationary phase of opposite charge cause the analyte to adsorb to the stationary phase. Ion exchange, often called *electrostatic interaction chromatography*, is described in greater detail in Chapter 5. A recent application that illustrates that ion binding and selective separation may take place by means other than electrostatic interactions was the use of an uncharged ligand, tetradecyl-16-crown-6, which complexes inorganic ions in the order $Ba^{+2} > Sr^{+2} > K^+ > Rb^+ > Ca^{+2} > NH_4^+ > Cs^+ > Na^+ > Li^+$. The crown ether ligand was coated onto a a Dionex MPIC resin, and the separations were performed using methanesulfonic acid as an eluent.[28] Depending on the degree of specificity of interaction between the stationary phase and a particular analyte, one might regard a separation of this kind to be a form of affinity chromatography.

1.4.6 Affinity chromatography

Affinity chromatography involves precisely the same kind of electrostatic, hydrophobic, dipolar, and hydrogen-bonding interactions described above, but the specificity of binding is extraordinarily high. Demands on the homogeneity of the stationary phase and on the rigidity of the support are often correspondingly low. Soft gels, such as agarose are often used as supports for affinity chromatography. An example of the use of *metal-interaction* affinity chromatography, often referred to as IMAC, was the use of Cu^{2+} and Ni^{2+} in the purification of cysteine-containing peptides that each had a free amino group, but lacked histidine or tryptophan residues.[29] The use of IMAC in preparative chromatography, is discussed in more detail in Chapter 3.

Extremely specific stationary-phase-analyte interactions, some of which are orders of magnitude more specific than the crown ether-inorganic ion complexation described above, are often described as *molecular recognition* to indicate that the stationary phase binds one compound to the virtual exclusion of all others. A recent review lists examples of ligands which have a highly specific affinity for certain analytes, including monoclonal antibodies against specific proteins, transition-state analogues for proteases, metal chelates for metal-binding proteins, and deoxythymidine 3'-phosphate,5-aminophenylphosphate for *Staphylococcus* nuclease.[30] As specific as affinity chromatography can be, the actual purification factor in separations is often comparable to that seen in a well designed separation on RPLC, HIC, IEC, or other conventional chromatographies. RPLC, HIC, and IEC have the capability of individuating many more peaks than affinity chromatography. For these reasons, affinity chromatography tends to be most useful as a preparative chromatography.

1.4.7   Chiral chromatography

The separation of enantiomers is particularly important in pharmaceutics, since many drug substances are chiral, and the biological activities of different enantiomers may be substantially dissimilar. In the absence of pre-chromatographic chemical modification to form diastereomers, enantiomer separation requires that a chiral selector be incorporated into the stationary phase or added to the mobile phase. The chiral selector, by interacting preferentially with one enantiomer, influences the migration of that enantiomer through the column. Preferential interaction is a phenomenon sometimes called *chiral recognition*, consonant with the phrase *molecular recognition*, above. Although many chiral phases have been designed to create three specific contact points between the analyte and the stationary phase, it has been pointed out that the achiral chromatographic support can serve as one of the contact points required for chiral recognition.[31]

There is a wide variety of commercially available chiral stationary phases and mobile phase additives.[32-34] Preparative scale separations have been performed on the gram scale.[32] Many stationary phases are based on chiral polymers such as cellulose or methacrylate, proteins such as human serum albumin or acid glycoprotein, Pirkle-type phases (often based on amino acids), or cyclodextrins. A typical application of a Pirkle phase column was the use of a N-(3,5-dinitrobenzyl)-α-amino phosphonate to synthesize several functionalized chiral stationary phases to separate enantiomers of β-blockers such as metoprolol, oxprenol, propanolol, pronethalol, pindolol, and bufarolol.[35] The β-blockers all have a hydrophobic aromatic ring, a basic secondary amino group, and a secondary hydroxyl group as functionalities. These interact, respectively, with the hydrophobic ring; the protonated carboxyl, phosphonate, or other electronegative group; and the amide functionality of the chiral stationary phase. Temperature control and mobile phase composition were useful to adjust retention.

The cyclodextrins are a family of basket-shaped oligosaccharides with a hydrophobic cavity. The α-cyclodextrin variant contains six glucose residues, β-cyclodextrin has seven, and γ-cyclodextrin has eight. Enantioselectivity is obtained generally only if the analyte has a hydrophobic portion to be included into the hydrophobic cyclodextrin cavity, which could mean that the analyte may require derivatization. Chiral ion-pairing agents such as quinine also have been used as mobile phase additives. Other chiral selectors include heparin and macrocyclic antibiotics such as rifamycins.[36] Some suppliers of chiral columns include Phenomenex (Torrance, CA; Chirex™), Daicel (Tokyo, Japan; ChiralPak® and Chiralcel®), J.T. Baker® (Phillipsburg, NJ), Regis Technical (Morton Grove, IL), and Macherey-Nagel (Düren, Germany).

1.4.8 Other chromatographic modes

Many other modes of chromatography have been described. One of these is chromatofocusing, in which a pH gradient is developed along the length of an ion exchange column using a complex mixture of buffers.[37,38] The pH gradient also can be generated in a column by means of an electric field, in which case the technique is called *isoelectric focusing*, a technique that is discussed in detail in Chapter 7. In either case, an analyte such as a protein will be eluted as the pH of the gradient approaches the point of minimum protein charge, known as its isoelectric point. Countercurrent chromatography is another technique in widespread use. A liquid stationary phase immiscible with the mobile phase is used for separation.[39]

1.4.9 Mixed-mode chromatographies and mixed-functionality resins

There is evidence of the participation of more than one mode of chromatography in many, if not most, separations. This can sometimes be exploited to customize a separation. The simplest means may be simply to couple two well defined columns together, as has been done for separations on a chiral column with pre-separation occurring on the reversed phase.[40] Another approach is to blend resins with different functionalities. Mixed-bed ion exchangers are a well established example of the utility of blending resins with different functionalities. Finally, different functionalities can be deliberately placed on the same bead to tailor column performance to a particular application. An example of a mixed-functionality, single-mode resin is the use of heterogeneous hydrophobic groups for direct injection of serum and plasma.[41] The strongly hydrophobic groups are sufficiently dilute such that adsorption of serum proteins is reversible, yet the phase is not so hydrophilic as to prevent the retention of low-molecular-weight components.

An interesting example of a mixed-mode chromatography is slalom chromatography. Slalom chromatography is a hydrodynamic or flow mode of separation; i.e., it is strongly dependent on the manner in which the column packing and flow rate interact with molecular size and shape to influence elution. A typical recent application was the mixed-mode separation of λ/HindIII fragments from endonuclease digestion of DNA on reversed phase columns such as the Shiseido (Tokyo, Japan) Capcell™-Pak C1 or Capcell™-Pak Phe or one of the Shandon Hypersil®-3 (phenyl, trimethylsilyl, cyanopropyl, dimethyloctyl, or octadecyl) packings (Hypersil, Ltd; Cheshire, U.K.).[42]

1.5   Detectors

1.5.1   Overview of detectors

Devices used for detection in column chromatography typically rely on differences in the physical or chemical properties of the eluent and the analyte. Alternatively, the solvent may be evaporated to allow detection by mass spectrometry, flame ionization, or other detection modes. Properties exploited in detection of analytes in the presence of mobile phase include absorbance of light in the infrared, visible, or ultraviolet range, fluorescence of absorbed light, light-scattering, light refraction, viscosity, conductivity, electrochemical reactivity, chemical reactivity, volatility, optical rotation, and dielectric constant. Of these detectors, by far the most common are those based on the absorbance of ultraviolet light and those based on the refraction of light. Post-column reactors can be connected in-line to introduce a reagent that will undergo a chemical reaction with the analyte, the product of which is then detected by light absorbance or fluorescence. Post-column reactors of this kind are in common use in amino acid analysis, clinical analysis, and other important areas. Detectors based on light-scattering, viscosity, and volatility are in wide use in the analysis of polymers. These detectors are detailed in a later chapter on gel permeation chromatography. Electrochemical detectors are popular in separations of sugars, while conductivity detectors are often used for detection of inorganic ions. These detectors are characterized in a chapter on ion chromatography. The present section focuses on the ultraviolet-visible absorbance (UV-VIS) detector, the fluorescence detector, and the differential refractive index (dRI) detector.

1.5.2   The UV-VIS detector

The ultraviolet-visible spectrophotometer is the most widely used detector for HPLC. The basis of UV-VIS detection is the difference in the absorbance of light by the analyte and the solvent. A number of functional groups absorb strongly in the ultraviolet, including aromatic compounds; carbonyl compounds, such as esters, ketones, and aldehydes; alkenes; and amides. It should be noted that a number of analytes of interest, particularly simple sugars, saturated hydrocarbons, alcohols, and polyethers absorb only very weakly above 200 nm, while many chromatographic solvents, such as toluene and acetone, absorb very strongly in the UV. Stabilizers commonly used in solvents such as tetrahydrofuran may also absorb in the UV range. Absorbance by the solvent or stabilizer interferes with detection of an analyte. Solvents that absorb only weakly in the UV range include water, methanol and other simple alcohols, and acetonitrile.

There are various designs for UV-VIS detectors. In the simplest design, UV-VIS is generated by a source such as a deuterium or mercury lamp. A slit monochromator is used to select a narrow band of UV-VIS light which is passed through the sample compartment and detected by a photomultiplier. Since some light may be scattered or re-emitted as fluorescence, a second monochromator may be placed between the sample and the photomultiplier. Since absorption of light heats the eluent containing the analyte, a thermal lens forms, distorting the light path. The effect is particularly pronounced with high-intensity sources, such as lasers. Thermal lensing has been used as a detection strategy,[43] but temperature control has long been recognized to be essential in high-sensitivity work.[44,45] A tapered cell has been proposed to minimize refractive index errors.[46]

Another strategy in UV-VIS design is to expose the sample to polychromatic light. The light then strikes a grating, which resolves the wavelengths of light much like a prism. Narrow ranges of wavelengths then strike discrete elements of an array of photodiodes.[47] Instruments based on this design are called *photodiode array detectors*. In a recent application illustrating the utility of the photodiode array detector, the chemical structure of procyanidins present in tannin could be inferred from the absorbance bands.[48] Peak homogeneity has been analyzed by an *ab initio* approach using a diode array detector.[49] In the absence of any assumptions about peak shape or the spectra of impurity or principal component, impurities in an unresolved peak could be detected at less than 1%.

The trend in the 1980s was to design more sensitive UV detectors. Although there are many applications for more sensitive detectors, HPLC is the technique of choice for assays that do not require high sensitivity but do require extremely high reproducibility. Several papers have described the compromise with respect to slit width between sensitivity and broad linear dynamic range.[50,51] For the maximum linear dynamic range necessary for highly accurate analyses, a narrow slit width is desired, while a wide slit width leads to a reduction in noise and improved sensitivity. For trace analyses, the critical relationship between bandwidth and linearity is illustrated in an experiment described by Pfeiffer et al.[51] using a liquid chromatography (LC) detector with an adjustable slit width. Benzoic acid, which has a relatively flat spectrum in the region of 254 nm, was used as the test compound (Figure 1). The linearity was tested at bandwidths of 2 and 16 nm at 1.0 AUFS (absorbance units full scale). As shown graphically in Figure 2, severe nonlinearity was obtained using the 16 nm slit width. Tests at intermediate slit widths showed nonlinear results when the slit width exceeded 4 nm.

Most instrument manufacturers evaluate linearity using a compound with a broad spectral band at the wavelength chosen for testing. However, for most multicomponent mixtures it is not possible to find a single wavelength where all components have an absorption maximum or broad spectral band. This led to the development of a more rigorous static test for evaluation of detector linearity using test compounds with changing UV spectra at the wavelengths chosen for testing.[51] The ultraviolet spectra of the test compounds, benzaldehyde and benzoic acid, are shown in Figure 1. Benzaldehyde was used to test detector linearity at 214 and 254 nm, while benzoic acid was used to test linearity at 280 nm. Using this static test, important differences between detectors from different manufacturers were observed. Some detectors were linear to 2.56 AUFS, while other detectors were very nonlinear well below 0.5 AUFS.[51]

A more recent paper suggests that the most complete assessment of detector performance may be obtained by testing detector linearity using both best- and worst-case conditions.[52] Dorschel et al.[52] also caution against using a least squares fit to data points using a nonzero intercept. Although the line may appear to be linear, substantial errors can result when the line (Figure 3) is used for quantitation of unknowns having concentrations at either extreme of the curve.[52] The article suggests that a linear model be used over a restricted concentration range or that a nonlinear model be applied over a broader range of concentration. A detailed comparison of five commercially available detectors showed the difficulty of absolute calibration.[53,54] Another article by Dose and Guiochon specifically discusses the nonlinearity observed with diode-array detectors due to polychromatic radiation.[55]

Another area of concern is wavelength accuracy. Wavelength inaccuracies of up to 10 nm have been observed not only between detectors from various manufacturers but also between detectors from the same manufacturer. This is particularly important at low wavelengths (<220 nm), a region commonly used to detect weak UV absorbers, where most organic compounds have rapidly changing absorption bands.[56] For these compounds, small changes in wavelength can result in large changes in detector response. A simple static test of wavelength accuracy applicable to the low-UV, mid-UV, and visible regions has been described in an earlier paper by Esquivel.[56] The test uses rare earth salts, which have sharp absorption bands in the UV-VIS region. Using this test, important differences in wavelength accuracy between UV detectors from various manufacturers were observed. The effects of refractive index lensing have been examined in detail.[57] A response resembling the derivative of a peak is observed, apparently due to focusing and defocusing of the beam image on the detector.

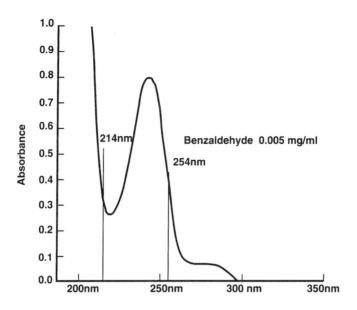

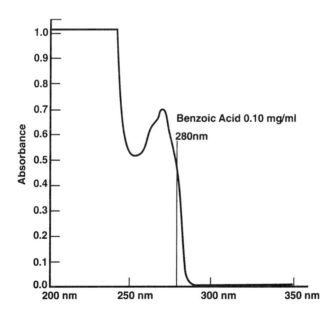

Figure 1 Ultraviolet spectra for benzaldehyde and benzoic acid: solvent, methanol; reference, methanol; cell, 1.0 cm. (From Pfeiffer, C. D., Larson, J. R., and Ryder, J. F., Linearity testing of ultraviolet detectors in liquid chromatography, *Anal. Chem.*, 54, 1622, 1983. Copyright American Chemical Society Publishers. With permission.)

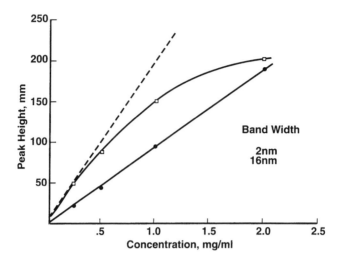

Figure 2   Effect of slit-width on linearity — benzoic acid at 254 nm, 1.0 AUFS. (From Pfeiffer, C. D., Larson, J. R., and Ryder, J. F., Linearity testing of ultraviolet detectors in liquid chromatography, *Anal. Chem.*, 54, 1622, 1983. Copyright American Chemical Society Publishers. With permission.)

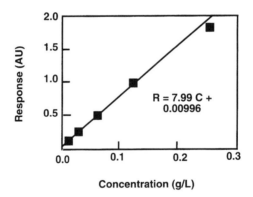

Figure 3   Least squares calibration line for photometric detector. (From Dorschel, C. A., Ekmanis, J. L., Oberholtzer, J. E., Warren, Jr., F. V., and Bidlingmeyer, B. A., LC detectors: evaluations and practical implications of linearity, *Anal. Chem.*, 61, 951A, 1989. Copyright American Chemical Society Publishers. With permission.)

Absorbance detectors are also commonly used in combination with post-column reactors. Here, most issues of detector linearity and detection limit have to do with optimization of the performance of the reactor. In a typical application, organophosphorus compounds with weak optical absorbances have been separated, photolyzed to orthophosphate, and reacted with molybdic acid, with measurement being performed by optical absorbance.[58]

1.5.3   The refractive index detector

The refractive index (RI) detector is described in more detail in Chapter 6. In this technique, the refractive index of the sample compartment is compared with that of a reference compartment, and the technique is usually called *differential* RI (dRI) detection. With improvement in sensitivity there has been a resurgence of interest in the RI detector in modes of chromatography other than GPC. Many compounds lack a strong absorbance in the UV, including alkanes, simple alcohols, ethers, sugars, and many ions; therefore, a UV-VIS detector alone is hardly universal. The refractive index detector is universal in the sense that for any given analyte, a solvent probably can be found such that the refractive index of the analyte differs enough to permit detection. The dRI detector is often useful in specialized applications, such as polymer analysis. The principal limitations that have made the use of RI detection less popular are sensitivity, variation in response with changes in temperature, peaks that may be negative or positive, sensitivity to variations in flow or dissolved gases, and incompatibility with gradient chromatography or analytes prepared in solvents different than the mobile phase. The problem of gradient incompatibility has been overcome to a considerable degree by a new design in refractive index detection.[59-61] Another approach has been to match the refractive indices of the components of the gradient,[62] for which rates of flow and gradient change were required.

Since the refractive index of the analyte may be either greater than or less than that of the solvent, peaks may be either positive or negative. Some data systems are incapable of properly integrating peaks of opposite polarities. Therefore, while it was possible to integrate negative peaks by switching the polarity of the detector leads, such data systems were of limited utility in separations involving peaks of both polarities. Increasingly, however, data systems are designed to accommodate negative peaks, and to process the area below the baseline as a positive quantity. The refractive index changes by about 10^{-3} RI units (1 mRI unit) per degree centigrade,[63,64] and since the limit of detection of modern RI detectors is approaching the nRI level (10^{-9} RI units), temperature control to 1 μdegree is required. This problem is usually finessed by using a reference cell so that the change in the solvent refractive index is compensated. This is not a rigorous solution for comparing data from run-to-run, however, since the refractive index of the analyte is not compensated for by the presence of the reference cell.

The problem of dissolved gases has been described at length.[65,66] Briefly, the refractive indices of degassed and air-saturated solvents differ by about 0.1 to 100 μRI units (10^{-7} to 10^{-4} RI units), with water exhibiting a small change on absorbing air, and organic solvents a large change. Similarly, as dry organic solvents absorb moisture from the atmosphere, the refractive index will change.[65] These changes in solvent composition result in drift. Artifactual peaks may be generated due to differences in composition between the sample and the mobile phase. Noise can result from using a binary system to mix the mobile phase.

There are a number of designs that have been used for RI detection.[65,66] A deflection RI detector measures the extent to which the incident beam is bent by traversing the sample.[67] A Fresnel detector measures the loss of intensity due to transmission vs. reflection at the interface between the flow cell and its contents.[65] A change of refractive index of about 90 nRI units can be detected. The interferometric detector, with a claimed detection limit of 10 nRI units,[66,68] splits a polarized light beam into two portions with orthogonal directions of polarization. Since the refractive index affects the speed of light, the sample and reference beams will be out of phase when the RI values of the two compartments are not identical. On recombination, interference due to mismatch of beam phase reduces the beam intensity. A Fabry-Perot interferometric design utilizing a He-Ne laser (634.2 nm) as the light source has been described.[69] This design exploits the proportionality of the change of refractive index to the change in wavelength for measurement. A detection limit of 15 nRI units was found. The linear range for RI detectors has been estimated to be about 1000 to 10000 times the minimum detectable limit,[66] although rigorous testing may suggest a more conservative estimate.[52]

1.5.4   The fluorescence detector

Another type of detector used in liquid chromatography is the fluorescence detector. Following absorption of light to form an excited electronic state, many events are possible. The excited molecule may chemically react. The absorbed light may be immediately re-radiated, which is known as light-scattering. When the excited molecule converts to a second, long-lived excited state, the re-emitted light is known as phosphorescence. The excited molecule may transfer all of its energy by long-range coupling to a distant molecule, a process known as Förster quenching.[70] The excitation may be dissipated entirely by vibrational motions, a process called "internal conversion". Stern-Volmer quenching describes de-excitation by collision with other molecules.[71] If, however, the excited state partially relaxes by vibration (internal conversion), then emits light at a lower wavelength than the exciting beam, the resultant emitted light is known as fluorescence.[72] It is common in fluorescence spectroscopy to observe a number of these effects and important to be able to differentiate them.

The absorption of light is an inefficient process. Typically, no more than 10% of the light passing through a chromophore is absorbed.[73] Internal conversion and quenching cause further losses in the yield of fluorescence per input quantum of energy. Reabsorption of emitted light, known as the inner filter effect, results in a further loss of fluorescence intensity. Fluorescence efficiency is a strong function of temperature due to collisional quenching of the fluorophore by chromatographically unresolved compounds and by solvent or dissolved gas molecules. Dissolved oxygen is a particularly potent quencher, as are sulfur-containing compounds.[74] Also, fluorescence may be affected indirectly by the column temperature in gradient chromatography, since fluorescence efficiency is strongly dependent on solvent composition. At high analyte concentrations, the inner filter effect may lead to a decrease in fluorescence, so concentrations should be kept as low as possible. High power sources, such as lasers, may cause photobleaching, a depletion of the fluorophore due to photochemical decomposition that may be an important source of detector non-ideality in microbore systems.[75] At low flow or high power, all of the analyte is photobleached, so one measures the total amount of analyte passing through the cell, while at high flow or low power, the fluorescence is proportional to concentration. To avoid scattering, solvent filtration is essential. Even so, column degradation fragments can introduce noise. The solvent alone may also scatter light. Finally, and most importantly, most analytes are not fluorescent and must be derivatized with fluorescent moieties such as anthracene, coumarin, phenanthrene, naphthalene sulfonates, fluorescamine, or *o*-phthalaldehyde.[76] Despite all of these disadvantages, fluorescence detection is still one of the most valuable techniques in trace analysis by HPLC.

The simplest fluorescence measurement is that of intensity of emission, and most on-line detectors are restricted to this capability. Fluorescence, however, has been used to measure a number of molecular properties. Shifts in the fluorescence spectrum may indicate changes in the hydrophobicity of the fluorophore environment. The lifetime of a fluorescent state is often related to the mobility of the fluorophore. If a polarized light source is used, the emitted light may retain some degree of polarization. If the molecular rotation is far faster than the lifetime of the excited state, all polarization will be lost. If rotation is slow, however, some polarization may be retained. The polarization can be related to the rate of macromolecular tumbling, which, in turn, is related to the molecular size. Time-resolved and polarized fluorescence detectors require special excitation systems and highly sensitive detection systems and have not been commonly adapted for on-line use.

Some very clever techniques have been used to optimize the performance of fluorescence detectors. It is possible to collect fluorescence at any angle relative to incidence, but interference from reflected or scattered light is strongest along the light path. Therefore, most fluorescence detectors collect light at 90° relative to incidence. The steradian cell used in the Applied Biosystems detector collects the fluorescence from 0° to ±90° relative to the incident beam, greatly increasing the sensitivity.[77] A combined UV-fluorescence detector was described in 1975.[78] Given the significant problems in calibrating fluorescence detectors, it is surprising that this design has not become generally available. Lasers have long been used as excitation sources for fluorescence.[79] They offer a number of advantages, including high intensity, narrow excitation wavelength, and plane polarization of the light source. Issues in cell design have been discussed.[80]

Another tactic in improving fluorescence detection is post-column derivatization. For example, post-column derivatization of sugars with guanidine has attained detection limits of 5 pmol, with linearity up to 1 nmol.[81] Post-column derivatization of sugars with benzamidine has also been used.[82] On-line, rapid-scanning fluorimetry is becoming increasingly availably.[83] Laser-induced fluorescence is becoming an increasingly important method for microbore systems, such as those in use in electrophoresis. Indirect fluorescence spectroscopy was used for the detection of oligosaccharides in the presence of the fluorophore, fluorescein.[84] The analytes, as they eluted, created a vacancy and therefore a negative peak, permitting detection at the picogram level. An alternative to photo-excitation to produce fluorescence is chemical excitation by a technique called "chemiluminescence".[85]

1.6   Chromatographic theory

In chromatography, a material to be analyzed is propelled along a pathway to a detector. Usually, the pathway is a column, but it can also be a plate, such as is used in thin layer chromatography, or a slab, such as is used in gel electrophoresis. The means of propelling the analyte is generally solvent flow, but it can also be the influence of an electric field, the chemical potential of a pH gradient, or the force field of a centrifuge. Two different analytes traveling at different rates are thereby separated. For the purposes of this section, it will be assumed that the analyte is separated in a column by means of solvent flow.

Gel permeation chromatography, described in Chapter 6, and capillary electrophoresis, described in Chapter 7, are examples of chromatographies in which there may be negligible interaction between the analyte and the stationary phase. In most chromatographies, however, there is an appreciable interaction between analyte and stationary phase, and the remainder of the chapter will focus on these *sorptive* chromatographies.[86] Some analyte distributes itself into the mobile phases, and some into the stationary phase. The interaction can be conceptualized as partition, as occurs between the two phases of a separatory funnel, or as adsorption, in which the stationary phase has discrete sites of interaction. As long as the interaction is relatively weak, elution can be accomplished under isocratic conditions. If, however, the interaction is strong, gradient elution may be required.

The practices of isocratic and gradient sorptive chromatography are very different. Isocratic chromatography tends to be very sensitive to the details of mobile phase preparation, temperature, pump speed, and sample composition. Gradient chromatography is usually more tolerant of small variations in these factors but may be extremely sensitive to column history, equilibration time, and gradient preparation.

The theoretical parameters of isocratic chromatography are often described using the *plate model*. One can imagine the analyte to be distributed

Table 3 Parameters in Separation

Symbol	Name	Formula	Note
t_o	Void time		Retention time of an unretained solute
t_n	Retention time		Retention time of analyte n
t'_n	Reduced retention time	$t_n - t_o$	
k'_n	Capacity factor	t'_n/t_o	
w_n	Peak width at half height		
N	Theoretical plates	$5.54(t_n/w_n)^2$	Other formulae are also used
N_{eff}	Effective plates	$N[k'_n/(1 + k'_n)]^2$	Useful for comparing columns
H	Plate height	L/N	L is column length
R_{n-m}	Resolution (between components n and m)	$2(t_n - t_m)/(w_n + w_m)$	Component n eluting after component m
α_{n-m}	Separation factor, selectivity	k'_n/k'_m	Component n eluting after component m

between the stationary phase and the flowing mobile phase. A given molecule is transferred into the stationary phase at a given point along the column. After a time, it is returned to the mobile phase and swept down the column a little way to a new site where the process is repeated. The locations at which the analyte is transferred into the stationary phase are conceptualized as *plates*, and the distance between them is conceptualized as the *plate height*.

The parameters of isocratic column performance are listed in Table 3. A substance n elutes at a characteristic retention time, t_n. An unretained solute has retention time t_o. Neglecting the solvent volume in the injector and tubing, the void volume in the column $V_o = Ft_o$, where F is the flow rate. As the solute plug traverses the column, it is broadened by turbulent and diffusional processes. Resistance to transfer between mobile phases and stationary phase also contributes to broadening. The perfection of column packing is estimated using the plate count, N, which is proportional to the square of the retention time divided by the peak width. Chromatographers variously use the peak width at half-height, at peak inflection points, or at the baseline. As long the procedure used to measure plates is consistent, any of these measures will serve to monitor column performance over time, which is the principal objective of making plate counts.

The quality of a separation is measured by the resolution. The resolution of two peaks, n and m, can be estimated as

$$R_{n-m} = 2(t_n - t_m)/(w_n + w_m)$$

where w_n and w_m are peak widths at half height. Under the assumption that the peaks are of equal width, this simplifies to

$$R_{n-m} = (t_n - t_m)/4\sigma$$

where $\sigma = 0.425w$. In Chapter 4, the discussion of isocratic column performance is extended to dispersive processes and peak moment theory. For the purposes of understanding basic theory of gradient separations, however, the only parameter needed is the capacity factor, $k' = (t_n - t_o)/t_o$. This measures how strongly an analyte is retained. The retention time depends greatly on mobile phase composition. In a gradient separation, the proportion of the strong solvent is continually increased. At the beginning of the separation, the composition of the mobile phase is such that the retention time of the analyte would be very long. As the composition of the mobile phase is programmed to include more strong solvent, the isocratic retention time of the analyte shortens. It is possible to understand gradient elution in terms of isocratic elution, using linear solvent strength theory.

The basis for linear solvent strength theory is the assertion that the logarithm of the capacity factor is linearly related to the solvent strength. This is expressed in the Equation:[87-90]

$$\log (k') = \log (k_w) - S\varphi$$

where k' is the capacity factor at a given solvent composition, k_w is the capacity factor when the strong solvent is absent, S is a constant for a given solvent, and φ is the solvent composition expressed as the volume fraction of organic solvent. This has been applied to IEC as well as to RPLC, although with less satisfactory results.[91] The implication is that one may measure the retention of peaks at any two values of solvent composition and then calculate the retention at any value of the solvent composition. If the solvent composition is varied linearly with time, such that $\varphi = \varphi_o + t \, \Delta\varphi/t_G$, where φ_o is the initial mobile phase composition, $\Delta\varphi$ is the change in mobile phase composition, t is time, and t_G is the gradient time, then the gradient steepness parameter, b, can be defined as $b = S\Delta\varphi t_o/t_G$. Ignoring non-ideal effects such as size exclusion, the retention time is

$$t_R = t_o + (t_o/b)\log(2.3k_o b+1)$$

where k_o is the value of k' extrapolated to initial solvent conditions. If k_o is large, then this equation can be solved by taking two or more isocratic measurements of retention time. Using the median value of k', where $\overline{k'} = 1/(1.15b)$, the value of b can be calculated by varying the flow rate or the gradient rate.[87,88]

The practical consequence of linear solvent strength theory is that, in principle, it is possible to estimate the retention times of all peaks at any

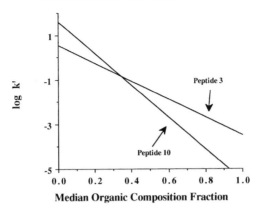

Figure 4 Crossover of peak retention times as a function of gradient rate in the separation of peptides. (Previously unpublished data are drawn from Swadesh, J. K., Tryptic fingerprinting on a poly(styrene-divinyl benzene) reversed phase column, *J. Chromatogr.*, 512, 3215, 1990.[92])

gradient rate or flow rate from just two chromatographic runs. Therefore, it is possible to optimize the resolution between any two peaks from just two measurements. In practice, it is perhaps most common to perform three or more gradient rate measurements while keeping the flow rate fixed.

In the analysis of the twelve peptides released from horse cytochrome c by the action of trypsin,[92] the gradient rate was varied. Then, log $\overline{k'}$ of two peptides was graphed against the median organic modifier composition $\overline{\varphi}$ to solve for the slope and intercept, as shown in Figure 4. Because of differences in slope, there was a crossover in retention when $\overline{\varphi}$ is about 0.3. At any other median solvent composition, those two peptides are better resolved. A resolution map can be constructed for all of the components, and a gradient rate and flow rate selected. A more extended discussion of this gradient separation, including the effects of temperature, is presented in Chapter 4.

1.7   Laboratory operations

1.7.1   Overview of laboratory operations

The conventional analytical process is comprised of sampling → sample preparation → analysis → calculation → approval of results → report → decision.[93] The introduction of productivity measurements to focus attention on continuous improvement and improving the reliability of assays to eliminate re-analysis can aid in re-engineering the process for greater efficiency.[93] Automation is another important aspect of improving efficiency.[94] The rate-limiting steps in many industrial laboratories, however, may precede or follow the conventional analytical process described above. Instead, assay development time, the discovery of assay limitations as industrial processes change, and the communication of results may be critical to the pace of industrial development. The present chapter, therefore, reviews in depth how analysis is planned and performed in the industrial laboratory.

The analytical chemist working on chromatographic procedures should consider

- Assay selection and design
- Sampling and sample handling
- Chromatographic optimization
- Experimental procedures to verify the validity of the assay
- Field testing of the assay
- Interpretation of results
- Presentation of findings

1.7.2   Assay selection

The professional analyst begins the design of a chromatographic assay by determining the *stringency* of the assay, i.e., the limit of tolerable error. In effect, the analyst must assess the tolerance of the overall project to errors in accuracy and precision. For example, in assaying for the concentration of a formulated drug with dose-dependent toxicity, considerable risks are incurred by error. Limits of a few percent error might be tolerable. By contrast, the risks of error in assaying a processed food for an off-flavor component are substantially less; more generous limits of error might be regarded as tolerable. The assay should be classified as a *concentration assay* (to determine the concentration of a major component), a *purity assay* (to determine the concentration of the major component relative to all components), or a *minor component assay* (to determine precisely the concentration or relative concentration of a minor, presumably undesirable, component). The goals of the assay should be explicitly established in these terms, since each kind of assay requires different levels of stringency in sample handling and preparation and in system performance.

The limit of tolerable error is generally smallest in a minor component assay. It will have been determined that a particular minor component must be at or below a threshold concentration for the product to be usable. Therefore, the decision to accept or reject an entire production batch may depend on the analytical result. Typical batches may contain the contaminant at a concentration very similar to the specification limit. In a minor component assay, the major component may be overloaded and out of the proper range of detection of the assay. Even so, the minor component may be at such low levels that assay noise interferes.

A purity assay is the analysis of multiple, often unknown minor components. Unless it has been determined separately that one or more of the components is deleterious to the product above a particular concentration, only an overall estimate of the total concentration of the minor components relative to the major component is required. It is rarely necessary to have a very precise determination of the concentration of the individual minor components, since errors in small values are still small. For example, a 10% uncertainty in the overall measurement of minor components in a material of purity 99.5% corresponds to an uncertainty range of 99.45 to 99.55%, which is within rounding error. Also, with the measurement of multiple components, there may be partial cancellation of overestimates of one component by underestimates of another.

Concentration assays are often the least demanding, since usually the component to be measured is abundant and minor components scarce. Even if resolution is poor or there is detector noise, accurate measurements of concentration can still be obtained. In concentration assays, the principal requirements are stringency in the precision of sample dilution and measurement of column losses of the major component. Detector calibration, another important issue in concentration assays, has been discussed above.

Having determined the limits of tolerable error, the complexity of the sample matrix should be assessed. The matrix for materials extracted from biological tissues or fluids is one of the most complex, while the matrix of a pure compound in solution is the least complex.

Assessing the complexity of the sample matrix is essential to determining the selectivity required for the assay. Under the best conditions, standard chromatographic methods are capable of separating mixtures containing no more than about 50 components. Peak overlap or co-elution are frequent occurrences. Changes or improvements in selectivity can be achieved by sample processing or by the use of specialized detection. Primary and secondary amines, for example, are easily quantitated with or without chromatographic separation by a number of derivatization chemistries discussed in Chapter 4. Aromatic compounds are readily quantitated in the presence of non-aromatic compounds with or without chromatographic separation by the selective detection afforded by UV.

The limitations of derivatization are that derivatization reactions only approach completion and never attain it, that the conditions of derivatization sometimes cause degradation, and that even very similar compounds are derivatized to different extents. Use of derivatization, therefore, requires a careful study of recovery of known components. A limitation common to the use of specialized detectors and derivatization is the response factor problem. The detector responds to different compounds to a greater or lesser extent. Measurement of correction factors to account for this is one of the most time-consuming aspects of analysis.

Once it has been determined that neither chemical processing nor selective detection are sufficient to assay the unfractionated sample within tolerable error and that separation will be required, several preliminary proposals for an assay should be considered. For example, if one wanted to conduct amino acid analysis, HCl and trifluoroacetic acid might be considered for hydrolysis; phenol, tryptamine, and sodium sulfite might be considered as antioxidants; IEC with post-column derivatization or RPLC with pre-column derivatization might be considered as separation methods; and fluorescence or UV-VIS spectrophotometry might be considered as detection methods. By establishing alternatives early in the process of assay development, the analyst generates backups in case method development shows that the primary assay choice is not sufficient for present or future needs.

Consideration should be given to future project needs and to other project requirements. Often, in the early stages of a project, sample is scarce, so sample-sparing assays are required. In later stages of development, greater sample consumption might be tolerated if greater precision can be obtained. A typical project requirement is that relatively complex assays devised in a research department must be simplified for quality control purposes. By designing an assay with future project needs in mind, one can avoid project delays.

Assessing the resources available for method development should also be done before beginning a project. The resources available include not only HPLCs, detectors, and columns, but also tools for sample preparation, data capture and analysis software, trained analysts, and especially samples representative of the ultimate analyte matrix. Also, it should be considered whether a fast, secondary method of analysis can be used to optimize sample preparation steps. Often, a simple colorimetric or fluorimetric assay, without separation, can be used for this purpose. A preliminary estimate of the required assay throughput will help to guide selection of methods.

1.7.3   Assay design

The design of an assay is, in large measure, prospective quality assurance. The factors that are likely to affect the results of the assay must be defined and controlled to the greatest extent possible. Once the general outlines of an assay have been established, key features should be examined, including optimization of sample preparation, sample stability, choice of standards, assay range, assay repeatability, optimization of separation, and optimization of detection.

A general feature of optimum sample preparation is that maximum recovery of the analyte is observed. Consider a graph of recovery vs. variation in one experimental condition. Figure 5 shows such a graph, with temperature as the experimental variable. The curve exhibits a maximum and a decline on either side of the maximum. The assay will be most reproducible at the point of zero slope, i.e., at the maximum recovery, because small variations in conditions will not affect the result. In hydrolysis of a protein to its constituent amino acids, for example, it will be found that at very high temperatures or long hydrolysis times, degradation of the product amino acids occurs, while at low temperatures or short hydrolysis times, the protein

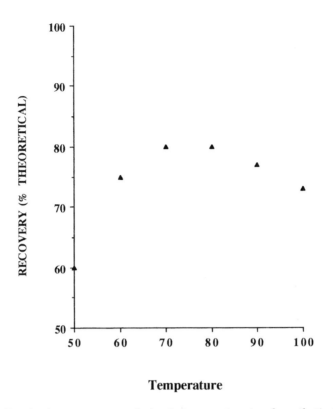

Figure 5 Graph of recovery vs. variation in temperature in a hypothetical chemical reaction.

is incompletely hydrolyzed. At the optimum temperature and time, the yield of amino acids is maximal.

The principle that an assay is most rugged when response changes slowly as a function of experimental variables can be applied to all aspects of assay optimization. There are, however, many response functions with multiple optima, and there may be cases in which the use of a local response minimum gives a more rugged assay. For example, using detection at an absorbance peak maximum is usually preferable to using detection at a wavelength at which the absorbance changes rapidly with wavelength. However, it may also be possible to use the local minimum of absorbance. Selecting the mobile phase composition such that retention times vary minimally with changes in composition gives ruggedness to an HPLC separation.

Standards are used for a number of purposes. An *external standard* contains a mixture of substances typically observed in the sample. Knowledge of the concentration of the standard substances allows calibration of the detector to compensate for run-to-run or day-to-day variability. External standards should not be subjected to hydrolysis or other sample processing steps, except as necessary for detection, since this would add other sources of variability to the result. An *internal standard* is added to a sample and to the external standard to correct for dilution. Internal standards should be chosen to exhibit no variation in recovery due to sample processing. The term *reference standard* is sometimes used to refer to a sample that has been characterized by a reputable vendor or institution. However, in industrial terms, a reference standard is a single lot of a material, representative of a production process, used to track the performance of an analytical assay over time. Unless material has been maintained as an integral unit or, at a minimum, subsamples stored under identical conditions, a sample cannot be regarded as a reference standard. By including the reference standard as a sample in every batch of sample analyses, one can monitor variation in recovery due to experimental conditions.

Also, a chromatographic profile or *fingerprint* of trace unknowns can be established and monitored, so that if product performance unexpectedly changes, there will be a starting point for troubleshooting. The effects of experimental variables on sample recoveries should be measured directly by controlled variation of an experimental factor, using the reference standard, or suitable external standards, or spiked addition of an external standard to the reference standard. A detailed example of the use of internal and external standards is presented in Chapter 4.

Considerable attention should be paid to obtaining samples truly representative of the production process early in method development. Production processes generate mixtures that are far more complex and variable than may be generally realized; therefore, a separation developed using an early production sample may prove to be inadequate for a later sample. Minor peaks observed in the chromatogram, whether known or unknown, serve as a record of the consistency of the process and can be used to monitor process changes. Sometimes a particular peak can be associated with a desirable or undesirable property of the product and used for controlled process optimization.

Sample stability and sample derivative stability are important questions that should to be resolved early in method development, particularly if derivatization is to be used. Some derivatives, such as *o*-phthalaldehyde/2-mercaptoethanol derivatives of amines are highly unstable in the absence of stabilizers and must be assayed immediately on preparation. Automatic sample preparation, which is available only on certain autoinjectors, is extremely helpful in controlling the reaction time and reaction-to-injection time. Some analytes, such as proteins, undergo associative or conformational equilibria on-column. Usually pH control or temperature control can be used to optimize separations. Samples of biological origin are often susceptible to microbial contamination and must be handled by clean or sterile techniques. By the simple experiment of injecting a freshly generated sample and an aged sample, one can avoid many of the pitfalls encountered in method development.

1.7.4 Sampling

Samples in solution are usually homogeneous, so samples drawn from different locations in the container are identical; however, sometimes even seemingly homogeneous solutions present sampling problems. When the solute is extremely dilute, when the sample size is extremely small, and when the requirements of sample-to-sample variation are stringent, solution inhomogeneity may be a problem. For solutions of picomolar concentration or greater and sample sizes of nanoliter or larger, inhomogeneity of sampling would not expected to be a serious problem.[95]

Supposedly identical samples in different containers often turn out not to be identical, and they should never be assumed to be so unless handling, preparation, storage, and the containers themselves are rigorously identical. Some samples, such as polymers, are extremely slow to dissolve and may generate artifactual chromatograms unless dissolution is allowed to proceed to completion. Solid samples, suspensions, and solutions capable of phase separation are intrinsically inhomogeneous.[96] Samples obtained from extremely large containers, such as reactors, also may be inhomogeneous. For inhomogeneous samples, a strategy for sampling should be devised. Samples can be drawn from different locations within the container and tested for equivalence.

1.7.5 Sample handling

For assays of stable materials with wide ranges of tolerable error, sample handling is of little concern. For assays of labile materials, especially assays for purity or for minor components, controlled sample handling procedures need to be established. There are three potential ways in which a sample may become contaminated, namely by the sampling tools, sample containers, and degradation on storage.

For solids, contamination by sampling tools or containers is rare. Analysis of trace metals, however, is one case in which special precautions must be taken. Plastic spatulas and plastic or acid-washed glass labware are often required for trace metals analysis. For liquids, contamination by sampling tools and containers occurs frequently. One common sampling tool for liquids is the automatic pipet. Liquid or vapor aspirated into the pipet piston during one sampling operation may contaminate a subsequent sampling. Before sampling from a bulk sample, cleaning the piston may prevent sample contamination. Sample containers are another common source of contamination of samples. Glass sample vials can be contaminated with dust and oil. The liners of caps are often glued in with resins that are soluble in organic vapors. The liners themselves may degrade when stored with corrosive solutions. Polyethylene and polypropylene containers are inert to most solvents.

Once a sample has been drawn, it should be provided with a unique label, noting the date at which it was separated from the bulk. Since samples may degrade with time and handling, the properties of one sample may differ from those drawn at a different time. Sample stability is a serious concern in pharmaceutics, since trace degradation can lead to batch rejection. Stability becomes a critical concern in biopharmaceutics, because biological drug substances are often prone to bacterial contamination. It cannot be assumed that a refrigerated or frozen sample is immune to degradation. Freezing and thawing, in particular, lead to denaturation and aggregation of biological materials. Phase separation can occur on refrigerated sample storage. Long-term storage of biological samples is often performed by freezing by stirring in liquid nitrogen and subsequently storing at –80°C. Thawing is performed by incubation with tepid water and agitation. These procedures do not guarantee the recovery of biological activity but often prevent aggregation.

Finally, developing a good labeling system is important to sample handling, particularly because samples may require long-term storage, sometimes under conditions of severe cold. In industrial production, samples are often identified by batch number. When a sample is drawn from the bulk, a new number should be assigned. Many organizations use sequential six-digit numbers to identify samples. Alternatively, by labeling the sample by notebook number and page number, a unique label can be established.

1.7.6   Chromatographic optimization

Some kinds of chromatography require relatively little optimization. In gel permeation chromatography, for example, once the pore size of the support and number of columns is selected, it is only rarely necessary to examine in depth factors such as solvent composition, temperature, and flow rate. Optimization of affinity chromatography is similarly straightforward. In RPLC or IEC, however, retention is a complex and sensitive function of mobile phase composition; column type, efficiency, and length; flow rate; gradient rate; and temperature.

Since retention in RPLC or IEC is a function of so many variables, it is advisable to structure the process of method development. The principal goal of optimization is to devise chromatographic conditions such that the peaks are roughly evenly distributed through the chromatogram and no peaks overlap. In the first phase of optimization, pH, temperature, gradient rate, and flow rate are fixed. Several columns may be surveyed under these standard conditions. In the later stages of optimization, linear solvent strength theory is useful to obtain precise solvent compositions, gradient rates, and flow rates.

In surveying a separation, a gradient from 100% weak solvent to 100% strong solvent is performed. In RPLC, the gradient might be from 100% water to 100% methanol or acetonitrile, while in IEC the gradient might be from water to 1 M NaCl. Some experimentalists prefer to begin the gradient from slightly less than 100% weak solvent, since mixing artifacts are sometimes observed. The solvent strength (i.e., the concentration of the strong solvent) at which each peak elutes is noted. Ideally, all components to be separated will elute as resolved peaks in a fairly narrow region of solvent strength in the range of 10 to 90% strong solvent. If these conditions obtain, isocratic separation may be possible.

If most peaks elute near the injection front or near the end of the gradient, it may be necessary to adjust the pH of the mobile phase before proceeding. At low pH, carboxylate anions will protonate to the neutral form, and amines will protonate to the cationic form. At high pH, carboxyl groups will deprotonate to the anionic form, and amines will deprotonate to the neutral form. At very high pH, hydroxyl groups may also deprotonate, a feature that is used in ion exchange separations of sugars as described in Chapter 5. By comparing separations performed with the mobile phase buffered at pH 2, pH 7, and (if permitted by the column packing) at pH 12, it is often easy to shift retention times that are too early or too late into the optimal region. In reversed phase chromatography, if components are still unresolved from the solvent front, it may be possible to include mobile phase additives to ion pair with the analyte, as is described in Chapter 4. Mobile phase additives and pH adjustment may also be used to improve peak shape.

Once the resolution has been optimized as a function of gradient rate, one can continue to fine-tune the separation, raising flow rate and temperature. In a study of temperature and flowrate variation on the separation of the tryptic peptides from rabbit cytochrome c, column performance doubled while analysis time was reduced by almost half using this strategy.[97] Commercially available software has been developed to aid in optimization. As a final note, in an industrial laboratory optimization is not completed until a separation has been shown to be rugged. It is a common experience to optimize a separation on one column, only to find that separation fails on a second column of identical type. Reproducibility and rigorous quality control in column manufacture remains a goal to be attained.

1.7.7   Data capture and analysis

The output of a detector is an electrical signal, the intensity of which is the *analog* form of the signal. For data processing and analysis, this must be converted into *digital* form. Chromatographic data systems and the principal issues of peak analysis, including peak shape assumptions, derivative techniques, signal filtering methods, and peak cutting techniques, have been reviewed.[98,99] Chapman presented a refreshing perspective on the mystical awe with which computer systems are reverenced; why it is not always a good idea to place such trust in what, after all, are fallible instruments; and the history of Food and Drug Administration (FDA) guidelines on software validation.[100] As software is incorporated ever more widely into HPLC controllers, detectors, data capture units, and data analysis systems, the chromatographer needs to be alert to the many, often insidious failures that can occur. A recent description of software validation in an industrial setting is instructive.[101] A Dionex signal generator, traceable to National Institute of Standards and Technology voltage patterns,[102] was used as a reference standard for digitally generated pseudo-chromatograms for analysis on a Hewlett-Packard® HP® 3350 LAS computer network. Peak deconvolution is becoming increasingly important, as sample complexity outstrips even the impressive peak fraction capacity of modern HPLC.[103,104]

1.7.8   Statistical analysis

Since most industrial laboratories operate under strict time constraints, it is a temptation to forego experimental design in favor of generating results quickly. It is almost always more cost-effective to re-analyze existing data than to generate new data. For re-analysis to be feasible, however, experimental conditions must be carefully controlled. Proper experimental design implies designing experiments with statistical analysis in mind. The first step in the process is to make a list of the possible variables, including suspected interactions between variables. This list constitutes an outline of the experimental design. From a list of variables and an estimate of assay precision, one can estimate the number of independent and the number of dependent replicates required for statistical analysis. Independent replicates are samples prepared, handled, and analyzed separately from one another. If a detector shows day-to-day variation in response, for example, one should plan to assay all samples on one day or to assay equal numbers of samples from each group on different days.

A central concept of statistical analysis is variance,[105] which is simply the average squared difference of deviations from the mean, or the square of the standard deviation. Since the analyst can only take a limited number n of samples, the variance is estimated as the squared difference of deviations from the mean, divided by n − 1. Analysis of variance asks the question whether groups of samples are drawn from the same overall population or from different populations.[105] The simplest example of analysis of variance is the F-test (and the closely related t-test) in which one takes the ratio of two variances and compares the result with tabular values to decide whether it is probable that the two samples came from the same population. Linear regression is also a form of analysis of variance, since one is asking the question whether the variance around the mean is equivalent to the variance around the least squares fit.

Take, as an example of an optimization problem, a derivatization reaction A + B → C in which it is desired to maximize the yield. The rate of formation of C might be dependent on pH, temperature, the concentrations of A and B, and the time allowed for the reaction to take place. A reasonable first step would be to perform an analysis of the reaction rate at fixed pH (say, pH 7), concentration, and temperature (say, 50°C) to estimate the time required for the reaction to reach completion. It is useful to follow the reaction well

Table 4   Example of ANOVA

	Temperature	
	50°C	70°C
pH 7	Group 1	Group 2
pH 9	Group 3	Group 4

beyond completion to see if decomposition of the product can occur. Also, by using a few replicates at various time points, it may be possible to estimate the standard deviation associated with the analysis.

Given an estimate of the time to completion and precision of the analysis, one can temporarily eliminate time as a variable and construct an analysis of variance (ANOVA) to examine the effects of pH and temperature. A simple ANOVA would consist of four groups, with several replicates in each group, as shown in Table 4.

Notice that the results from one of the time points of the kinetic analysis may be used for the Group 1 values, assuming that day-to-day and sample-to-sample variability is minimal and that the time used for the remaining groups is the same as that of the Group 1 sample. From the ANOVA will be obtained not only information about the best temperature and pH, but also an indication of whether pH and temperature interact with one another. Further ANOVA or kinetic analysis can help to pinpoint the optimal yield as a function of all of the variables.

A full investigation of a five-factor problem such as the reaction above would probably require greater resources than most industrial analytical laboratories would normally care to expend. By arbitrarily defining a limited number of values for the variables and carefully controlling the sources of variance, the problem can be analyzed according to ANOVA of lower dimensionality. There are many other optimization strategies. The advantage of designing the experiment using an ANOVA model is that the results are structured. All of the known variables are controlled, and many groups are directly comparable to other groups.

1.7.9   Presentation of results

Preparing reports and presentations is a major function of an industrial analyst. Unlike academia, in which graphics are used primarily as an adjunct to finished publications, in the industrial environment, presentations of inter-mediate data are frequently used to guide decision-making on the fly. Since those making the decisions are rarely analytical experts, graphic presentation is an art of its own. In effect, the analyst needs to construct a picture so simple that even a non-expert can understand the implications. Some of the formats that are extremely useful include chromatographic overlays for sample comparison or comparison of sample with blank; statistical quality control charts and trend analysis, demonstrating the change of a value over time; and the substitution of tabular data with graphic presentations.

Of the various methods of data presentation, the one with which starting analysts may be least familiar is trend analysis and statistical quality control. In an industrial environment, analysis is often centered around the production of batches of material. The properties of those batches may change over time due to random effects or to subtle changes in the production process. In either case, the quality of the product may change. Analysis is used to track the change in the properties of batches over time. Industrial analytical methods, therefore, need to be extremely rugged. Millions of dollars may depend on the analyst's judgment as to batch equivalence.

The statistical control chart is a graph of a value measured for samples over time.[96,106] One usually graphs the property measured, but one can also graph the standard deviation of replicates within a run. By analyzing a reference sample each time that a series of unknown samples are run, control of the measurement process can be demonstrated. It is wise to bracket groups of unknowns with reference samples so that if there is a failure of the method, analytical control can be demonstrated for the groups that are bracketed. Given a sufficient number of measurements of the control sample, an estimate of the true mean value and standard deviation of the measurement around that mean can be obtained. Of the values subsequently obtained for the standard, 95% should lie within no more 2 standard deviations of the mean, and any values outside of 3 standard deviations should be regarded as clearly indicative of method failure.

One can apply a similar approach to samples drawn from a process over time to determine whether a process is in control (stable) or out of control (unstable). For both kinds of control chart, it may be desirable to obtain estimates of the mean and standard deviation over a range of concentrations. The precision of an HPLC method is frequently lower at concentrations much higher or lower than the midrange of measurement. The act of drawing the control chart often helps to identify variability in the method and, given that variability in the method is less than that of the process, the control chart can help to identify variability in the process. Trends can be observed as sequences of points above or below the mean, as a non-zero slope of the least squares fit of the mean vs. batch number, or by means of autocorrelation.[106]

1.8   The role of laboratory management

1.8.1   Overview

Analytics plays a critical and sometimes underestimated role in the pace and direction of industrial projects. As laboratories are increasingly automated, the management skills of scientists are increasingly relied upon to coordinate a complex and dynamic workplace.[107] There are many aspects of project management which are not within the discretion of an analytical manager. The exigencies of project timing and institutional culture, as well as the nature and volume of the sample stream, influence the flow and methodology of laboratory work. The manager of the analytical laboratory, however, can provide leadership essential to the success of an enterprise.

There are many kinds of analytical laboratories, ranging from clinical laboratories handling vast throughput by invariant methods to research support facilities with widely variant sample streams. Factors that are taken into consideration for management style include minimization of cost and turnaround time and maximization of throughput, precision, assay repeatability over time, flexibility, and sensitivity of detection. It is impossible to optimize all of these desirable characteristics simultaneously. To achieve certain goals, others will be sacrificed. In addition to instruments, columns, detectors, and general scientific procedure, wise scientific management is an essential element of the production of quality industrial data.

One leading industrial scientist said about the process of drug development:[108]

> "The 'managers' must ... be strong leaders, accomplished and respected scientists themselves, who must exhibit broad vision, long-term perspective, trust in other professionals, and the ability to inspire others... The public and the ethical industry are best served by decisions based on good science, adherence to high standards, and independent, expert review... If the industry starts with high quality science, effective analyses, and honest, responsive presentations, its regulatory problems will be few."

This is, more generally, a good overview of what is required in analytical management. While the primary emphasis must be on good science — high technical skill, careful experimental design, application of efficient analytical tools, thorough statistical analysis, and effective management of data — the "broad vision, long-term perspective, trust in other professionals, and the ability to inspire others" listed by Cuatresecas above are skills essential to the laboratory of the future.[107] Furthermore, these leadership skills are definable and learnable. Noting that leadership in the laboratory is directly related to bottom-line financial performance, some institutions are actively adopting these principles, while others lag. This section will expand on the concepts implicit in Cuatresecas' outline, beginning with "broad vision", progressing through "good science", and closing with some practical suggestions on budgeting and equipment acquisition.

1.8.2 The laboratory vision

Given the same goals and resources, two equally qualified individuals often end up with very different degrees of success. To some extent, the different outcomes are attributable to the relative levels of motivation. The personal outlook that defines motivation is what is meant by "vision". Leaders are often required to take risks and to sacrifice, and to persuade others to do likewise. Vision is the ability to look beyond the present and calculate the future, permitting one to see that the necessary risks and sacrifice are worthwhile. Effective leaders are able to communicate this vision to others, motivating the entire team.

Defining a vision in an industrial analytical laboratory is difficult because of the pressures to operate within the short-term time horizon often associated with product development. If the laboratory environment requires extremely high sample throughput, it is tempting to abandon efforts to develop the staff. Given the pace of change in the modern laboratory, however, it is both more humane and more productive to automate repetitive tasks and to develop a work environment of high learning.[107] The managerial vision increasingly will need to focus on the human needs of the work force in order to obtain a flexible and productive laboratory and keep the staff engaged and learning.

Scientific workers tend to be better educated, more independent, and more highly motivated than nontechnical workers. Managers of technical laboratories, who often come from a career in science, need to recognize that management of technical employees is a new profession, distinct from the conduct of technical work. All workers, technical and nontechnical, value the following in descending order of importance:[109]

- Interesting work
- Full appreciation of work done
- The feeling of being in on things
- Job security
- High wages
- Promotion
- Good working conditions
- Personal loyalty of the supervisor
- Tactful discipline, and
- Help with personal problems

Some practical means to accomplish these goals are

- Define achievement by projects advanced or questions answered, rather than by the number of samples run. Let the analyst know what the results of an analysis will mean to a specific project. Ensure that each analyst receives public credit for work well done.
- Make a list of basic skills useful in analytical chemistry, such as public speaking, writing skills, statistics, electronics, and computer science. Determine where each analyst is strong or weak, and make a commitment to further training. Reserve several hours each week for special training in basic skills. Presentation by junior analysts of a published paper to a group of peers is one useful format.

- Work with those who submit samples to limit submissions to those likely to yield useful information. In a typical sample queue, some are outside of the parameters required for a standardized assay, having a concentration too low for the precision required or containing matrix components incompatible with the assay. Others may be low on the list of project priorities. Analytical morale gains if every sample run leads to results that are valued.
- Periodically — weekly, if possible — review progress and project future work. Junior analysts will usually tolerate periods of overload as long as the periods are anticipated, finite, and clearly related to defined project goals. Recognition of extra effort through thanks may be sufficient if the periods of overtime are brief. If sustained periods of overtime are required, compensatory leave or financial compensation are essential to maintain good will and limit turnover.
- Plan ahead for those brief periods of slack time, and use them productively. In many laboratories, instrument maintenance, operator training, and staff development are in chronic deficit. During slack time, those deficits can be addressed.
- Establish a program to monitor the publications, patents, and announcements of competitors and communicate these to the analytical group. If there is a genuine commitment in the organization to "be the best", a good first step is to recognize the very skilled competition to be faced.
- If possible, collaborate with research departments of local universities on publishable work, even if that requires working outside of business hours. Publication builds training and morale. It is also a way of ensuring that junior analysts will have proof of their accomplishments in the event of corporate reorganization, restructuring, or downsizing.
- Rotate the work that is least well loved. If possible, include yourself in the rotation. There are benefits to taking part in laboratory work, even if only for a few days per year. It is a form of "managing by walking around", helping a manager stay current with laboratory operations. Also, it emphasizes the notion that no work that advances projects is unworthy. Further, it helps the manager maintain a sense of how long it takes to accomplish tasks. Finally, by taking part in laboratory operations, the manager gains a sense of which tasks are repetitive and can with advantage be automated.

1.8.3 Serving the public

An analyst will handle many tens of thousands of samples over a career. The speed and complexity of sample analysis and reporting is rising and will rise further still. The quality and reproducibility demanded of sample analysis is also rising. Analyses may become numbingly repetitive and results may not receive much appreciation. In analytics, the rewards for work well done are rarely proportionate to the penalties for work done late or

THE SCIENCE MOB
Fraud, complacency, and secrecy
in the scientific establishment

Figure 6   "The Science Mob". (Reproduced with permission from *The New Republic* and Mr. Robert Grossman.[110])

work discovered to be erroneous in a regulatory audit. The low rate of fraud or error in science is a powerful testament to the fundamental dedication and decency of the overwhelming majority of scientists.

Still, some members of the lay public are clearly skeptical of science and scientists, as indicated by the view of the profession presented on the cover of a national magazine in Figure 6.[110] Although we may be assured that scientists as a group are no more deficient in morals than, say, cartoonists, it is important to understand the basis of public perceptions and to examine the practical implications for laboratory management. Industrial products are very widely distributed and many are used intimately. There are real consequences that can arise from a serious error in manufacture or design, as illustrated by the case of L-tryptophan.

L-tryptophan is a naturally-produced, natural amino acid used as a dietary supplement. Samples from one manufacturer were found to be contaminated by trace amounts of another naturally occurring compound — 1,1'-ethylidene *bis*-[L-tryptophan]. The latter compound appears to have been responsible for causing a number of deaths and additional cases of chronic eosinophilia-myalgia,[111] some of which might have been prevented had there been a fuller portfolio of analytical data. The public record indicates no dereliction of science in this episode. Rather, the producer's apparent strong focus on improving the fermentation yield overshadowed full characterization of the minor components or the process of their removal.

Industrial scientists, therefore, hold a very high public trust. There is a need to go beyond the requirements of scientific canons and demonstrate to the lay public that its trust has been upheld. A number of industrial firms have recognized and responded to this need. Wilder, of Eastman Chemical, for example, describes a proactive pollution prevention policy developed in partnership with the Environmental Protection Agency.[112] This work describes the key role of analytics in helping chemical manufacturers take leadership roles in redesigning processes for waste prevention. Since the analytical laboratory is a principal referee in the production process, there is always a risk that time pressures will tempt some members of the production team to "work the referee". An important function of the chain of command of the analytical laboratory, therefore, is establishing a clear operational policy to ensure that any such pressure does not reach the bench level:

- Define the turnaround time for each kind of routine sample analysis and ensure that sample submitters understand requirements for sample size, concentration, and matrix.
- Ensure that all reported results are reviewed prior to release, even if by a peer or a junior.
- Allow sufficient development time for new assays so that deadlines can be comfortably met in all but exceptional cases. When assays require new instrumentation or new personnel, build in a reasonable learning period.
- Make it clear that the laboratory is devoted to good science.

1.8.4 Trust

The level of trust within a working group is probably the single most important predictor of success.[107] Teamwork is predicated on trust, and trust is built not by words but by actions. Particularly essential in establishing trust are managerial actions. A recent case history provides an instructive counterexample.[113] The point which emerges from this case study is that when each is striving for his or her own benefit to the detriment of others, the security and well being of all are eroded.

Organizations thrive or fail on trust. Trust is a precious commodity impossible to quantitate on a bottom line, fragile, and, when damaged, difficult to repair. When there is trust in an organization, resources and information are shared freely and efficiently. Ironically, during the "efficiency drives" of the recent past, the employment security that once was the cornerstone of trust in scientific organizations was eliminated. The analytical manager must therefore establish trust within his operation despite the uncertain employment environment.

The essence of trust is a sense of predictability and the security that follows. The most predictable things are those which one directly knows and controls. A good analytical work environment is characterized by a leader who:

- Directs only to the degree required by an individual's own limits, involving each worker in as many decisions as possible and delegating to each worker considerable control.
- Has a genuine interest in the personal and professional wellbeing of each group member and therefore treats each as a unique individual.
- Respects each worker as a responsible, well intentioned adult, striving to do his or her best.
- Is ethical, fair, and consistent, communicating honestly and openly about events in the organization.
- Recognizes accomplishment both publicly and privately and encourages continual worker education and growth.
- Serves as a facilitator, rather than as an overseer.

and by a staff that:

- Has a sense of purpose and an understanding of its role in the corporate mission.
- Feels an internal commitment among staff members to each other and to the organization.
- Cooperates, rather than competes.
- Shares the tedious work.
- Voluntarily trains new arrivals.

Trust, which implies respect, is the fundamental source of high morale and productivity, loyalty, and enthusiasm. A common corporate aphorism is that "no one is irreplaceable". In fact, the loss of even the least efficient member of one's team to some degree scars the whole. While everyone may be replaceable, one must ask whether replacing people is worth the price in trust lost. There may be cases in which an employee proves incapable of performing at a minimal level of competence. In such cases, transfer or termination should be performed promptly for the benefit of the group, the corporation, and the employee. The recent trend toward solving corporate performance problems by dismissal of employees, however, is unwholesome.

1.8.5   Good science

A description of how large academic laboratories are administered, illustrating how different management styles can be successful, has appeared in the literature.[114] There is great diversity in laboratory operations in large laboratories, but clearly contact between those who plan the projects, those who have high skills in project operations, and those who execute the more routine work is regarded as important. Contact between the different groups allows review of the details of how work is done and redefinition of the goals of the work. Review is one of the most important features of good science.

Good science requires good records. Complete records require the identification of each sample with a unique number, traceable to written records describing donor and recipient, date of receipt, the sample composition, the notebook page describing sample preparation, sample hazards, assays requested, and the date after which the sample should be discarded. The sample label should include the name of one analyst responsible for storing and distributing the sample, safety information, the date of receipt, and the expiration date. The sample label should be resistant to conditions of storage. If a sample is transferred from one container to another, either in process of distribution or storage, a new number should be assigned and the aliquot should be regarded as a new sample. For such aliquots of original samples, labeling by initials, notebook number, page number, and sample number is usually sufficient. Some additional suggestions:

- Validate routine methods, i.e., define the conditions under which the assay results are meaningful.[115] To do that, one must select samples that are truly representative of the product stream. This may be a difficult task when the process is still under development and the product stream variable. The linearity of detector response should be defined over a range much broader than that expected to be encountered. Interference from the sample matrix and bias from analyte loss in preparation or separation often can be inferred from studies of linearity. Explicit detection or quantitation limits should be established. The precision (run-to-run repeatability) and accuracy (comparison with known standards) can be estimated with standards. Sample stability should be explored and storage conditions defined.
- Establish detailed written protocols. A suitable test of the quality of the protocol is that a naive operator should be able to analyze accurately blind samples with no further instruction.
- Perform periodic unannounced audits of operator performance by submitting blind samples of known composition. Comparison of observed vs. expected results may reveal bias due to differences in methodology or flaws in technique. The results of the tests should not be made the basis of punishment or reward.
- Establish control charts of instrumental performance. Day-to-day variations in pump flow rate, relative response factors, absolute response to a standard, column plate counts, and standard retention times or capacity factors are all useful monitors of the performance of a system. By requiring that operators maintain control charts, troubleshooting is made much easier. The maintenance of control charts should be limited to a few minutes per day.

- When appropriate, require that operators be capable of performing mechanical repairs of their own instruments. This is the simplest way of teaching instrumental function and greatly improves operator self-confidence. With the highly-computerized systems of today, it is likely that supplemental assistance from the manufacturer will be necessary for instrumental maintenance.

When the sample is injected onto the HPLC, the chromatographic run should also be supplied with a unique number. Since HPLC systems are becoming increasingly complex and components are swapped in and out, it may save time to make a drawing of the system complete with component serial numbers and insert a photocopy into the log book. This is called the system definition and should be updated routinely. To make it impossible to tamper with a photocopy (a requirement in laboratories subject to audit), the copy should be mounted permanently into the notebook with glue and tape. If standard methods are used, the parameters also should be recorded into the logbook. The analysis parameters used to calculate results should be recorded. Although there is presently an inclination to require that hard-copy printouts of all data be retained, the rise in the volume of data that a single analyst can generate is such that retaining hard-copy is becoming impractical. In a high-throughput laboratory, a single analyst may generate over 10,000 pages of output over a year. An audit may require selecting a single analysis, performed years ago, from the output of the entire division. Clearly, management of data is a critical component of managing an analytical laboratory.

Thorough validation of a method is a major step in ensuring the validity of results. By understanding the analytical issues that arise due to variability in the sample stream and errors in sample manipulation, one can build safeguards into the procedure. By defining the limits of detection and the expected variation in system performance, one sets limits on interpretation of data. It is probably better practice to report an observed value, even if it falls below the official level of detection, rather than listing the result as less than a threshold or as zero. Be forewarned that reporting a value below the limit of detection is an eternal source of perplexity to nonanalysts and may lead to attempts to draw conclusions from insignificant differences or demands to lower the official limit of detection.

Standards and blanks are the usual controls used in analytical HPLC. Standards are usually interspersed with samples to demonstrate system performance over the course of a batch run. The successful run of standards before beginning analysis demonstrates that the system is suitable to use. In this way, no samples are run until the system is working well. Typically, standards are used to calculate column plate heights, capacity factors, and relative response factors. If day-to-day variability has been established by validation, the chromatographic system can be demonstrated to be within established control limits. One characteristic of good science is that samples are run once only and all data is used in final reporting. In this way, an auditor can be more certain that there has been no tampering with results. Multilevel calibration is used as a proof of linearity. The inclusion of a reference sample, i.e., a well characterized sample representative of the process stream, is used as a control for sample handling. Results from the reference sample are often used as part of the acceptance criteria in determining whether a batch run was successful.

The use of internal standards is somewhat controversial.[115] There is agreement that an internal standard may be used as a correction for injection volume or to correct for pipetting errors. If an internal standard is included before sample hydrolysis or derivatization, it must be verified that the recovery of the internal standard peak is highly predictable. Ideally, the internal standard is unaffected by sample handling. Using an internal standard to correct for adsorptive or chemical losses is not generally approved, since the concentration of the standard may be altered by the conditions of sample preparation. An example of internal vs. external standards is given in Chapter 4.

Statistical analysis is a very important element of good science, and it is underused in most analytical laboratories. In designing methods, factor analysis is useful for identifying factors that are likely to affect results. Correlation coefficients, while useful, are often overused or misused. The density of data points selected for a regression should be evenly distributed, rather than clustered, to prove linearity by means of a correlation coefficient. It should be recognized that linearity is an idealization of detector performance. The upper and lower limits of linearity should be established. While many analysts are familiar with regression, relatively few use the related method, analysis of variance, to determine the actual significance level of a correlation. Analysis of variance can sometimes be used in correlating the sizes of minor peaks to process changes and is therefore useful in process validation and process monitoring. In the course of monitoring a production process over time, one develops the very large sample populations necessary to use statistics to explore unsuspected relationships. Propagation of error is another method of great utility in analytical chemistry. By understanding the contributions of individual variables to the uncertainty in a measured value, one can systematically improve the precision of an assay.

Maintenance of instrument logbooks and detector calibration are important issues. A logbook should include a full description of the instrument, including serial and model numbers. Each instrumental user should be instructed to look for specific signs of impending failure, such as changes in pump flow rate and the level of medium-term baseline noise. It should be explained that recording these in the logbook may help to pinpoint when an instrumental failure occurred, thereby limiting the amount of data that may need to be re-checked. Column serial numbers and day-to-day performance characteristics, such as plate count, should be logged for the same reason. Developing a test procedure for a detector can be somewhat involved. For the multiple wavelength UV detector, one can scan a known sample over a wavelength range to determine accuracy of wavelength and measure absorbance of different concentrations at a fixed wavelength to determine accuracy of photomultiplier response. Other detectors may be much more difficult to calibrate. The performance characteristics of fluorimetric and electrochemical detectors, for example, tend to exhibit much greater day-to-day variability than UV detectors. If possible, methods involving these detectors should not rely on absolute response, but rather on relative response.

1.8.6　Budgeting

Cost-benefit analysis is a part of managing a laboratory that many laboratory managers dread, yet it can be made relatively painless. The first step in doing cost-benefit analysis for instrumental acquisition is to determine how many hours an analysis or an analysis step presently requires and how many hours would be saved by acquiring a new instrument. If the anticipated time savings over the lifetime of an instrument, calculated from salary plus overhead plus benefits, is greater than the instrumental cost, there is a strong case for for purchasing the instrument. For example, if the cost of labor is estimated at \$100,000/year, and a technician would save 5 minutes per day by purchasing an automatic pipettor expected to last three years, then about 60 hours (equivalent to \$3000) would be saved over the lifetime of the pipettor. If the pipettor costs \$300, the cost-benefit ratio is 0.1, and the pipettor is a real bargain. From that simple calculation, it should be clear why automation is cost effective in most laboratories and why making every minute productive is an important part of corporate success. Scientists, trained in the methods of exact calculation, may find it vexing to make the approximations necessary for cost-benefit analysis. Cost-benefit analysis, however, is an important part of communicating scientific needs to financial managers.

For acquisitions of instruments, cost-benefit analysis is usually only the first step. It is much more difficult to write a justification for an instrument based on project utilization. In some companies, it may be difficult to get an instrument unless one can go outside the company to get data to prove the benefits of an instrument. That scientific literature demonstrating the utility of an instrument on related compounds is not an acceptable substitute to data on the company's compound can be distressing; it can be difficult to get the necessary permissions to take a sample out of the company to a university or to an instrumental applications laboratory, and it can be awkward to present data to a conference of senior managers who have little understanding of the science involved in the presentation. However, the present system often makes it incumbent on the laboratory director to do these things to get the instruments necessary to complete projects on time.

One of the most difficult decisions that good managers face is the decision to request additional personnel. Adding a person to the group is a commitment of your time and training to ensure that the new hire will be doing valuable work well. Adding a new person is a significant commitment of corporate resources. In addition to salary, benefits, and overhead, each new hire requires instrumentation and supplies. Also, it is not unusual for a new hire to become fully productive only after a period of many months. The decision to add personnel should come only after making a full study of project timelines over as long a period as possible. Consideration should be given to automation, contracting work out, or temporary hires, and the addition of personnel should be reserved as a last resort. Employing recent graduates is one way to bring to the corporation new technology and new ways of looking at problems. It may, however, be more cost effective as a general strategy to systematically train existing employees.

Minimally, research and development (R&D) upper management must communicate general project priorities to R&D personnel, and then request, often through Project Management, appropriate time commitments and personnel needs from each department. The decline of the ranks of middle management, however, has damaged lines of communication. One sometimes even finds that information that is freely available through such public domain sources as scientific publications, corporate press releases, or shareholder reports does not circulate to those who need it. However, assuming that approximate project deadlines are known, one can calculate manpower needs fairly accurately. A good starting point is to determine what characteristics of the compound under development are likely to be most important in keeping development on track. For a polymer, the molecular weight or the presence of additives might be expected to be important, while for a biopharmaceutical, the presence of residual cell culture media might be important. Required detection limits should be estimated. After compiling a list of which characteristics are likely to be important to define, the next step is to propose an assay, including approximate assay time, separation technique, and detector. Staff discussion at this stage may reveal potential weaknesses in the assay and suggest a backup. Since staff are now aware of what work is likely to be coming up, they may be able to assemble literature and materials. As the project start time approaches, preliminary method development can be done. Finally, one needs to estimate the production rate, i.e., the number of batches per year, that will require assay. At this point, estimating staff requirements is fairly straightforward.

1.8.7 Technology transfer

Technology transfer may involve the transfer of an analytical method from a research group to an analytical group, the transfer of a method from one analytical laboratory to another analytical laboratory or to a quality control unit, the transfer of a separation method from an analytical laboratory to a process development group, or the transfer of a production method from one facility to another. Therefore, the skills of the analytical laboratory manager in defining and communicating technical issues are critical to maintaining the pace of development. Often, there are differences in culture between the corporate divisions that lead to misinterpretation and conflict. In the words of a researcher in the area of conflict:[116]

> "The pressures of rapidly bringing high quality new products to market have increased managerial awareness that cross-functional integration is necessary to success. ... [T]he 1980s are littered with the detritus of failed ventures or merged ones that could not create their strategic syncretism of science and business in time."

Perhaps the most important step in ensuring the smooth transfer of technology is the production of a written method. As well as serving as a technical guide, the written method serves to limit conflict by providing a record of who said what when. A written method should begin with an introduction in plain, nontechnical language that describes the function and limitations of a method. Issues that may arise during the course of sample handling, particularly dissolution, detection, and stability, should be described briefly. Specification of standards, solvent grade, and HPLC column should be established. Methods of calculation, including an example, should be included. A method should also include a system suitability test, i.e., a test that can be run with standards to determine whether the system used by the laboratory receiving the new method is capable of performing the analysis. A method should also include acceptance criteria, i.e., criteria to determine whether the test results are likely to be valid. Examples of elements of a system suitability test are the measurement of column plate height, the calculation of separation factors, and the sizes of the standard peaks relative to full scale. An element of acceptance criteria could be the size of the peak of the principal component. Brief methods are more likely to be read in detail than long methods.

In addition to a written method, transfer should include several samples of well-characterized materials, portions of which are retained by the original laboratory for troubleshooting purposes. For distant laboratories, the availability of data transfer by modem can be extremely helpful in troubleshooting. Often, comparison of chromatograms transmitted by mail or by FAX is unsatisfactory for rapid troubleshooting. Senior management should be briefed on the advantages of installing uniform data systems throughout the organization, including the importance of data systems capable of transmission by modem.

1.8.8 Conclusions

Managing an analytical laboratory requires knowledge of how to manage people, instruments, data analysis, data systems, and project schedules. A background in science does not necessarily provide one with the tools required to manage effectively. The analytical manager should review the literature on social science and management,[117-119] recognizing that the human component is extremely important to successful operations. As one writer put it, middle management, though much maligned, serves as a stabilizing force to maximize system efficiency and aid the implementation of technological change:[120]

> "Fortunately, American culture produces many strong people who can be extraordinarily good at coming up with creative solutions. ... Middle managers ... have to... recognize that most technical and people problems are inseparable. [T]op management will have to start developing and rewarding this activist role."

Operating an analytical laboratory also requires considerable technical depth. In the selection of staff for the analytical laboratory, some thought should be given to this need. A background in organic chemistry is useful in optimizing derivatization reactions. Knowledge of statistical methodology is necessary in experimental design and in interpreting data. Spectroscopy and electrochemistry are important in understanding detectors. Computer and electronics skills find application in integrating HPLC systems and in management and reporting of data. The coming years promise to be some of the most productive years of science — for those that have the skills.

Acknowledgments

Many thanks are due to Russell Bernard and to Frank Dubinskas for discussion and references on sociocultural studies of organizations. Professor Chickery Kasouf of the Department of Management at Worcester Polytechnic Institute is also thanked for useful commentary.

References

1. McNair, H. M., HPLC equipment update — VII, *J. Chromatogr. Sci.*, 25, 564, 1987.
2. Stevenson, R.W., The world of separation science: PITTCON® '95. Mixed currents along the Big Muddy, *Am. Lab.*, 27 (May), 26C, 1995.
3. Ishii, D. and Takeuchi, T., Open tubular capillary LC, *J. Chromatogr. Sci.*, 18, 462, 1980.
4. Knox, J. H., Theoretical aspects of LC with packed and open small-bore columns, *J. Chromatogr. Sci.*, 18, 453, 1980.

5. Scott, R. P. W., Microbore columns in liquid chromatography, *J. Chromatogr. Sci.*, 18, 49, 1980.

6. St. Claire, III, R. L. and Jorgenson, J. W., Characterization of an on-column electrochemical detector for open-tubular liquid chromatography, *J. Chromatogr. Sci.*, 23, 186, 1985.

7. Sagliano, Jr., N., Shih-Hsien, H., Floyd, T. R., Raglione, T. V., and Hartwick, R. A., Aspects of small-bore column technology, *J. Chromatogr. Sci.*, 23, 238, 1985.

8. Ishii, D., Asai, K., Hibi, K., Jonokuchi, T., and Nagaya, M., A study of micro-high-performance liquid chromatography. I. Development of technique for miniaturization of high-performance liquid chromatography, *J. Chromatogr.*, 144, 157, 1977.

9. Tsuda, T. and Novotny, M., Packed microcapillary columns in high performance liquid chromatography, *Anal. Chem.*, 50, 271, 1978.

10. McGuffin, V. and Zare, R. N., Femtomole analysis of prostaglandin pharmaceuticals, *Proc. Natl. Acad. Sci. U.S.A.*, 82, 8315, 1985.

11. Renn, C. N. and Synovec, R. E., Packed microcapillary columns in high performance liquid chromatography, *Anal. Chem.*, 63, 568, 1991.

12. Schwartz, H. E., Karger, B. L., and Kucera, P., Gradient elution chromatography with microbore columns, *Anal. Chem.*, 55, 1752, 1983.

13. Majors, R. E., Recent advances in HPLC columns and packings, *J. Chromatogr. Sci.*, 18, 488, 1980.

14. Majors, R. E., New chromatography columns and accessories at the 1992 Pittsburgh Conference, Part I, *LC-GC*, 10, 188, 1992.

15. Ehwald, R., Fuhr, G., Olbrich, M., Göring, H., Knösche, R., and Kleine, R., Chromatography based on membrane separation with vesicular packing material, *Chromatographia*, 28, 561, 1989.

16. Gorowek, J., Characterization of mesopores using temperature-programmed desorption, *Am. Lab.*, 27 (May), 16, 1995.

17. Karger, B. L., Snyder, R. L., and Horvath, Cs., *An Introduction to Separation Science*, Wiley-Interscience, New York, 1973, 33–46.

18. Majors, R. R., High performance liquid chromatography on silica gel, *Anal. Chem.*, 44, 1722, 1972.

19. Martin, O. C. and Pagano, R. E., Normal- and reverse-phase separation of fluorescent (NBD) lipids, *Anal. Biochem.*, 159, 101, 1986.

20. Rosolowsky, M. and Campbell, W. B., Synthesis of hydroxyeicosatetraenoic (HETES) and epoxyeicosatrienoic acids (EETS) by cultured bovine coronary artery endothelial cells, *Biochim. Biophys. Acta*, 1299, 267, 1996.

21. Previati, M., Bertolaso, L., Tramarin, M., Bertagnolo, V., and Capitani, S., Low nanogram range quantitation of diglycerides and ceramide by high-performance liquid chromatography, *Anal. Biochem.*, 233, 108, 1996.

22. Mori, S., Size exclusion chromatography and nonexclusion liquid chromatography for characterization of styrene copolymers, *Adv. Chem.*, 247, 211, 1995.

23. Okamoto, M., Kakamu, H., Nobuhara, K. and Ishii, D., Effect of silver-modified silica on retention and selectivity in normal-phase liquid chromatography, *J. Chromatogr. A*, 722, 81, 1996.

24. Lee, S. T. and Olesik, S. V., Normal-phase high-performance liquid chromatography using enhanced fluidity liquid mobile phases, *J. Chromatogr. A*, 707, 217, 1995.

25. Monde, T., Kamiusuki, T., Kuroda, T., Mikumo, K., Ohkawa, T., and Fukube, H., High-performance liquid chromatographic separation of phenols on a fluorocarbon-bonded silica gel column, *J. Chromatogr. A*, 722, 273, 1996.

26. Baillet, A., Corbeau, L., Rafidson, P., and Ferrier, D., Separation of isomeric compounds by reversed-phase high-performance liquid chromatography using Ag+ complexation. Application to *cis-trans* fatty acid methyl esters and retinoic acid photoisomers, *J. Chromatogr.*, 634, 251, 1993.

27. Cacia, J., Keck, R., Presta L. G., and Frenz, J., Isomerization of an aspartic acid residue in the complementarity-determining regions of a recombinant antibody to human IgE: identification and effect on binding affinity, *Biochemistry*, 35, 1897, 1996.

28. Edwards, B. R., Giaque, A. P., and Lamb, J. D., Macrocycle-based column for the separation of inorganic cations by ion chromatography, *J. Chromatogr. A*, 706, 69, 1995.

29. Hansen, P., Andersson, L., and Lindeberg, G., Purification of cysteine-containing synthetic peptides via the selective binding of the α-amino group to immobilised Cu^{2+} and Ni^{2+} ions, *J. Chromatogr. A*, 723, 51, 1996.

30. Jones, C., Patel, A., Griffin, S., Martin, J., Young, P., O'Donnell, K., Silverman, C., Porter, T., and Chaiken, I., Current trends in molecular recognition and bioseparation, *J. Chromatogr. A*, 707, 3, 1995.

31. Davankov, V. A., Separation of enantiomeric compounds using chiral HPLC systems. A brief review of general principles, advances, and development trends, *Chromatographia*, 27, 475, 1989.

32. Bui, K. H., Liquid chromatographic resolution of enantiomers of pharmaceutical interest, in *HPLC in the Pharmaceutical Industry*, Fong, G. W. and Lam, S. K., Eds., Marcel Dekker, New York, 1991, chap. 10.

33. Szepesi, G., *HPLC in Pharmaceutical Analysis*. Vol. I. *General Considerations*, CRC Press, Boca Raton, FL, 1990, chap. 8.

34. Francotte, E. and Junker-Buchheit, A., Preparative chromatographic separation of enantiomers, *J. Chromatogr.*, 576, 1, 1992.

35. Pirkle, W. H. and Burke, III, J. A., Chiral stationary phase designed for β-blockers, *J. Chromatogr.*, 557, 173, 1991.

36. Ward, T. J., Chiral media for capillary electrophoresis, *Anal. Chem.*, 66, 634A, 1994.

37. Hutchens, T. W., Li, C. M., and Besch, P. K., Development of focusing buffer systems for generation of wide-range pH gradients during high-performance chromatofocusing, *J. Chromatogr.*, 359, 157, 1986.

38. Hutchens, T. W., Li, C. M., and Paige, P. K., Performance evaluation of a focusing buffer developed for chromatofocusing on high-performance anion-exchange columns, *J. Chromatogr.*, 359, 169, 1986.

39. Berthod, A., Separation with a liquid stationary phase: the countercurrent chromatography technique, *Instr. Sci. Technol.*, 23(2), 75, 1995.

40. Rizzi, A. M. and Plank, C., Coupled column chromatography in chiral separations: systems employing β-cyclodextrin phases for chiral separation, *J. Chromatogr.*, 557, 199, 1991.

41. Kanda, T., Shirota, O., Ohtsu, Y., and Yamaguchi, M., Synthesis and characterization of polymer-coated mixed-functional stationary phases with several different hydrophobic groups for direct analysis of biological samples by liquid chromatography, *J. Chromatogr. A*, 722, 115, 1996.

42. Hirabayashi, J. and Kasai, K.-I., Applied slalom chromatography. Improved DNA separations by the use of columns developed for reversed-phase chromatography, *J. Chromatogr. A*, 722, 135, 1996.

43. Skogerboe, K. J. and Yeung, E. S., Single laser thermal lens detector for microbore liquid chromatography based on high-frequency modulation, *Anal. Chem.*, 58, 1014, 1986.

44. Callmer, K. and Nilsson, O., Modification of a Varian liquid chromatography UV-detector for high-sensitivity measurements, *Chromatographia*, 6, 517, 1973.

45. Peck, K. and Morris, M. D., Optical errors in a liquid chromatography absorbance cell, *J. Chromatography*, 448, 193, 1988.

46. Li, J., Hillier, E., and Cotter, R., A new programmable multiwavelength HPLC detector, *J. Chromatogr. Sci.*, 23, 446, 1985.

47. Alfredson, T. A. and Sheehan, T., Recent developments in multichannel, photodiode-array, optical LC detection, *J. Chromatogr. Sci.*, 24, 473, 1986.

48. Bartolomé, B., Hernández, T., Bengoechea, M. L., Quesada, C., Gómez-Cordovés, C., and Estrella, I., Determination of some structural features of procyanidins and related compounds by photodiode-array detection, *J. Chromatogr. A*, 723, 19, 1996.

49. Keller, H. R., Kiechle, P., Erni, F., Massart, D. L., and Excoffier, J. L., Assessment of peak homogeneity in liquid chromatography using multivariate chemometric techniques, *J. Chromatogr. A*, 641, 1, 1993.

50. Stewart, J. E., Spectral-bandwidth effects of variable-wavelength absorption detectors in liquid chromatography, *J. Chromatogr.*, 174, 283, 1979.

51. Pfeiffer, C. D., Larson, J. R., and Ryder, J. F., Linearity testing of ultraviolet detectors in liquid chromatography, *Anal. Chem.*, 54, 1622, 1983.

52. Dorschel, C. A., Ekmanis, J. L., Oberholtzer, J. E., Warren, Jr., F. V., and Bidlingmeyer, B. A., LC detectors: evaluations and practical implications of linearity, *Anal. Chem.*, 61, 951A, 1989.

53. Torsi, G., Chiavari, G., Laghi, C., and Asmudsdottir, A., Responses of different UV-visible detectors in high-performance liquid chromatographic measurements when the absolute number of moles of an analyte is measured, *J. Chromatogr.*, 518, 135, 1990.

54. Yokoyama, Y. and Sato, H., Performance of some commercial UV detectors for indirect photometric chromatography, *J. Chromatogr. Sci.*, 26, 561, 1988.

55. Dose, E. V. and Guiochon, G., Bias and nonlinearity of ultraviolet calibration curves measured using diode-array detectors, *Anal. Chem.*, 61, 2571, 1989.

56. Esquivel, J. B., Wavelength accuracy testing of UV-visible detectors in liquid chromatography, *Chromatographia*, 26, 321, 1988.

57. Evans, C. E., Shabushnigg, J. G., and McGuffin, V. L., Experimental and theoretical model of refractive index artifacts in absorbance detection, *J. Chromatogr.*, 459, 119, 1988.

58. Schieffer, G. W., Preliminary examination of a new post-column photolysis-molybdate reaction detection system for the determination of organophosphorus compounds by high performance liquid chromatography, *Instr. Sci. Technol.*, 23, 255, 1995.

59. Renn, C. N. and Synovec, R. E., Refractive index gradient detection of biopolymers separated by high-temperature liquid chromatography, *J. Chromatogr.*, 536, 289, 1991.

60. Hancock, D. O. and Synovec, R. E., Refractive index gradient detection of femtomole quantities of polymers by microbore size-exclusion chromatography, *Anal. Chem.*, 60, 1915, 1988.
61. Hancock, D. O. and Synovec, R. E., Rapid characterization of linear and star-branched polymers by concentration gradient detection, *Anal. Chem.*, 60, 2812, 1988.
62. Mhatre, R. and Krull, I. S., Determination of on-line differential refractive index and molecular weight via gradient HPLC interfaced with low-angle laser light-scattering, ultraviolet, and refractive index detection, *Anal. Chem.*, 65, 283, 1993.
63. Stevenson, R. L., UV-VIS absorption detectors for HPLC, in *Liquid Chromatography Detectors*, Vol. 23, Vickrey, T. M., ed., Marcel Dekker, New York, 1983, 33.
64. Colin, H., Jaulmes, A. Guichon, G., Corno, J., and Simon, J, Construction and performance of an improved differential refractometer detector for liquid chromatography, *J. Chromatogr. Sci.*, 17, 485, 1979.
65. Munk, M., Refractive index detection, in *A Practical Guide to HPLC Detection*, Parriott, D., Ed., Academic Press, San Diego, 1993, chap. 2.
66. Munk, M., Refractive index detectors, in *Liquid Chromatography Detectors*, Vol. 23, Vickrey, T. M., Ed., Marcel Dekker, New York, 1983, chap. 5.
67. Dark, W. A., UV and dRI detectors in liquid chromatography: the workhorse detectors, *J. Chromatogr. Sci.*, 24, 495, 1986.
68. Hagel, L., Interferometric concentration determination of dextran after gel chromatography, *Anal. Chem.*, 50, 569, 1978.
69. Woodruff, S. D. and Yeung, E. S., Refractive index and absorption detector for liquid chromatography based on Fabry-Perot interferometry, *Anal Chem.*, 54, 1174, 1982.
70. Förster, Th., 10th Spiers Memorial Lecture. Transfer mechanisms of electronic excitation, *Discuss. Faraday Soc.*, 27, 7, 1959.
71. Stern, O. and Volmer, M., The extinction period of fluorescence, *Physik Z.*, 20, 183, 1919.
72. Lakowicz, J. R., *Principles of Fluorescence Spectroscopy*, Plenum Press, New York, 1983, chap. 7.
73. Cantor, C. R. and Schimmel, P. R., *Biophysical Chemistry, Part II*, W.H. Freeman, San Francisco, 1980.
74. Swadesh, J. K., Mui, P. W., and Scheraga, H. A., Thermodynamics of the quenching of tyrosyl fluorescence by dithiothreitol, *Biochemistry*, 26, 5761, 1987.
75. Sugarman, J. H. and Prud'homme, R. K., Effect of photobleaching on the output of an on-column laser fluorescence detector, *Ind. Eng. Chem. Res.*, 26, 1449, 1987.
76. Lingeman, H., Underberg, W. J. M., Takadate, A., and Hulshoff, A., Fluorescence detection in high performance liquid chromatography, *J. Liq. Chromatogr.*, 8, 789, 1985.
77. Weinberger, R., Fluorescence detection for liquid chromatography, *Lab. Pract.*, 36, 65, 1987.
78. Steichen, J. C., A dual-purpose absorbance-fluorescence detector for high-pressure liquid chromatography, *J. Chromatogr.*, 104, 39, 1975.
79. Van Den Beld, C. M. B. and Lingeman, H., Laser-based detection in liquid chromatography with emphasis on laser-induced fluorescence detection, *Pract. Spectrosc.*, 12, 237, 1991.

80. Shelly, D. C. and Warner, I. M., Fluorescence detectors in high-performance liquid chromatography, in *Liquid Chromatography Detectors,* Vol. 23, Vickrey, T. M., ed., Marcel Dekker, New York, 1983, chap. 3.

81. Yamauchi, S., Nakai, C., Nimura, N., Kinoshita, T., and Hanai, T., Development of a highly sensitive fluorescence reaction detection system for liquid chromatographic analysis of reducing carbohydrates, *Analyst,* 118, 773, 1993.

82. Coquet, A., Veuthey, J.-L., and Haerdi, W., Comparison of post-column fluorescence derivatization and evaporative light-scattering detection to analyse saccharides selectively by LC, *Chromatographia,* 34, 651, 1992.

83. Retzik, M. and Froehlich, P., Extending the capability of luminescence spectroscopy with a rapid-scanning fluorescence spectrophotometer, *Am. Lab.,* March, 68, 1992.

84. Richmond, M. D. and Yeung, E. S., Development of a laser-excited indirect fluorescence detection for high-molecular weight polysaccharides in capillary electrophoresis, *Anal. Biochem.,* 210, 245, 1993.

85. Brinkman, U. A. Th., de Jong, G. J., and Gooijer, C., Use of luminescence techniques for sensitive and selective determinations in HPLC (high-performance liquid chromatography), *Pure Appl. Chem.,* 59, 625, 1987.

86. Karger, B. L., Snyder, L. R., and Hórvath, Cs., *An Introduction to Separation Science,* John Wiley & Sons, New York, 1973.

87. Schoenmakers, P. J., Billiet, H. A. H., Tijssen, R., and de Galan, L., Gradient selection in reversed-phase liquid chromatography, *J. Chromatogr.,* 149, 519, 1978.

88. Snyder, L. R., Dolan, J. W., and Gant, J. R., Gradient elution in high-performance liquid chromatography. I. Theoretical basis for reversed-phase systems, *J. Chromatogr.,* 165, 3, 1979.

89. Stadalius, M. A., Gold, H. S., and Snyder, L. R., Optimization model for the gradient elution separation of peptide mixtures by reversed-phase high-performance liquid chromatography. Verification of retention relationships, *J. Chromatogr.,* 296, 31, 1984.

90. Aguilar, M. I., Hodder, A. N., and Hearn, M. T. W., High-performance liquid chromatography of amino acids, peptides, and proteins. LXV. Studies on the optimisation of the reversed-phase gradient elution of polypeptides. Evaluation of retention relationships with β-endorphin-related polypeptides, *J. Chromatogr.,* 327, 115, 1985.

91. Hearn, M. T. W., Hodder, A. N., and Aguilar, M. I., High-performance liquid chromatography of amino acids, peptides, and proteins. LXXXVII. Comparison of retention and bandwidth properties of proteins eluted by gradient and isocratic anion-exchange chromatography, *J. Chromatogr.,* 458, 27, 1988.

92. Swadesh, J. K., Tryptic fingerprinting on a poly(styrene-divinyl benzene) reversed-phase column, *J. Chromatogr.,* 512, 315, 1990.

93. McDowall, R. D., Change? I've always worked this way, *LC-GC,* 1, 5, 1995.

94. Majors, R. E., Trends in sample preparation and automation — what the experts are saying, *LC-GC,* 13(9), 742, 1995.

95. Laitinen, H. A. and Harris, W. E., *Chemical Analysis,* McGraw-Hill, New York, 1975, 565–582.

96. Taylor, J. K., *Quality Assurance of Chemical Measurements,* Lewis Publishers, Chelsea, MI, 1987.

97. Swadesh, J. K., High-speed tryptic fingerprinting, *Peptide Res.,* 3, 282, 1990.

98. Papas, A. N., Chromatographic data systems: a critical review, *CRC Crit. Rev. Anal. Chem.*, 20, 359, 1989.

99. Hunt, R. J., Assessment of the results from data processing systems using a digital chromatogram simulator, *J. HRC &CC*, 8, 347, 1985.

100. Chapman, K., Challenges, transitions, and paradigm shifts for information technology, *Pharm. Technol.*, 19, 120, 1995.

101. Hayes, T. L., Kohne, J. W., and Miller, T. L., The functional testing of chromatographic software as part of computer validation, *LC-GC*, 13, 960, 1995.

102. McConnell, M., Canales, M., and Lawler, G., A validation protocol for analog-to-digital interfaces in chromatographic data systems, *LC-GC*, 9, 486, 1991.

103. Nelson, T. J., Deconvolution method for accurate determination of overlapping peak areas in chromatograms, *J. Chromatogr.*, 587, 129, 1991.

104. Economou, A., Fielden, P. R., and Packham, A. J., Deconvolution of overlapping chromatographic peaks by means of fast Fourier and Hartley transforms, *Analyst*, 121, 97, 1996.

105. Sokal, R. R. and Rohlf, F. J., *Biometry*, W.H. Freeman, San Francisco, 1969.

106. Howarth, R. J., Quality control for the analytical laboratory. Part 1. Univariate methods: a review, *Analyst*, 120, 1851, 1995.

107. Swadesh, J. K. and Kasouf, C. J., Industrial analytical operations: organizational implications of automation, *Crit. Rev. Analyt. Chem.*, 25, 195, 1995.

108. Cuatresecas, P., Preface, in *Drug Development*, 2nd ed., Hammer, C. E., Ed., CRC Press, Boca Raton, FL, 1990.

109. Kovach, K. A., What motivates employees? Workers and supervisors give different answers, *Bus. Horizons*, Sept.–Oct., 58, 1987.

110. Grossman, R., The Science Gang, *New Republic,* May (cover), 1992.

111. McCormick, D., L-Trp's lesson: the process is the product, *Biotechnology,* 10, 5, 1992.

112. Wilder, D. R., New directions in industrial environmental analytical chemistry: beyond compliance testing, *Crit. Rev. Anal. Chem.*, 25, 77, 1989.

113. Connors, J. L. and Romberg, T. A., Middle management and quality control, strategies for obstructionism, *Human Org.,* 50, 61, 1991.

114. Barinaga, M., Labstyles of the famous and well funded, *Science*, 252, 1776, 1991.

115. Karnes, H. T., Shiu, G., and Shah, V., Validation of bioanalytical methods, *Pharm. Res.*, 8, 421, 1991.

116. Dubinskas, F. A. Culture and conflict: the cultural roots of discord, in *Hidden Conflict in Organizations: Uncovering Behind-the Scenes Disputes,* Kolb, D. and Bartunek, J., Eds., Sage Press, Thousand Oaks, CA, 1992, 188–209.

117. Tingstad, J. E., *How to Manage the R&D Staff: A Looking-Glass World,* AMACOM Books, New York, 1991.

118. Dodgson, M., *The Management of Technological Learning: Lessons from a Biotechnology Company,* de Gruter, Berlin, 1991.

119. Elmes, M. B. and Kasouf, C. J., Managerial and Organizational Cognition Interest Group for the Academy of Management, Annual Meeting, Vancouver, BC, 1995.

120. Sayles, L. R., Middle managers can rescue business, *The New York Times,* p. 11, February 14, 1993.

chapter two

Automated sampling in the process environment

Jeffrey R. Larson

0-8493-2682-6/97/$0.00+$.50
© 1997 by CRC Press, Inc.

2.1   Issues in process HPLC

2.1.1   Overview

The present section discusses issues involved in the process laboratory, drawing samples for chromatographic analysis from a reactor or any other component of a production process. These issues can be categorized as

- Safety
- Durability
- Sampling site and frequency
- Sample preparation, especially the removal of particulates
- Speed of analysis

Sterility, an important issue in biopharmaceutical manufacturing, is not covered in this section. Briefly, in sampling from a biopharmaceutical process, the sample port should not compromise the physical integrity of the process step. In other words, the port should be designed so that flow is unidirectional out of the process stream, with no possibility of flow back into the process. Also, the port should not constitute a stagnant pool in the flow path. Stagnant pools do not give representative samples, are potential locations for the growth of adventitious organisms, and are difficult to clean between batches. There are similar concerns in process scale manufacturing of biopharmaceuticals. In particular, the fittings must be designed to preserve the physical integrity of the system. This requirement places restrictions on the flow cell used in detectors.

The first commercially available process scale instrument was introduced in 1974 by Applied Automation.[1] Since then, other instrument companies, such as Waters™, Dionex, Siemens, and Knauer, have also developed instruments. Larger particle size column packings (50 to 100 μ) are commonly used to reduce back pressure and maintenance on the liquid handling system.[1] Resolution is generally sacrificed for throughput, as is discussed in the Chapter 3. An analytical high performance liquid chromatograph (HPLC) may be coupled to a process liquid chromatograph (LC) as a monitor for the separation. Although chromatography is used for production of high-value-added materials such as pharmaceutics, HPLC applications at the pilot and production scale more typically use the HPLC for automated sampling and analysis. Therefore, analytical applications will be the focus of this chapter.

There are several advantages of the use of HPLC for process monitoring. First, HPLC provides both qualitative and quantitative information about a process. At the research or pilot reactor stage of development, real time monitoring increases research efficiency and provides the data for process optimization. Second, because HPLC permits continuous real-time monitoring of reactors or other process components, process upsets that might go unnoticed by manual sampling are detected promptly. Production batches that might require reprocessing or even rejection in the absence of continuous monitoring can be salvaged by timely intervention. Also, continuous monitoring provides a database for statistical process control.[2]

2.1.2   Safety

A primary concern in the installation of temporary on-line or permanent process HPLC instruments in the process environment has to do with flammable vapors. For operation in a hazardous environment, the National Fire Protection Association (NFPA) has established a standard (NFPA 496) for enclosures of electrical equipment. These enclosures are purged with clean air or inert gas at sufficient flow and pressure to reduce the concentration of any flammable gas or vapor to a safe level. There are two degrees of hazard classifications, Division 1 (normally hazardous) and Division 2 (hazardous only under abnormal conditions). Division 1 includes locations in which flammable gases are normally present and Division 2 includes locations in which flammable gases are normally confined but could escape if failure or breakdown of containers occurs. The purge requirements for the first case are more stringent and require Type X purging, while Division 2 situations require Type Z purging. In both cases an internal pressure of 25 Pa (0.1 in H_2O) must be maintained in the enclosure. For Type Z purging, an alarm or indicator is required to detect failure of purging, but safety interlocks are not required. For Type X purging, power must automatically be removed if purge is lost. In addition, an interlock must also be provided to cut power if the enclosure door is opened. For HPLC enclosures, flammable limit detectors are generally used to detect solvent spills inside the box.

A diagram of a typical instrument is shown in Plate 1A.[*2] These instruments are all designed to meet the requirements of NFPA 496 and are suitable for use in hazardous environments. Conventional HPLC equipment must be housed in a purged enclosure suitable for use in Division 1 and 2 areas. As shown in Plate 1B,[**] the enclosure is similar in size to the commercial instrument shown earlier. The entire enclosure is air purged. The enclosure has an explosion proof switch to cut power to all devices in the enclosure. Air must be flowing into the enclosures before power can be turned on and loss of air purge cuts power to the instruments. Two Rexnord flammable gas detectors are used to detect the presence of any flammable vapors inside the cabinet. The enclosures contains several shelves. The lower shelves are used for mounting of up to three pumps. A shelf on the right side above the pumps is used for mobile phase. Several air-actuated Rheodyne valves are mounted on a shelf on the left side of the cabinet. A heated HPLC column enclosure is also provided. The detectors are mounted on the top shelf, and all electrical components are mounted above the rear of the top shelf.

[*]   Plate 1A follows p. 178.
[**] Plate 1B follows p. 178.

2.1.3   Durability

Process LC instruments have higher requirements for ruggedness than analytical instruments. Process down time is extremely expensive. Also, bring-
ing service personnel into a production area is undesirable from the standpoint of confidentiality. Removing equipment from a pharmaceutical
production area for service requires considerable time for re-installation.
Even in a process not subject to regulation, re-installation is time consuming;
therefore, process hardware should be reliable, durable, and simple to service. For example, most process LC units use gas-driven pumping mechanisms or syringe pumps rather than piston-type reciprocating pumps, which
are used most often in laboratory work.

The decision to select a permanent or a temporary installation of HPLC
for on-line analysis depends on the environment. As shown in Figure 1, a
permanent HPLC installation must be rugged, reliable, reproducible, and
capable of being operated and maintained by personnel with limited technical training. A primary goal is to monitor upsets. A temporary on-line
installation is more flexible but less rugged and reproducible. A primary
goal may be to identify which components are most important to monitor
on a continuous basis. The training level required of the operator of a temporary installation is much higher.

2.1.4   Sampling site and frequency

Obtaining a representative sample from a reactor is often not so straightforward as it might seem. Large vessels tend to have some degree of inhomogeneity, even with stirring. There may be a time lag in sampling from effluent
or feed lines. In developing an on-line system, therefore, samples should be
drawn from several locations to ensure that the sample drawn from the
primary sampling port is representative of process status.

2.1.5   Sample handling

Sample preparation, injection, calibration, and data collection, must be automated for process analysis. Methods used for flow injection analysis (FIA)
are also useful for reliable sampling for process LC systems.[1] Dynamic dilution is a technique that is used extensively in FIA.[1,3] In this technique, sample
from a loop or slot of a valve is diluted as it is transferred to a HPLC injection
valve for analysis. As the diluted sample plug passes through the HPLC
valve it is switched and the sample is injected onto the HPLC column for
separation. The sample transfer time typically is determined with a refractive
index detector and valve switching, which can be controlled by an integrator
or computer. The transfer time is very reproducible. Calibration is typically
done by external standardization using normalization by response factor.
Internal standardization has also been used. To detect upsets or for process
optimization, absolute numbers are not always needed. An alternative to

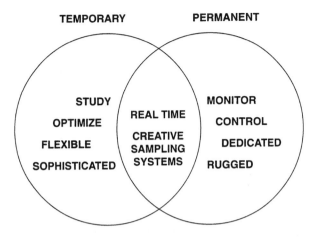

TEMPORARY   PERMANENT

STUDY   MONITOR

OPTIMIZE   REAL TIME   CONTROL

CREATIVE

FLEXIBLE   SAMPLING   DEDICATED

SYSTEMS

SOPHISTICATED   RUGGED

**ON-LINE
MONITORING**

Figure 1   Goals of temporary vs. permanent on-line HPLC analysis.

dynamic dilution is the use of a stirred mixing chamber in which a known volume of sample is diluted to a known concentration. The use of the stirred mixing chamber for FIA and HPLC has been described.[4]

For permanent process HPLC instruments, sampling systems are usually designed for each application. Manufacturers of process HPLC will design sampling systems at additional cost. Concerns about confidentiality may surface if the vendor requires more knowledge of a particular process than the industrial customer cares to give. Sampling is a key parameter for a successful on-line project. Errors in sampling can have serious consequences to process-scale projects, as work at the production level is so expensive; therefore, careful attention to sampling is critical.

Filtration of the process sample is a major concern in many applications. An agitated ultrafiltration cell has been used for a continuously pumped fermentation broth stream.[5] A flow-through process filter housing equipped with a 0.45-μ Durapore microporous filter has been used for monitoring the isomerization of glucose to fructose.[6] This system was very reliable, since the filter only had to be replaced on a monthly basis. The use of tangential flow filtration or cross-flow filtration has increased the maintenance-free lifetime of process analyzers dramatically. Tangential flow filters are available in several different geometries and can be constructed of various materials. In tangential flow filtration, only a portion of the liquid feed passes through the filter, while most of the liquid feed flows tangentially to the filter membrane. This tangential flow then sweeps away solid particles, so the filter is self cleaning. Tangential flow filters are available in several forms including tubular cross-flow, pleated sheet, plate and frame, and spiral wound filters.[7]

2.1.6   Speed of analysis

One key factor in designing a method of chromatographic analysis for process monitoring is matching the frequency of sampling the process to the run time of the analytical method. High-speed separations now permit run times of one minute or less in favorable cases. By automating sample preparation, injection, and analysis, it is possible to monitor a process in real time. One important application of process monitoring is the use of analytical HPLC to monitor a process scale chromatographic run. The fractions to be blended are determined by the analytical results. In order to automate fraction collection of the process stream, the speed of the analysis must be on the order of 100 to 1000 times faster than the process chromatography. In general, sampling frequency should be sufficiently high so that a process variable is sampled at least 10 to 20 times during that period of most rapid change, e.g., during elution of a chromatographic peak.

2.2   Case histories

Several applications of on-line HPLC used as temporary installations for process monitoring are presented in this section. These case histories include

- Catalytic amidation/cyclization step in a pesticide intermediate
- Microbore HPLC of a catalytic hydrogenation process
- Analysis of a bromination reaction
- Measurement of peak molecular weight of polystyrene

These installations of on-line process analytical equipment operated over extended periods of time with the high degree of dependability required in the process environment. Considerable cost savings in process development time were achieved through automation of analysis.

2.2.1   Catalytic amidation/cyclization step in a pesticide intermediate

This application was performed on a lab-scale reactor and later in a mini-plant-scale reactor. The reaction studied was the vapor phase catalytic amidation/cyclization step in a pesticide process. As shown in Figure 2, two reactions are taking place on the catalyst bed.

The equipment consisted of two Waters™ (Waters Corp.; Milford, MA) M-45 pumps, a Waters™ 481 UV detector, a six-port Valco sampling valve (A2L6P) with 0.08" holes in the valve body and rotor, a Rheodyne Model 7413 injection valve with a 1-μl loop, a valve interface box, and a Digital Equipment LSI-11/23-based microcomputer system. The microcomputer was used to control all valves, collect raw data from the UV detector, integrate the chromatogram, and store and plot results.

A diagram of the sampling system is shown in Figure 3. The two starting materials were pumped to the catalyst bed using a dual piston Eldex® (Napa,

AMIDATION

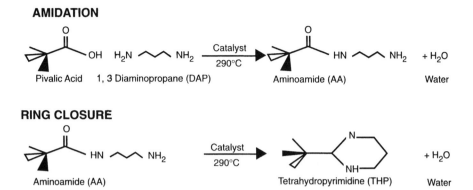

RING CLOSURE

Figure 2 Vapor phase catalytic amidation/cyclization reaction in a pesticide process monitored by on-line HPLC.

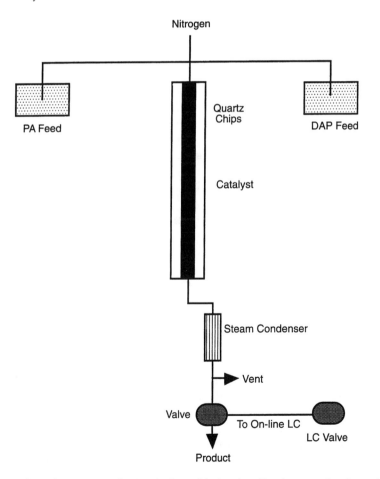

Figure 3 Sampling system for catalytic amidation/cyclization reaction in pesticide intermediate.

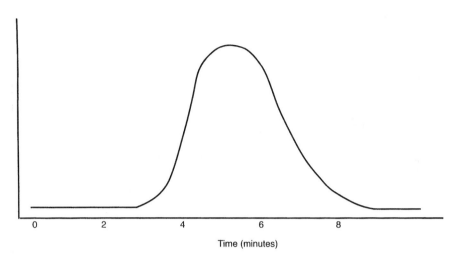

Time (minutes)

Figure 4 Dynamic dilution of sample for HPLC analysis of amidation/cyclization reaction. Transfer solvent: methoxyethanol; flow rate 1 ml/min. Detection: refractive index.

CA) pump. The catalyst was packed in a 3/4" × 2-ft glass column, which was placed inside a three-zoned furnace. A preheater zone consisting of quartz chips was packed on top of the catalyst. Nitrogen gas was used as a carrier for the vapor phase reaction. The vapors from the catalyst bed were condensed and the nitrogen passed through a tee in the line to a trap. The 1/4" exit tube of the condenser was reduced to 1/8" using a stainless steel reducing union and a graphite ferrule on the glass condenser. A short piece of the 1/8" Teflon® tubing was used to connect the condenser to the modified six-port Valco valve. This valve had an 0.08" bore to minimize restriction. Typically, Valco valves have 0.01" or 0.03" bores. The liquid sample was fed by gravity through a 75-μl loop to a collection bottle. Every 20 minutes the Valco valve was air actuated and the 75-μl sample transferred through 1/16" × 0.02" i.d. stainless steel tubing using a Waters™ M-45 pumping methoxyethanol at a flow of 1 ml/min. By this means, the sample was transferred to a Rheodyne 7413 LC injection valve located about 50 ft away. Due to space limitations in the lab, the LC could not be located closer to the reactor.

The 1/6" × 0.02" i.d. transfer line also functioned as a sample dilution device; in other applications, a stainless steel column packed with glass beads has been found to be useful for dilution. This simple dynamic dilution technique has been used extensively in flow injection analysis.[3] A refractive index detector is typically used to measure the sample transfer time. As shown in Figure 4, approximately 5 minutes is required to transfer the sample plug to the Rheodyne valve. As the apex of the sample band passes though the Rheodyne valve, the valve is activated and 1 μl injected onto the liquid chromatographic column. The sample transfer time was checked periodically over 1 year of operation and found to be stable.

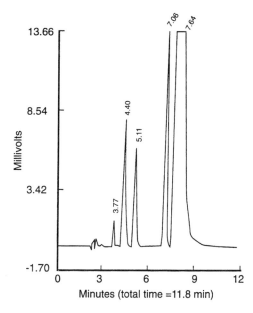

Figure 5 Separation of reactor product of amidation/cyclization reaction. Columns: 4.6 × 150-mm Zorbax® ODS coupled to a 4.6 × 250-mm Partisil® SCX. Mobile phase: acetonitrile — water (30:70) with 50 mM monobasic potassium phosphate per 10 mM phosphoric acid buffer. Flow rate: 2 ml/min. Detection: UV at 210 nm. Injection: 1 μl.

The sample drawn from the reactor consisted of an acid, several amines, and a neutral species. Two of the components were not resolved to baseline by reversed phase LC, so a dual column reversed phase ion-suppression/ion-exchange technique was used. A chromatogram of the separation is shown in Figure 5.

This temporary on-line HPLC analysis, providing continuous monitoring of catalyst activity, not only resulted in significant savings in analytical time, but also accelerated the optimization of process parameters such as reactor bed temperature and flow rate. The plot shown in Figure 6 shows experiments run to determine optimum bed temperature. Figure 6 is a real-time plot, displayed continuously on the video monitor and updated automatically after each sample analysis. The figure shows that conversion to the desired product increases with reactor bed temperature up to 325°C. Above that temperature, competing reactions occur, as evidenced by the appearance of several unknown peaks in the chromatogram and a decline in the consumption of one of the starting materials. In less than 12 hours, the reactor bed temperature was optimized, with the optimum occurring at 300°C.

To model the amidation and cyclization reactions, an internal standard was weighed into the feed, and the feed placed on a balance for continuous weighing during the reaction. The HPLC system was calibrated using units of mol/kg, so that density effects could be neglected. A series of experiments

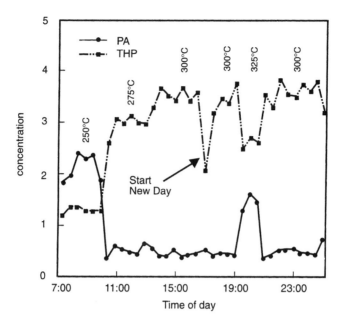

Figure 6 Optimization of catalyst bed temperatures using on-line HPLC.

was then run using four catalyst bed lengths at two different temperatures and four different flow rates. The mass balance showed 97% recovery on average for the 32 experiments. Estimated rate constants and a predictive model were obtained, completing the optimization and process design.

2.2.2 On-line microbore HPLC monitoring of a catalytic hydrogenation process

The process under investigation, as shown in Figure 7, was the reduction by hydrogen of a polysulfide feed to a mercaptophenol. Manual sampling was impossible, since that would have caused reactor upset. The reaction, occurring on a continuous fixed bed reactor catalyst, was analyzed and optimized for parameters such as catalyst type and loading, bed temperature, and flow rate. The temporary installation of an on-line HPLC monitor at the benchtop level (1" diameter reactor) was so successful in helping to control the reaction that the pilot plant reactor (6" diameter) was designed with a built-in sample cup.

$$Ar \; S_n \; Ar \quad \xrightarrow{H_2/Cat} \quad 2ArSH + (n-2) \; H_2S$$

$$n = 2, 3, 4, 5$$

Figure 7 Catalytic hydrogenation of polysulfide feed to form a mercaptophenol.

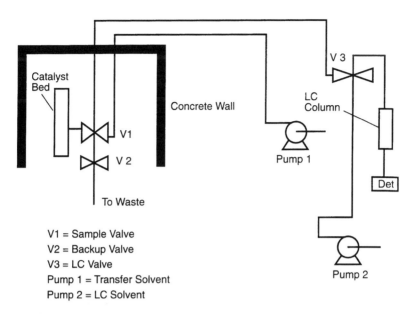

Figure 8 Sampling system for catalytic hydrogenation step at pilot plant.

The equipment used in this application included two Waters™ M-45 pumps, a Waters™ 481 UV detector with microbore cell, an air-actuated Rheodyne 7413 injection valve with a 1-µl injection loop, an air-actuated Valco four-port sampling valve (A2CI4UW2) with no groove in the injection entry ports, an air-actuated Valco three-port switching valve (AC3W), and a Digital Equipment LSI-11/23 microcomputer. The LC system was located in a purged cabinet suitable for use in Class I, division 2 areas. The cabinet was in a heated room about 40 feet from the reactor column. The two Valco valves were mounted next to the reactor column, while the microcomputer was located in the control room.

A diagram of the sampling system is shown in Figure 8. The liquid leaving the catalyst bed passed through a flange in the process pipe into a 1-ounce capacity cup inserted between two V-shaped plates. One of the plates had a hole drilled through it for insertion of 1/16" stainless steel tubing. A 1"-to-1/16" pipe-to-tube reducing union was located on the side of the reactor to allow connection of the 1/16" × 0.03" i.d. sample line, which was inserted through the hole in the steel plate into the liquid in the sample cup. The other end of the sample line was connected to a 0.5-µ in-line filter, which was connected directly to a Valco valve.

Normally the Valco valve was in the inject position. In this position, the reactor feed was blocked from entry, since the valve rotor had no grooves. To sample the reactor, the valve was switched for 1.2 seconds to the load position, using the 750-psi reactor pressure to fill a 2-µl slot in the valve rotor from the sample collected in the cup in the reactor. Upon switching back to

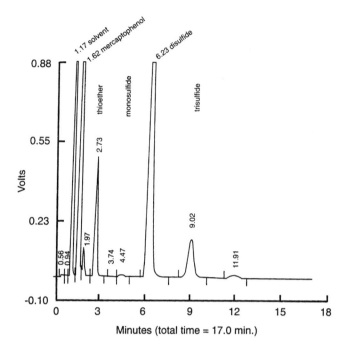

Figure 9 Typical microbore HPLC chromatogram of catalytic hydrogenation process. Column: 1 × 250-mm Partisil® ODS (Whatman). Mobile phase: acetonitrile:water:tetrahydrofuran, 85:15:2 (v:v:v). Flow rate: 0.2 ml/min. Detection: UV at 214 nm. Injection: 1 μl.

the inject position, the sample was transferred through 1/16" × 0.02" i.d. stainless steel tubing to the Rheodyne valve located in the LC cabinet about 60 feet away. Acetonitrile, pumped at 1 ml/min, was used as the transfer solvent. Transfer, as in the preceding case history, accomplished dynamic sample dilution. Injection onto the microbore column was performed as the apex of the sample band passed through the Rheodyne 7413 valve. The Valco valves and sample line were enclosed in a Styrofoam® plastic foam box with SR-1 heat tracing at 60°C to prevent freezing. The three-port Valco valve served as the safety valve in the unlikely event of the sample valve sticking in the load position.

The LC system consisted of a Waters™ M-45 pump with a micro-flow module, a Waters™ Model 481 UV detector equipped with a microbore cell, and a 1 × 250 mm Partisil® ODS reversed phase column. A typical chromatogram obtained under these conditions is shown in Figure 9. The response was calibrated by external standardization.

Parameters such as feed rate, catalyst bed temperature, and reaction pressure were optimized by use of the temporary on-line LC installation. Reactor upsets could also be monitored. Figure 10 demonstrates how continuous monitoring can aid in detection of an upset. Due to a problem with a level control valve, the reactor filled with liquid, preventing the reaction

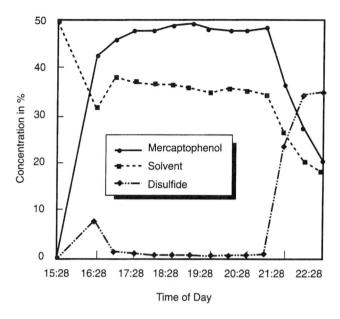

Figure 10 Real-time plot of reactor component concentrations in catalytic hydrogenation step illustrating early detection of baseline upset by on-line micro-HPLC.

from occurring. The on-line LC system readily detected the fault. In 1 hour, the concentration of the starting material increased from 0.5 to 24% and the product decreased from 48 to 36%. This partially converted batch was fully salvaged.

In another experiment, the optimum feed rate was determined by slowly increasing the pump rate while monitoring the concentration of starting material. As shown in Figure 11, the concentration of starting material increased as the feed rate was increased, and decreased to the desired level of 1 to 2% level as the pump speed was reduced.

2.2.3 Liquid chromatographic analysis of a bromination reaction

As illustrated in Figure 12, the reaction mixture contains mono-, di-, and tribrominated glycols, hydrobromic acid, and water. The mixture is extremely corrosive, and the reactor is operated at a temperature just above the freezing point of the product. The key to successfully sampling this mixture was the use of a corrosion-resistant tantalum sampling system. In addition, the sample line was continuously flushed with reactor solvent except during sampling.

A diagram of the sampling system is shown in Figure 13. The sample was drawn through tubing 1/16" o.d. × 0.03" i.d. × 6" in length. The tip of the tube was inserted through a piece of Kynar® tubing, which was placed inside a Teflon®-coated stainless steel dip pipe to prevent the tube from being bent during agitation of the reactor. The other end of the tube was connected to a four-port Valco tantalum valve mounted on top of the reactor. The

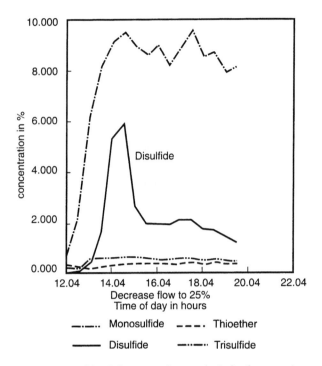

Figure 11   Optimization of feed flow rate for catalytic hydrogenation using temporary on-line LC.

$$C(CH_2OH)_4 + HBr \xrightarrow[\text{5-10 PSI}]{\substack{110°C \\ \text{Solvent}}} C(CH_2OH)_{4-n}(CH_2Br)_n + H_2O$$
$$n = 1,2,3$$

Figure 12   Bromination of pentaerythritol.

sample exit line from the four-port valve consisted of a short piece of tantalum tubing (1/16" o.d. × 0.03" i.d.) connected to about 15 feet of 1/8" Teflon® PFA tubing. The Teflon® tubing was connected to a four-port Valco stainless steel switching valve mounted on the side of the cabinet housing the HPLC equipment. The tantalum valve and sampling line were steam traced, insulated, and maintained at 115°C. As an additional precaution to prevent plugging, the sample line was flushed with reactor solvent at 1 ml/min.

    To obtain a sample, the reactor was manually pressurized with nitrogen to 10 to 12 psi, and the four-port switching valve was activated to divert the flow of reactor solvent to recycle. The pressure forced about 4 ml of sample through the 5-μl slot of the tantalum valve. The valve was then switched to transfer the sample to the HPLC located about 15 feet away. The solvent for transfer and dynamic dilution was ethoxyethanol, with a boiling point of

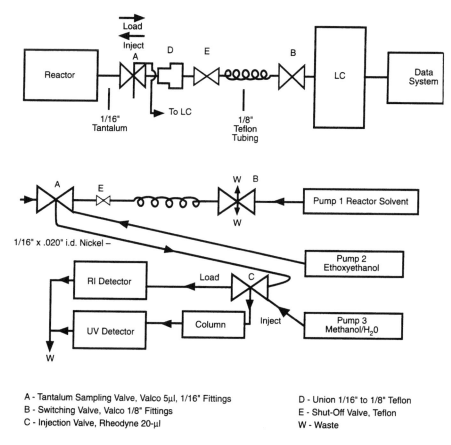

A - Tantalum Sampling Valve, Valco 5µl, 1/16" Fittings
B - Switching Valve, Valco 1/8" Fittings
C - Injection Valve, Rheodyne 20-µl

D - Union 1/16" to 1/8" Teflon
E - Shut-Off Valve, Teflon
W - Waste

Figure 13   Diagram of sampling system for brominated glycols.

130°C. The Rheodyne Model 7010 injection valve, equipped with a 20-µl loop, was switched to injection at the apex of the sample band, as observed on the refractive index detector. The complex kinetics of the production of mono-, di-, and tri-brominated glycols is shown in Figure 14. Optimization of parameters such as the flow rate of acid resulted in a 15% reduction in batch cycle time and eliminated the need for manual analysis and intervention to obtain a desired endpoint composition.

2.2.4   On-line GPC technique for measurement of peak molecular weight of polystyrene

Polymers, which are slow to dissolve, represent a significant problem for conventional sampling. Recirculation of sample and solvent is one approach, but excessive sample dilution may occur. In the system described in the present section, a flow-through valve was modified to contain an internal dissolution chamber, as is shown in Figure 15. A known volume of solvent

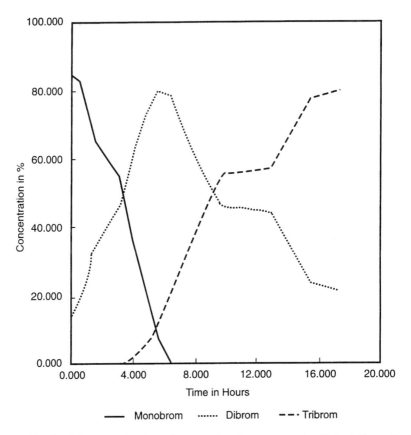

Figure 14   Real-time monitoring of conversion from mono- to di- to tri-brominated product.

(1 ml) is transferred to the 2-ml internal chamber, then a few microliters of sample at 100°C are transferred from the process line to the internal chamber with a grooved rod of the kind used to determine sample volume in process gas chromatographs. A slow flow of air provided mixing. Since the valve was maintained at 100°C, dissolution was rapid. The sample was then transferred to the liquid chromatograph using controlled-flow air.

A diagram of the sampling system is shown in Figure 16. A narrow molecular weight standard was added to the diluent to correct retention times for drift due to variations in flow rate. The dilution volume was fixed by filling a loop on a six-port valve, labeled T3 in the diagram. A Porter flow controller was used to deliver air at 4 ml/min to the six-port valve, transferring the 1 ml of diluent to the internal chamber on valve T4. A four-port valve, labeled T6 in Figure 16, was installed above the sampling valve to provide a second source of flow-controlled air for transfer of the dissolved sample to the liquid chromatograph. A second four-port valve (T5) was also installed between the sampling valve and the diluent valve (T3) for diversion

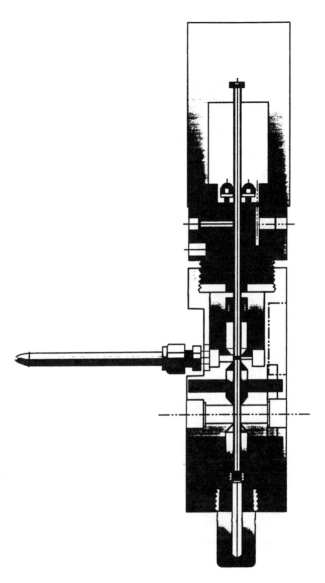

Figure 15 Diagram of polymer sampling valve.

of the sample stream to the LC located in a purged cabinet about 10 feet away. One minute was required to transfer the sample to the LC. Separation was done by gel permeation chromatography using a Polymer Laboratories PL-Gel™ 5-µ mixed-bed column with tetrahydrofuran as the mobile phase. The flow rate was 0.9 ml/min, and UV detection at 254 nm was used for measuring the molecular weight distribution. A Spectra-Physics 4270 integrator was used to control the entire system.

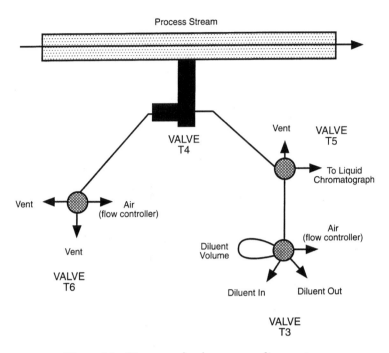

Figure 16   Diagram of polymer sampling system.

2.3   Future directions for on-line HPLC process monitors

2.3.1   General

HPLC is extremely useful in monitoring and optimizing industrial processes. Conventional process monitors measure only bulk properties, such as the temperature and pressure of a reactor, while HPLC permits continuous real-time monitoring of consumption of starting materials, product composition, and impurity profile. There are a number of new initiatives relevant to HPLC for process monitoring, including sample preparation, automation, miniaturization, and specialized detectors.

2.3.2   Automated sample processing and process monitoring

Many process mixtures, notably fermentations, require sample preconcentration, microdialysis, microfiltration, or ultrafiltration prior to analysis. A capillary mixer has been used as a sample preparation and enrichment technique in microchromatography of polycyclic aromatic hydrocarbons in water.[8] Microdialysis to remove protein has been coupled to reversed phase chromatography to follow the pharmacokinetics of the metabolism of acetaminophen into acetaminophen-4-O-sulfate and acetaminophen-4-O-glucuronide.[9] On-line ultrafiltration was used in a process monitor for *Aspergillus niger* fermentation.[10]

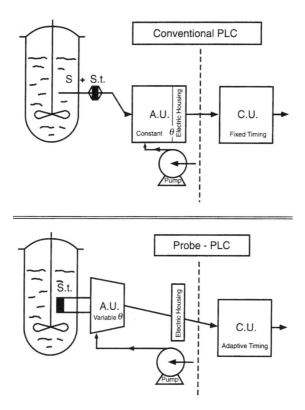

Figure 17 Diagram of probe-process LC. (From Guillemin, G. L., *J. Chromatogr.*, 441, 1, 1988. Copyright Elsevier Publishers. With permission.)

If simple sample pretreatment procedures are insufficient to simplify the complex matrix often observed in process mixtures, multidimensional chromatography may be required. Manual fraction collection from one separation mode and re-injection into a second mode are impractical, so automatic collection and reinjection techniques are preferred. For example, a programmed temperature vaporizer has been used to transfer fractions of sterols such as cholesterol and stigmasterol from a reversed phase HPLC system to a gas chromatographic system.[11] Interfacing gel permeation HPLC and supercritical fluid chromatography is useful for nonvolatile or thermally unstable analytes and was demonstrated to be extremely useful for separation of compounds such as pentaerythritol tetrastearate and a C_{36} hydrocarbon standard.[12]

Automation of sample handling and instrumental reliability in the past have been significant limitations to the use of HPLC for process monitoring. Guillemin[13,14] has described an approach known as probe-process liquid chromatography (PLC). As shown in Figure 17, the sample handling, sampling line, sample conditioning, and chromatographic valves and columns are housed in a single unit, which can be immersed in the mixture to be analyzed. The detector, pump, and electronics are housed in a separate, temperature-controlled unit. The microprocessor-controlled instrument uses the deferred standard concept[15,16] to characterize peaks despite changing retention times, to readjust automatically the time functions of the analysis, and to quantitate results. The deferred standard procedure consists of injecting a reference compound during each analysis sequence. The software recognizes the standard by its area and retention time, and automatically adjusts time functions such as the start time of the next sample. An automated process monitor to perform system calibration and sample preparation and separation continuously for up to a week has been reported for environmental and process monitoring.[17] Illustrations included separations of metal cyanides from plating, mono- and dichloramine from drinking water, chromium, copper and lead, and aluminum.

Since reproducibility of the flow system is critical to obtaining reproducibility, one approach has been to substitute lower-performance columns (50- to 100-μ packings) operated at higher temperatures.[1] Often, improvements in detection and data reduction can substitute for resolution. Chemometric principles are a way to sacrifice chromatographic efficiency but still obtain the desired chemical information. An example of how meaningful information can be derived indirectly from chromatographic separation is the use of system or vacancy peaks to monitor chemical reactions such as the titration of aniline and the hydrolysis of aspirin to salicylic acid.[18]

2.3.3   Miniaturization

One of the major practical problems to the installation of HPLC as a permanent process monitor is the need to replace solvent. A large solvent reservoir may present problems both in terms of size and safety. One solution is the use of packed capillary columns, which consume much less solvent than conventional columns, as the comparison (at constant linear velocity) in Table 1 shows.

Table 1   Solvent Consumption as a Function of Column Diameter

LC column (Flow rate, ml/min)	Solvent consumption (ml/day)
Conventional, 5-mm diameter (2 ml/min)	2880
Small bore, 1.6-mm diameter (0.2 ml/min)	288
Micro-LC, 0.3-mm diameter (7 μl/min)	10

Another advantage of the micro-LC approach is that the required sample size is minimal, so the sample can be drawn from a 1-l laboratory scale reactor without influencing the reactor composition. The ISCO μLC-500 microflow syringe pump has proven to be reliable and reproducible in evaluations in our laboratory. Capillary liquid columns have been fabricated on planar devices such as silicon to form a miniaturized separation device.[19]

Column switching in electrophoretic separations on miniaturized planar devices is remarkably easy. The same approach of etching planar chips has been developed in column chromatography to produce a miniaturized and eventually portable screening device.[20] A photometer using radiolumines-cence from irradiated inorganic or diamond crystals was useful as a source for a miniaturized UV detector, with effective detection limits equivalent to about 100 mAU on a conventional detector.[21] Verzele et al.[22] discuss other advantages of micro-LC, including higher efficiency and easier interfacing with techniques such as mass spectroscopy. Despite the greater stringency of requirements for sample handling and column treatment, the considerable advantages of micro-LC make it attractive in process monitoring.

2.3.4   Specialized detectors

Because process mixtures are complex, specialized detectors may substitute for separation efficiency. One specialized detector is the array amperometric detector, which allows selective detection of electrochemically active com-pounds.[23] Electrochemical array detectors are discussed in greater detail in Chapter 5. Many pharmaceutical compounds are chiral, so a detector capable of determining optical purity would be extremely useful in monitoring syn-thetic reactions. A double-beam circular dichroism detector using a laser as the source was used for the selective detection of chiral cobalt compounds.[24] The double-beam, single-source construction reduces the limitations of flicker noise. Chemiluminescence of an ozonized mixture was used as the principle for a sulfur-selective detector used to analyze pesticides, proteins, and blood thiols from rat plasma.[25] Chemiluminescence using *bis* (2,4, 6-trichlorophenyl) oxalate was used for the selective detection of catalytically reduced nitrated polycyclic aromatic hydrocarbons from diesel exhaust.[26]

Nuclear magnetic resonance, using ^{13}C dynamic nuclear polarization stimulated with a nitroxide radical immobilized on silica, has been used as an on-line technique in normal-phase chromatography of halogenated hydrocarbons.[27] Flow splitting was used to obtain simultaneous ^1H NMR and mass spectrometric data from an on-line separation of an isocratic reversed-phase system of the anti-fungal agent fluconazole and two related triazoles.[28] On-line mass spectrometry has proved especially important in characterizing recombinant proteins. Mass spectrometry has been used for the characterization of intact and tryptic peptides of recombinant single-chain tissue-plasminogen activator of the vampire bat, Desmodus rotun-dus.[29] Electrospray mass spectrometry has been used for other tryptic digests, such as cytochrome c.[30] Laser-induced infrared thermal lensing was used for both direct and indirect detection of environmentally important halophenol herbicides separated by isocratic normal phase chromatography.[31]

A final aspect of process analytical chemistry is the vulnerability of the sensitive detector components to the harsh conditions sometimes encoun-tered in process sampling. It may be possible to physically separate sensitive components, especially the electronics, from the sampling site. Fiber optics have been used in a UV detector to allow signal collection at high temperature, while the detector electronics were at a site remote from the chromatographic apparatus.[32]

2.4   Summary

The operation of HPLC equipment in a process environment requires special considerations. One of the most important concerns is safety. For operation in hazardous production plant areas instrumentation must meet the requirements of NFPA 496, which is the standard for purged and pressurized enclosures for electrical equipment. In addition, process HPLC hardware must be reliable, durable, and easy to service. Sampling is another key consideration. Obtaining a representative sample, particularly from an inhomogeneous reactor, requires ingenuity. Automation of sample processing is essential for continuous process monitoring. The use of tangential flow filters for sample preparation has increased the reliability and decreased the maintenance requirements on process HPLC equipment dramatically. The speed of analysis should be high enough to permit a sampling frequency much more rapid than the change of the process variable of interest. Microbore HPLC is useful to reduce solvent consumption, an important issue in the process environment.

HPLC as a purification technique and as a tool for process monitoring has become increasingly attractive and will find many new applications in the future. Low pressure LC, probe LC, and micro-LC are techniques important to the future of process chromatography. Specialized detectors and multidimensional chromatographic approaches are also of increasing use. Additional literature is available.[22,33-36]

References

1. Synovec, R. E., Moore, L. K., Renn, C. N., and Hancock, D. O., New directions in process liquid chromatography, *Am. Lab.*, 21, 82, 1989.
2. Cotter, R. L. and Li, J. B., The advantages of at-line sampling for process monitoring by HPLC, *Lab. Rob. Autom.*, 1, 251, 1989.
3. Ruzicka, J. and Hansen, E. H., *Flow Injection Analysis*, 2nd ed., Wiley, New York, 1988.
4. Garn, M. B., Gisin, M., Gross, H., King, P., Schmidt, W. and Thommen, C., Extensive flow-injection dilution for in-line sample pre-treatment, *Anal. Chim. Acta*, 207, 225, 1988.
5. Gressin, J. C., Online process chromatography: a useful tool in biochemical processes, *Biotech. Forum*, 5, 38, 1988.
6. Fallick, G. J., Online high performance liquid chromatography, *Anal. Proc. (London)*, 24, 108, 1987.
7. Michaels, S., Crossflow microfilters: the ins and outs, *Chem. Eng.*, Jan. 16, 84, 1989.
8. Doložel, P., Krejčí, M., and Kahle, V., Enrichment technique in an automated liquid microchromatograph with a capillary mixer, *J. Chromatogr. A*, 675, 47, 1994.

9. Chen, A. and Lunte, C. E., Microdialysis sampling coupled on-line to fast microbore liquid chromatography, *J. Chromatogr. A*, 691, 29, 1995.

10. van de Merbel, N. C., Ruijter, G. L. G., Lingeman, H., Brinkman, U. A. Th., and Visser, J., An automated monitoring system using on-line ultrafiltration and column liquid chromatography for *Aspergillus niger* fermentations, *Appl. Microbiol. Biotechnol.*, 41, 658, 1994.

11. Señorans, F. J., Reglero, G., and Herraiz, M., Use of a programmed temperature injector for on-line reversed-phase liquid chromatography-capillary gas chromatography, *J. Chromatogr. Sci.*, 33, 446, 1995.

12. Cortes, H. J., Campbell, R. M., Himes, R. P., and Pfeiffer, C. D., On-line coupled liquid chromatography and capillary supercritical fluid chromatography: large-volume injection system for capillary SFC, *J. Microcol. Sep.*, 4, 239, 1992.

13. Guillemin, C. L., New concept in chromatographic instrumentation. The probe-process liquid chromatograph, *J. Chromatogr.*, 441, 1, 1988.

14. Guillemin, C. L., Problems and proposed solutions for on-line process liquid chromatography: the software-driven chromatograph, *Proc. Cont. Qual.*, 1, 23, 1990.

15. Guillemin, C. L., Calibrating chromatographs on-line. Deferred standard method, *Instrum. Technol.*, 22(4), 43, 1975.

16. Guillemin, C. L.,The deferred standard, *Proc. Cont. Qual.*, 6(1), 9, 1994.

17. Barisci, J. N. and Wallace, G. G., Development of an improved on-line chromatographic monitor with new methods for environmental and process control, *Anal. Chim. Acta*, 310, 79, 1995.

18. Mizrotsky, N. and Grushka, E., Use of system peaks in liquid chromatography for continuous on-line monitoring of chemical reactions, *Anal. Chem.*, 67, 1737, 1995.

19. Manz, A., Verpoorte, E., Effenhauser, C. S., Burggraf, N., Raymond, D. E., Harrison, D. J., and Widmer, H. M., Miniaturization of separation techniques using planar chip technology, *J. High Resol. Chromatogr.*, 16, 433, 1993.

20. Cowen, S. and Craston, D. H., An on-chip miniature liquid chromatography system: design, construction and characterization, in *Micro Total Analysis Systems*, van den Berg and Bergveld, Eds., Kluwer Academic, The Netherlands, 1995, 295.

21. Rinke, G. and Hartig, C., UV photometer based on radioluminescence with applications in HPLC and process control, *Anal. Chem.*, 67, 2308, 1995.

22. Verzele, M., Dewaele, C., and De Weerdt, M., Micro-LC: what are its chances?, *LC-GC*, 6, 966, 1988.

23. McReedy, T. and Fielden, P. R., Amperometric detector for high-performance liquid chromatography, featuring a glassy carbon working electrode array in the wall-jet configuration, *Analyst*, 120, 2343, 1995.

24. Rosenzweig, Z. and Yeung, E. S., Laser-based double-beam circular dichroism detector for liquid chromatography, *Appl. Spectr.*, 47, 207, 1993.

25. Ryerson, T. B., Dunham, A. J., Barkley, R. M., and Sievers, R. E., Sulfur-selective detector for liquid chromatography based on sulfur monoxide-ozone chemiluminescence, *Anal. Chem.*, 66, 2841, 1994.

26. Li, H. and Westerholm, R., Determination of mono- and di-nitro polycyclic aromatic hydrocarbons by on-line reduction and high-performance liquid chromatography with chemiluminescence detection, *J. Chromatogr. A*, 664, 177, 1994.

27. Stevenson, S. and Dorn, H. C., ^{13}C dynamic nuclear polarization: a detector for continuous-flow, on-line chromatography, *Anal. Chem.*, 66, 2993, 1994.

28. Pullen, F. S., Swanson, A. G., Newman, M. J. and Richards, D. S., 'On-line' liquid chromatography/nuclear magnetic resonance spectrometry — a powerful spectroscopic tool for the analysis of mixtures of pharmaceutical interest, *Rapid Comm. Mass. Spectr.*, 9, 1003, 1995.

29. Apffel, A., Chakel, J., Udiavar, S., Hancock, W. S., Souders, C., and Pungor, Jr., E., Application of capillary electrophoresis, high-performance liquid chromatography, on-line electrospray mass spectrometry and matrix-assisted laser desorption ionization-time of flight mass spectrometry to the characterization of single-chain plasminogen activator, *J. Chromatogr. A*, 717, 41, 1995.

30. Davis, M. T., Stahl, D. C., Hefta, S. A., and Lee, T. D., A microscale electrospray interface for on-line, capillary liquid chromatography/tandem mass spectrometry of complex peptide mixtures, *Anal. Chem.*, 67, 4549, 1995.

31. Tran, C. D., Huang, G., and Grishko, V. I., Direct and indirect detection of liquid chromatography by infrared thermal lens spectrometry, *Anal. Chim. Acta*, 299, 361, 1995.

32. Svensson, L. M. and Markides, K. E., Fiber optic-based UV-absorption detector cell for high-temperature open tubular column liquid chromatography, *J. Microcol. Separation*, 6, 409, 1994.

33. Cotter, R. L. and Li, J. B., An at-line approach for HPLC process monitoring, *Adv. Instrum. Control, Part 1*, 44, 313, 1989.

34. Cotter, R. L., Limpert, R. J., and Deluski, C., Rapid and reliable polymer analysis by on-line liquid chromatography, *Am. Lab.*, 18, 50, 1987.

35. Haak, K. K., Carson, S., and Lee., G., Monitoring of high purity water using on-line ion chromatography, *Am. Lab.*, 18, 50, 1986.

36. Mowery, R. A., Online process liquid chromatography, *Chem. Eng.*, 88, 145, 1981.

chapter three

Chromatography in process development

Patricia Puma

0-8493-2682-6/97/$0.00+$.50

3.1 Scope of process chromatography

3.1.1 Introduction

Most industrial processes involve processing very large lots of materials of relatively low unit cost. The numbers and types of contaminants to be removed from these systems do not require chromatographic separations, which would increase unit cost significantly without providing a corresponding increase in purity. In biopharmaceutics, the lot sizes are relatively small and the unit costs are relatively high. In addition, the purification challenges presented can be formidable and can only be addressed chromatographically. Use of other techniques such as precipitation and recrystallization may not be possible. It has been estimated that more than half the costs of processing a recombinant protein arise from downstream processing rather than fermentation.[1] Fortunately, since the potency of biopharmaceuticals is relatively high, with dosages in the milligram range, expensive chromatographic materials and instrumentation are still compatible with a cost-effective process. This chapter will focus on the use of chromatography in the purification of biological materials for pharmaceutical applications. From the earliest stages of development, the process must be designed in such a way that quality is built into the system, in order to ensure that the finished biopharmaceutical product meet essential criteria of purity, stability, safety, and lot-to-lot uniformity. To be successful as an industrial process, it must also be rugged and flexible enough to be scalable, yet sensitive and well defined in order to be validatable.

Preparative liquid chromatography (PLC) is focused on purifying relatively large amounts of a specific product at the lowest cost, as opposed to analytical high performance liquid chromatography, where the emphasis is on highest resolution independent of cost. Typically, PLC has been done in glass columns containing packings with larger particles, and these columns are run at relatively low pressures (< 50 psi). More recently, two variations of PLC have gained increasing acceptance. The first, which is closer in many ways to analytical chromatography than to conventional PLC, involves use of stainless steel columns packed with relatively small particle sized packings (10 to 30 μ) and run at medium pressures of 1000 to 1500 psi.[2] The second variation involves running small particle sized packings in glass columns at low pressures. With the commercial availability of rigid polymer-based packings with small particle sizes and small particle size distributions, operation in this mode can provide a combination of resolution and throughput at low pressure, a combination not thought possible a few years ago.

Purification of biopharmaceuticals often involves the removal of materials with physical characteristics very similar to the desired product, such as failure sequences from DNA synthesis or misfolded proteins from bacterial fermentations. The contaminants, however, may have biological characteristics very different from the desired product, including different antigenicities, bioactivities, and specificities. There are even systems in which the desired biopharmaceutical can be viewed as a trace contaminant of the feedstock. An example of this would be isolation of a nonrecombinant protein from a mammalian culture. In this case, the protein of interest is a relatively small percentage of the total protein present. Added to these challenges is that of designing a process chromatography system that meets current good manufacturing practices (cGMPs). The chromatographic media, hardware, and overall process must be well documented and capable of being validated.

Lastly, considerations of process scale are a very critical point that must be kept in mind even at very early stages of process development.[3] Approval is given for a particular product produced by a very specific process. While changes are permitted in the IND (investigational new drug) stage of biopharmaceutical product development, the process becomes more and more fixed as the project nears the Phase III clinical stage. Once regulatory approval for the product is in progress, the purification process is fixed. The rationale for process design and optimization of purity, throughput, and cost must begin very early on.

3.1.2   Biopharmaceutical materials

3.1.2.1   Overview
While proteins and peptides continue to be the biopharmaceuticals of most widespread interest, new drugs based on carbohydrates and synthetic DNA require industry to develop new separation strategies for process scale purification. Each type of molecule presents its own unique set of challenges in purification with regard to contaminants, breakdown products, altered sequences, misfolding, and, in the case of proteins, immunological variants. The following review highlights some of the more significant problems posed by biopharmaceuticals produced in cell culture or by fermentation (e.g., recombinant proteins) and those produced by chemical synthesis (e.g., oligonucleotides).

3.1.2.2   Biopharmaceuticals from cell culture or fermentation: proteins
While recombinant proteins can be manufactured in *Escherichia coli* in a soluble form, they are most often made in systems in which they accumulate as large insoluble complexes in particles called "inclusion bodies".[4,5] Prior to purification, inclusion bodies must first be solubilized, often using denaturants such as guanidine-HCl. The recombinant protein must then be refolded (or in a more correct sense, folded for the first time) to obtain the desired activity (e.g., enzymatic, immunological, hormonal). The processing of inclusion bodies is extensive and involves a number of nonchromatographic purification steps.[6] These include cell disruption and removal of cell debris by centrifugation or filtration, followed by denaturation/solubilization of the inclusion bodies and refolding of the protein of interest.[7,8] This is the point in purification where difficulties in separation increase sharply, and it is here that chromatographic techniques are usually introduced. In

some cases, the protein may be refolded after initial chromatographic isolation. In these cases, the first chromatographic purification step may be run
in the presence of a denaturant. The types and number of contaminants in
these systems are large and complex and include host organism proteins,
lipids, nucleic acids, and, in the case of many nonbacterial systems, carbohydrates.

With regard to denaturants, it should be mentioned that while the cha-
otrope urea was commonly employed as a denaturant in early industrial
processes (its lack of charge lends itself very well to ion-exchange, for example) and is still seen in the research literature, today it is seldom chosen for
process scale. Even the highest purity, commercially available ureas potentially are contaminated with cyanates. Cyanate ion contamination of concentrated urea solutions[8] can lead to carbamylation of lysine residues, a modification with very serious implications for protein drugs, since significant
changes in antigenicities could result from such modifications. These would
be extremely difficult to detect. Guanidine hydrochloride (GuHCl) is the
most favored chaotropic agent used for denaturation and solubilization.
Unfortunately, GuHCl is highly charged and this limits the choices open to
the process chromatographer for downstream processing. Ion-exchange separations in the presence of 7 M GuHCl, for example, are impossible. Affinity
chromatography techniques are also of limited usefulness in the presence of
GuHCl, since GuHCl is a very effective elution buffer for immunoaffinity
chromatography.

While glycosylated and produced in a soluble, active form, recombinant
proteins produced in mammalian cell lines are not produced in as great a
concentration as in bacterial systems. They are also more labile to proteolytic
degradation from naturally occurring proteases in mammalian cells. In addition, the glycosylation of proteins produced in mammalian culture is significantly affected by cell culture age and culture conditions.[9] Different glycosylation can seriously alter protein activities *in vivo*. Contamination with
viruses is of particular importance when isolating products produced in
mammalian cell culture.[10] Recombinant proteins produced in yeast also accumulate in particulate form in 30- to 60-nm particles,[11] and the glycosylation
patterns of proteins produced in yeast may be very different from those seen
in the naturally occurring mammalian protein. Differences in glycosylation
affect the antigenicity of recombinant proteins, and these effects are potentially of great significance in the clearance of protein drugs.

Proteins, whether recombinant or naturally produced, also present other
special challenges to purification due to misfolding (often involving disulfide
bond scrambling), denaturation, and modified sequences. In addition, a
number of other biological and chemical modifications occur, including
formylation, acetylation, phosphorylation, sulfation, glycosylation, deamidation, and proteolytic cleavage.[12] These modifications are often very difficult to detect and present serious challenges in separation. They are potentially serious and may affect protein products by adversely affecting stability,
solubility, and bioactivity. These variants also possess antigenicities differing

from the desired protein, which can affect clearance rates and even produce serious immunological effects when given to patients.

While the chromatographic techniques available for use with proteins include ion-exchange, hydrophobic interaction, affinity, reversed phase, and gel filtration,[13] certain design preferences exist. In an analysis of 100 papers from scientific journals on protein purification, it was found that ion-exchange chromatography was the most common purification technique, followed closely by affinity chromatography.[14] The chromatographic sequence used most often was ion exchange (IEC), followed by affinity, followed by gel filtration. This approach still reflects that taken by many researchers in biotechnology, particularly in the early stages of research or process development, when there is a need to rapidly purify larger amounts of material. There are relatively few detailed published procedures for complete large-scale purifications of specific recombinant proteins. One that is available is that for purification of recombinant insulin from *E. coli*.[15] The process includes ion-exchange, followed by reversed phase high performance PLC, followed by size exclusion chromatography, and was used successfully at production scale.

Affinity chromatography has serious limitations of scale in process chromatography. Experts are wary of this approach and have found that an early commitment to use affinity chromatography for the sake of convenience or expediency more often than not results in serious long-term limitations. Affinity supports are relatively expensive, since they often entail the use of costly ligands or coupling of the ligands to an activated support. Making an affinity support in-house, either in its entirety or by performing a coupling reaction to a commercially available resin (e.g., coupling proteins to CNBr-activated agarose), puts the burden of the quality control work involved in manufacturing the affinity packing on the shoulders of the user, rather than on those of a packing manufacturer. Generation of the affinity support can become a rate-limiting step in production. In one case in the author's experience, the scale-up of the chemical coupling reactions involved in making the support was itself a major research project. In another case, the production process for a Phase I clinical study incorporated an immunoaffinity support using an antibody ligand that was in very limited supply. Affinity chromatography was abandoned when it became obvious that it could not be scaled to produce a commercially viable process. This necessitated a complete redesign of the purification, with considerable loss of time and money.

While use of affinity chromatography can present problems in large-scale purification, it does have undeniable advantages when it comes to its ability to yield maximum purification on one step. There are promising approaches being taken to provide purification schemes that will ensure smooth transfer to larger scale. One possible approach which ensures availability of a ligand or for use in situations where there are no suitable affinity ligands available is to bioengineer into the protein of interest specific binding sites for which a suitable antibody is readily available. The antigenic determinant on the specific binding sequence can be used as the binding site for purification. This approach was taken to purify fusion proteins for which suitable monoclonal antibodies were not easily made.[16] If the antigenic determinant is attached to the fusion protein by a cleavable linker, linker and binding site can be removed at some later stage of purification. An inherent problem with this approach is that enzymatic or chemical cleavage of the linker will increase the probability of heterogeneity in the product. Affinity chromatography has potential uses for very early stage processing to remove products from mammalian cell culture as they are produced. This is done by integrating the affinity support into a contained cell culture system.[9] Since highly purified product was isolated very early in the process, proteolytic degradation was kept to a minimum. Use of an affinity support in culture was made possible by the development of a stable, nontoxic, autoclavable affinity support: Actigel-ALD. The activated supports generally used in purification are not autoclavable or chemically inert, with toxic groups from the support poisoning cells in culture.

One subset of affinity chromatography that shows promise is immobilized metal-chelate affinity chromatography (IMAC), which takes advantage of the fact that many proteins, perhaps one third, bind to transition metals.[17] These metals are chelated to chromatography supports. Activated supports have become commercially available and the ligands (metals) are relatively inexpensive. One recent variant of IMAC is chelating peptide immobilized metal ion affinity chromatography, or CP-IMAC.[18] Metal binding sites are engineered into proteins that do not naturally possess them to enable purification by IMAC. This is done by genetically engineering the expression of a covalently attached small peptide which possesses metal affinity. Small peptides are preferred since they are less likely to interfere with the desired activity of the protein of interest and could potentially be left on after purification. IMAC on nickel affinity columns has been successfully used to purify clinical-grade malaria vaccine candidates.[19] The recombinant proteins were bound by polyhistidine affinity tails. The tails were left in place after purification, as no biological effect such as autoantibodies was found to result from the polyhistidine tails in animal models. It should be borne in mind, however, that use of metals for purification is not feasible with many proteins. Metals can catalyze disulfide rearrangement and affect protein activity and stability.

Hydrophobic interaction chromatograph (HIC), while very attractive in principle, has proved difficult to scale up for processing. A recent series of articles explores some of the unique problems associated with process-scale HIC. Load sample preparation[20] must be carefully examined to prevent protein aggregate formation in the presence of the relatively high salt concentrations used in this technique. Successful scale-up also requires the setting of wide specifications to accomodate routine variations in the feed.[21] The effect of the salt concentration on capacity may be somewhat more involved than in other chromatographic modes and is a key parameter in obtaining reproducible success as the scale changes.[22]

3.1.2.3   Biopharmaceuticals from organic synthesis: oligonucleotides

Recent interest in antisense DNA drugs has intensified work in the area of oligonucleotide chromatographic purification on a preparative and process scale. Purification of synthetic nucleic acids which have been synthesized on solid supports presents a different type of challenge to the process chromatographer. The types of contaminants are not as varied as in the case of a recombinant protein, and some problems such as misfolding due to disulfide bond scrambling are nonexistent. The contaminants which are present, however, are exceptionally difficult to separate. Automated synthesizers produce oligonucleotides in a stepwise manner, adding one monomer at a time to the growing oligonucleotide chain. From 1 to 3% of the reactions fail during each cycle, however, producing a heterogeneous mixture of oligonucleotides containing failure sequences consisting of oligonucleotides of varying length and possibly even branched DNA.[23] The additive effect of these failures during synthesis of, for example, a 20-mer, will result in purities of between 50 and 60%. In addition to problems due to inefficiencies in coupling, postsynthetic problems such as depurination and cleavage reactions occur and result in additional failure sequences.

Other problems arise when modified oligonucleotides are synthesized. Oligonucleotides are most commonly synthesized today for pharmaceutical purposes in the form of phosphorothioates (PS), in which sulfurization of the phosphodiester bond has taken place (Figure 1).

Phosphodiester   **Phosphorothioate**

Figure 1   Chemical structure of phosphodiester and phosphorothioate.

Sulfur is substituted for oxygen in order to decrease susceptibility of the oligonucleotide to nuclease digestion.[24] The presence of intra and extracellular nucleases is a serious limitation to the use of oligonucleotides as therapeutics. A certain small percentage of the PS oligonucleotide, however, accumulates in the phosphodiester (PO) form (i.e., with oxygen replacing sulfur). Removal of PO from the phosphorothioate preparation is very difficult. The up-front costs involved in synthesis (particularly raw material costs such as the cost of phosphoamidites and solid support matrices such as controlled pore glass) mean that synthetic DNAs are very high-value-added products — purification yields must be even higher than those required in protein purifications in order for the process to be cost effective.

Purification steps prior to chromatography also raise problems with using many of the chromatographic packings used for other biopharmaceuticals. In solid-phase synthesis, the completed DNA chain must first be cleaved from the solid support, and several blocking groups must be removed. This is most often done by incubation in concentrated ammonium hydroxide. Working with ammonium hydroxide in and of itself presents major problems in purification with regard to corrosion and safety. Classically, reversed phase liquid chromatography (LC) on bonded silica phases has been used to separate full-length oligonucleotides containing dimethoxytrityl (DMT), protecting groups from those lacking the trityl group (DMT-off failure sequences), although reverse phase LC cannot separate from one another the DMT-on sequences of varying lengths.[25] DMT-on oligonucleotides are more hydrophobic than the DMT-off failure sequences and are eluted in mobile phases with higher levels of organic solvent. In order to use silica, however, the pH must first be reduced. Removal of ammonium hydroxide prior to chromatography is often done by use of a nonchromatographic method such as evaporation under reduced pressure, which is in itself a difficult technique to scale. As soon as the pH of the unpurified oligonucleotide mixture decreases, the DMT group begins to cleave from the oligonucleotide, eliminating the "handle" used for the primary step in upstream purification. The advent of polymer supports has simplified this somewhat. The ammonium hydroxide solutions can be loaded directly onto polymer-based reversed phase supports and washed away prior to elution with organic solvent. Further purification after removal of the DMT group is relatively straightforward in the case of the phosphodiester oligonucleotides. Nucleic acid separation can be achieved by ion exchange chromatography, mainly due to electrostatic interactions between the phosphate groups of the nucleic acid and the positively charged groups of the ion-exchange medium.[26] Resolution of the n-mer from the (n − 1)-mer can be achieved relatively easily on anion ion-exchange supports. Purification of the more commonly used phosphorothioates, on the other hand, cannot be done as easily. The phosphorothioates are highly charged and bind very tightly to anion ion-exchange columns.[27,28] High salt concentrations, in the range of 2 to 3 M salt, must be used for elution. There are greater similarities in the charges of the n, n − 1, and oxygenated species of oligonucleotides than, for example, between proteins of differing length or sequence. Metelev et al.[29] were able to separate oligoribonucleotides on ion-exchange HPLC, with retention being directly related to length. These methods, however, were primarily analytical methods done with HPLC and are not easily transferred to a preparative or process scale.

More recently, it was shown that addition of dextran sulfate as a displacer improved recovery of phosphorothioate oligonucleotides during ion exchange chromatography.[30] Careful use of overload chromatographic conditions, even in the absence of added displacers, can turn the potential disadvantages of working with phosphorothiotes into very real advantages at process scale, as will be seen in the case history at end of this chapter.

3.1.3   Scale-up considerations

3.1.3.1   Development at small scale

This section will discuss some of the adjustable parameters in method development that are critical to successful process development. As a general approach these parameters are best examined on analytical scale before increasing the scale to something approaching production size for several reasons:

- One has the freedom to examine a multiplicity of variables without sacrificing large amounts of expensive product.
- The process can be scaled relatively quickly in a few steps, e.g., five-fold, tenfold, 100-fold.
- Even in the best designed systems, reoptimization is unavoidable, but it is best to limit reoptimization to problems truly related to scale.

For example, during transfer to process scale, new equipment may be required (e.g., a process chromatography skid with, for example, a different pump design) and methodologies will evolve to compensate for limitations at large scale (e.g., the method of pH adjustment during buffer preparation will change as buffer lots go from 1 liter to several hundred liters). A solid database at small scale provides a solid framework with which to separate issues that more likely are a consequence of increasing scale (new equipment, introduction of alternate suppliers of raw materials, new operators, problems due to changes in mass transfer) from issues that are more closely related to the science involved in process design (choice of optimum separation medium and separation mode for the product, effect of variations in feedstock on purification, interrelationships between capacity/flow/column configuration).

3.1.3.2   Choosing a packing material

The basic approach to choosing a packing for process scale is similar in some respects to that for choosing a packing at an analytical or small preparative scale. By the time one is faced with the project of designing a purification at process scale, characterization of the sample to be purified and contaminants to be removed will be far enough along to provide a good indication of which mode(s) of chromatography to use, and the major contaminants will have been identified. In the process of doing analytical workup and preparing small amounts of product for evaluation in *in vitro* cell culture work and *in vivo* animal studies, potential chromatography supports will be identified and even evaluated in an analytical particle and/or column size. While small-scale analytical work can provide a good scientific basis for designing a process purification system, significant optimization involving choice of particle size, size distribution, and composition of the support remain to be done. Special considerations must be given to the rigidity and chemical composition of the support, including choice of functionalized or nonfunctionalized supports and the use of bonded or coated chemical supports. When purifying biopharma-ceuticals, questions of leachables, resulting from both chemical degradation and loss of particle integrity, are paramount. Cleanability and its relation to the useful life of the resin are also crucial from a regulatory as well as an economic perspective.

3.1.3.2.1   Particle size and size distribution.   Until a few years ago, the general approach in choosing a process-scale chromatography packing was to use media with larger particles and relatively broad particle size distri-butions. This approach was dictated in part by the packings that were com-mercially available and was often limited to the rigid silica-based packings of HPLC and relatively soft dextran-based packings of LC.[31] From a theoret-ical perspective, it is preferable to use a packing with smaller particles having a relatively small size distribution range. The particle distribution range *dp* for the particles should vary by no more than 1.5× from the smallest to largest particles.[31] While smaller particles increase efficiency, they are also respon-sible for decreased permeability and increased pressure. The narrower the particle size distribution, the less likely a packing is to give a significant pressure drop due to the smaller particle components.

In a theoretical evaluation of the use of smaller sized particles (5 to 45 μ) for HPLC, Golshan-Shirazi and Guiochon[32] concluded that, in designing a process and for any particular particle size, the correct combination of col-umn length and run conditions produces similar percentage yields. Calcu-lations indicated that the optimum particle size for maximum percentage yield was approximately 20 μ. Until relatively recently, however, it was not possible to readily obtain smaller packings. Packings of >100 μ were used, as they had better permeability characteristics and in some cases improved capacity, even though they also produce somewhat lower column efficien-cies. Today, however, there are many excellent packing materials available in sizes < 100 μ. The more recent trend in preparative and process-scale chromatography has been to use smaller particle sizes (20 to 30 μ or 10 to 20 μ) with relatively narrow size distribution ranges,[33] in contrast to the 100 μ and above sizes and relatively wider particle size distributions (e.g., 55 to 105 μ) once generally associated with preparative packings. Use of smaller particles, however, has put new demands on column hardware design and packing procedures (see below).

3.1.3.2.2   Chemical composition of packings.   Today, a wider variety of different support materials is available from which to choose. Silica is still widely used, though preparative grades often possess a relatively wide particle size distribution as compared to polymer-based supports. One seri-ous limitation of silica-based supports is the low stability of silicas to alkaline pH conditions, which limits use of caustic solutions in sanitization and depyrogenation. Polymer-based supports, which include poly(styrene-divi-nyl benzene)- or methacrylate-based materials, are widely available and have gained increased acceptance and use. Nonfunctionalized poly(styrene-divinyl benzene) particles themselves serve as an excellent reversed phase packing. These can be functionalized to yield ion-exchange, hydrophobic interaction, and affinity packings. Polymer-based packings are rigid and base stable, generally offer a fairly narrow particle-size distribution range, and often have excellent flow properties. They can yield very good separations at relatively high linear velocities as compared to silica packings or the softer dextrans and agarose supports. It has been this author's observation that in many cases performance of a wide variety of these polymer-based supports improves as flow increases. Dextran-based packings (e.g., Sephadex™ from Pharmacia; Uppsala, Sweden) and agarose-based supports, while considerably less rigid, are still used very often in biopharmaceutical purification. Agarose-based packings have been improved considerably (e.g., Sepharose™ fast flow resins from Pharmacia) and offer improved rigidity and flow properties over what was available several years ago. The cross-linked dextrans and agaroses are also quite stable during treatment with caustic cleaning solutions.

3.1.3.3   Column configuration and flow

Problems occur in translating a purification from an analytical or small preparative scale to a process scale in choosing the correct column configuration (dimensions). The simplest approach to scale-up involves increasing column diameter while keeping column configuration (defined here as dimensions in terms of ratio of height to column diameter) and linear velocity unchanged. This direct approach greatly increases chances of success. It should be remembered that short, wide columns permit maximum flow and are preferable to longer columns. Larger column diameters often result in decreased stability of the packed bed. Columns that are longer rather than wider generally show improved efficiency and may provide improved resolution. Obviously, a compromise between these considerations must be reached, and this compromise is specific to a particular separation.

There is a limited variety of column hardware sizes available for process applications with regard to column diameter. Small-scale experiments done with a view towards predicting conditions at a large scale must be planned with this in mind (see Table 1). For example, when a separation developed on a 2.2-cm diameter × 4-cm height column bed is scaled tenfold with regard to column load, the new bed volume required involves small changes in configuration. These changes, while relatively minor, may have an impact on a separation in some systems. Some changes are substantial and may necessitate extensive re-optimization of the purification methodology. Optimization experiments should be done at a relatively small scale. For example, to evaluate the effect of the 3:1 configuration on the separation, it could be run at small scale in a 5 × 15-cm column.

The linear velocity at which a separation is run is another critical parameter to be optimized in scaleup. When scaling up a process, many chromatographers feel it is advisable, at least initially, to keep linear velocity and column height constant, since this will ensure that the feedstock has the same

Table 1   Changes in Configuration during Scale-Up

Relative scale	Bed volume	Column Height (cm)	Diameter (cm)	Configuration (height:diameter)
1	15.2 ml	4.0	2.2	1.8:1
10	152 ml	7.8	5.0	1.6:1
100	1520 ml	24.0	9.0	2.3:1
1000	15.2 l	21.5	30.0	0.7:1
10,000	152 l	53.8	60.0	0.9:1
10,000	152 l	23.9	90.0	0.3:1

residence time in the packing, with the same chance to interact with packing materials, at both scales. It is this author's observation that the configuration of the column should be kept constant as well, at least until the point where the relative importance of configuration has been studied and is thoroughly understood.

In general, it is preferable to optimize a purification scheme such that it uses the highest linear velocity possible. This increases throughput and, in some cases (such as with polymer-based supports), improves performance. As with other parameters, velocity can be optimized at some smaller scale. It should be noted, however, that if other parameters, such as column configuration or sample load change after scale-up, some optimization work may have to be done for flow as well. From a theoretical perspective, high flow is not an impediment to purification. A theoretical treatment of the effect of flow rate on recovery in overload conditions demonstrated that production rates pass through a maximum which is reached at relatively very high flow.[34]

A very serious problem in working with large column diameters is a decrease in bed stability. Unstable regions are produced in the bed during packing; these can be described as bridges, or arches, surrounded by air spaces.[35] The bridges collapse due to shear forces resulting from solvent flow. Particles can then redistribute themselves and eventually produce a void, with accompanying loss in efficiency. In larger diameter columns, the wall is farther away from most of the packing and the resulting loss of wall support increases instability.

3.1.3.4   Column capacity

Specifications for chromatographic packings often describe maximum loading in terms of the maximum capacity of a unit quantity of the packing to bind some analyte, often a well characterized protein such as bovine serum albumin. The static loading capacity[35] is very different from functional capacity,[2] which is the maximum amount of a particular feedstock that can be loaded and still achieve acceptable purification and recovery. Functional capacity is determined empirically for each type of load and associated set of equilibration and elution conditions and is the capacity definition that is of most concern in developing a process separation.

Preparative-scale and process-scale columns are often described as being run most efficiently in a state of column overload. This concept of overload or nonlinear chromatography arose from a comparison of the optimal conditions identified in analytical-scale HPLC. These conditions are referred to as linear chromatography. There has been an increased emphasis over the last few years on understanding the theory and practical applications of overload chromatography. This activity has been fueled both by the explosive growth of biopharmaceutical chromatography and the availability of new packing materials and column hardware to address the specific needs of this industry.

3.1.3.4.1   Linear vs. nonlinear isotherms.   Analytical scale chromatography relies primarily on linear isotherm retention, in which sufficiently small sample sizes are loaded such that any increase in sample amount results in increases in peak height only, with no alteration in peak retention time.[31] In analytical separations, loads are kept at <1 mg/g packing. As a result of this, sample retention (k') and peak separation (α factor) are independent of mobile phase velocity (u), while column efficiency (N) is affected by u. One approach to preparative chromatographic separation, often used in early stage development to prepare larger amounts of pure product, involves use of larger diameter columns loaded at <1 mg/g and run essentially like large analytical columns. Retention is determined by a linear isotherm. This approach is impractical in process scale chromatography, where the goal is to obtain the highest yield of pure compound with the smallest outlay of time and cost. The approach used in process scale systems involves application of loads of >1 mg/g in order to maximize throughput, and these are defined as being overload conditions. Mazsaroff and Regnier[36] define column overload as occurring when peak width increases more than 10% beyond that gotten for an analytical load. In overload conditions, k' and the separation factor α decrease significantly, and the effect of u on N is also reduced. Industrial columns are routinely run in such overloaded conditions.

The challenge of finding a suitable framework from which to predict behavior of sample bands in nonlinear chromatography has received increased attention as use of chromatography in the biopharmaceutical and biotechnology industries increases. In overload conditions, peaks are highly non-Gaussian, the usual definition of peak resolution is not particularly useful, and columns exhibit low efficiencies as measured by conventional methods.[37] One can describe three kinds of band patterns with regard to resolution.[38] When a small sample size is used, elution bands are completely separate. In lightly overloaded situations, touching bands are obtained; in heavily overloaded situations, overlapping bands are seen. Relatively empirical approaches to predicting band behavior have been suggested in the literature, in which the plate number is redefined and measured in a different manner from that done in analytical separations.[39] More theoretical treatments were attempted by Golshan-Shirazi and Guiochon.[40,41] These authors first derived equations that could be used to predict band profiles of large sample loads based on experimental conditions using Langmuir isotherms. The experimental system used was an extremely simple one when compared to the complex systems used in preparative chromatography. The second study involved a somewhat more complex system with binary mobile phase separations of simple solutes in reverse phase and normal phase systems. While predicted and experimental results agreed well, these systems were also simple as compared to a real-life situation. In a computer simulation study, an approach was proposed for designing preparative scale separations.[38] Key experiments were outlined that would be done using small sample loads under isocratic and gradient conditions. These could be used to design optimum conditions for overload separations.

More recently, there have been attempts to study band patterns as they are affected by shock layers in nonlinear chromatography.[42] Shock layers are steep boundaries that develop when the boundary front of an elution band becomes very steep and self-sharpening at high concentrations. While comparison of predicted and experimental data was promising, this study, like the others mentioned above, was done with single-component samples and awaits further analysis with the kinds of multi-component feeds more frequently encountered in process purifications.

3.1.3.4.2 Displacement chromatography. A form of chromatography which holds great promise but has made relatively small inroads into actual use at a process scale is displacement chromatography. Displacement chromatography is a form of overload chromatography in which a solution of a substance called the "displacer" is loaded behind the feed. The displacer is chosen such that it has a higher affinity for the stationary phase than any of the feed components. Thus, the displacer competes for adsorption to the column and in this manner drives the feed components through the column while they mutually displace each other and separate into component bands.

Horváth[43] presented an excellent review of the history of displacement chromatography and its potential use in process-scale separations. He noted that while the displacement phenomenon had been noted as early as 1906 and displacement chromatography used by Tiselius and others for biochemicals, the method was eclipsed by linear elution chromatography for many years, largely due to the lack of adequate chromatographic supports and instrumentation to support displacement and the rapid development of linear chromatography, in both theory and practice. Since the mid-1980s there have been increasing numbers of reports in the literature of displacement chromatography being used successfully for peptides[44,45] and proteins.[46-48]

Currently, one of the main roadblocks to utilizing this technique in purification of biopharmaceuticals is a lack of commercially available displacers deemed safe for use in pharmaceutical purifications. A review of some of displacers used in the literature[43] indicates why this is a problem: tetrabutylammonium bromide for dipeptide separation, *n*-butanol in separation of adenosine monophosphates from adenosine, octyldodecyldimethylammonium chloride for purification of polymyxin B. Though in theory the displacer front and product front are separate, in reality there is always potential for some contamination of product with a displacer, which itself may be very difficult to separate from the product. More recently, proteins have been displaced using less toxic, more easily separable and possibly more pharmaceutically acceptable displacers including chondroitin sulfate,[49] dextran sulfate,[50] carboxymethyl starches,[51] and protamines.[52] Imidazole, N-protected histidines, and tryptophan were also found to have utility as displacers in metal affinity chromatography systems.[53]

An additional problem exists in which impurities in the displacer itself complicate separation.[54] Also, that the displacer itself must be removed from the column lengthens regeneration time and can adversely affect throughput. Ironically, while the difficulties involved in identifying displacers and in column regeneration have retarded use of displacement as a preparative method, there has been renewed interest in using displacement chromatography in analytical and semi-preparative applications for enrichment of trace compounds.[55,56]

An associated phenomenon may also occur in overload chromatography of some biomolecules which can be referred to as a displacement effect. In a computer simulation study of overlapping bands in overload chromatography, it was noted that gradient elution could result in high recovery of very pure product, and the results were compared to previous reports in the literature of similar success obtained with displacement chromatography.[38] Golshan-Shirazi and Guiochon[57] discussed this phenomenon in some detail as it relates to nonlinear elution. In unresolved bands, containing several components, the proportion of each component in the stationary phase is lower than for unmixed components, resulting in a larger velocity for each. The first component band is less retained than when alone, and it is pushed, or displaced, forward by the second component. This phenomenon can be exploited very successfully at process scale to achieve good yields from very difficult separations, as is described in the case history at the end of this chapter.

3.1.4 Special packing material requirements

One factor of critical importance in choosing a packing for process-scale purification of a biopharmaceutical is its cleanability. It must be periodically determined that material is not carried over from cycle to cycle and that there is adequate control of bioburden and endotoxin loads. Operating under current good manufacturing practice (cGMP) involves developing and validating suitable methodologies for cleaning the packing that may include sterilization, sanitization, and/or depyrogenation. The packing should also be able to withstand a reasonable number of cleaning cycles. As mentioned previously, silica-based packings are not stable to the caustic cleaning solutions favored for removal of pyrogens, although they can be sanitized with acidic alcohol solutions. Cleaning must be validated under cGMP requirements, and validation is usually best performed at process scale, e.g., by taking measurements during an actual production run.[10]

The stability of the packing is also of prime importance with regard to chemical leachables from the support. These should be at minimum amounts under the normal operating conditions as well as the cleaning cycles to which the packing is subjected. The time to determine this is during the early process development stage, before significant expenditures of time and money have been spent on a process that can never produce a safe product or operate in a GMP environment. An excellent place to begin is with the manufacturer. Suppliers who provide process-scale supports for the pharmaceutical and biotechnology industries are increasingly aware of regulatory requirements and often maintain Drug Master Files on their products which are available for examination by the Food and Drug Administration (FDA). In some circumstances, and under special arrangements with the vendor, customers may have access to some of this information as well. The Parenteral Drug Association has set up a model program for vendor certification to help with this process.[58]

Determination of the useful lifetime of a resin occupies much effort in any process development program. In fact, because of validation considerations, it is perhaps the major cost factor. A very useful model has been presented for determining the number of cycles for which a given resin can be used.[59] This program involves activities in six main areas:

- Resin evaluation of both new and used resins (titration of total binding sites, total protein capacity, flow vs. pressure, particle size distribution, total organic carbon removed by cleaning procedures, and microbial and endotoxin analysis)
- Chemical challenge of the resin with the harshest conditions encountered during the process, including accelerated cycling studies
- Evaluation of production column resins (periodic sampling of resin in process columns)
- Model cycling of the first column in the purification stream (this usually is subjected to the harshest cleaning conditions and sees the dirtiest feedstocks), process monitoring of production columns (yield and purity of product, HETP measurements, pressure-vs.-flow tests)
- Demonstrations of resin cleaning (e.g., by sampling resin storage solution)
- Demonstration of removal of leachates (review of manufacturer's regulatory support file, actual measurement of possible leachates in the lab).

3.1.5   Economic considerations

Mazsaroff and Regnier[36] reviewed the key parameters involved in designing a preparative chromatographic process. They identified throughput as the single most important variable in preparative chromatography. In a series of equations, they set up an economic model for evaluating the cost-effectiveness of a purification system, in which column efficiency is evaluated in terms of feed, product purity, product yield, ability of the column to be cleaned, and throughput and operating costs. They noted that even expensive instrumentation and column hardware can be cost effective in the long run, since they will last for many years. The overall conclusion of this analysis was that even very expensive columns can be cost effective, if the cycle time is sufficiently high to maximize throughput. This means that even process-scale HPLC is an option for particular situations.

Colin[2] came to a similar conclusion in a review of this subject area. He emphasizes that it is important to distinguish early on the difference between purification costs (e.g., equipment, solvents, packing material) and production costs (purification and cost of making the crude sample). He noted that a crude sample resulting from a multistep synthesis can itself be very expensive and will enable one to tolerate much higher purification costs. This is indeed the case in purification of synthetic oligonucleotides, where even very steep purification costs are a fraction of the costs of even the raw materials required for synthesis, let alone the total cost of synthesis.

The different areas outlined above for scale-up considerations all provide approaches to reducing cycle time and maximizing throughput:

- Being able to run columns in overload conditions
- Optimizing flow, load, and configuration to run shorter columns at maximum flow and thereby minimize processing time
- Utilizing packings with smaller particle sizes and narrower size distributions to obtain the highest efficiencies
- Operating when possible in low pressure conditions to minimize equipment and column hardware costs
- Choosing rigid, base-stable packings that can withstand the rigors of sanitization/cleaning cycles and be turned around relatively quickly

One interesting concept mentioned in the literature to reduce cycle time is that of backflushing.[60] In this method, the column flow is reversed just after the product of interest has eluted. After a delay time, the next injection is made at the opposite end of the column from the first. This is designed to minimize the time spent eluting strongly retained contaminants.

3.2   Special hardware requirements

Process scale columns generally are constructed essentially the same as analytical columns. With larger column diameters, however, this type of construction does not allow for adequate stabilization of the packing bed.[2] A somewhat different column design that is particularly useful at process scale is the compression column. Compression columns are increasingly popular for use at the process scale. Compression can be static (applied only temporarily during packing) or dynamic (applied continuously). Radial compression is a system, primarily available at the semi-preparative scale, in which compression is placed along the entire wall of the column by immersion of the column in a pressurized fluid. An example of radial compression is the Waters™ radial compression cartridges (Waters Corp; Milford, MA).[61] In dynamic axial compression (DAC), a piston is used to compress the column bed, and this compression is maintained during use. This technique has taken on particular importance with the increasing use of smaller particles for process-scale work. Columns packed using DAC have very good reduced plate heights, between two and three times the particle diameter, for smaller sized silica packings with several particle size distributions, including 8 to 13 μ, 10 μ, 10 to 20 μ, 12 to 25 μ, and 12 to 30 μ.[35]

Another column design that has received increased attention is one that enables radial-flow chromatography. Columns typically use axial separation of samples; i.e., components travel down the length of a vertical column. In a radial flow column, sample is introduced into a central channel of the column and elutes outward from that to an outer channel. Because of the minimal bed depth involved and the large cross-sectional area in the outer channel, pressures under 20 psi are encountered at relatively high flow rates.[62] Successful purifications with radial flow columns have been reported for purification of recombinant proteins,[63,64] with significant advantages offered by radial flow technology in terms of speed, pressure drop, and overall throughput.

Before releasing a process column for chromatography, it is advisable to perform some test to measure efficiency, such as calculating height equivalent theoretical plates (HETP), both to forestall any problems in the column bed and to provide a benchmark by which to measure column reproducibility and predict degradation of the bed or material. Examples of compounds that are relatively innocuous for use in pharmaceutical applications are 1% NaCl (for gel filtration), concentrated buffer solutions (for ion exchange), and benzyl alcohol and parabens for reverse phase LC.[10]

The piping, valves, and fittings used in a biopharmaceutical process chromatography skid also have unique requirements due to the need to be able to validate their cleanability. Sanitary fittings are the preferred design for use in pharmaceutical manufacturing today. These fittings are sold under various trade names (e.g., Sanitech™ and Triclover™) and are designed to eliminate use of glues and allow for easier and more direct hookups and installation. Their primary advantage from a regulatory aspect, however, is that this design eliminates dead spaces and thus potential areas that can serve as traps for microbial contamination. Piping, particularly with regard to the design and installation of joints, must also be planned in order to

eliminate stagnant pools that trap microorganisms. The installation qualification (IQ) phase of validation of a water-for-injection system, for example,
involves extensive documentation and inspection of all welds in the system.
Welding must be done by certified technicians using approved materials.
With regard to piping components, especially plastic components, stringent
requirements exist and must be adhered to when designing a purification
system.[61]

Use of organic solvents in elution buffers at a process scale involves
safety issues not normally encountered at analytical or semi-preparative
scales. A large-scale reverse phase system requires explosion-proof chromatography skids. This adds considerably not only to the cost of the skid itself
(e.g., a nonexplosion-proof unit may cost $100,000 vs. $130,000 for the explosion-proof version), but to many of the accessory costs, as well. For example,
the explosion-proof cabinet must be purged constantly by clean, water- and
oil-free instrument air. Supplying this air requires installation of a relatively
large, costly compressor system. If plastic piping is to be used with organic
solvents at high flow rates, static discharge can develop. This potential
hazard necessitates that the piping must be extensively grounded by wrap-
ping it with with copper mesh or copper wiring.

3.3   Validation considerations

Validation activity is the crucial end-stage in the process development of a
biopharmaceutical chromatographic purification and one which young companies often underestimate. Regarding process validation, the FDA issued
a guideline in 1986 which states:

> "Process validation is establishing documented evi-
> dence which provides a high degree of assurance that
> a specific process will consistently produce a product
> meeting its predetermined specification and quality
> attributes."

"Validation" is not "optimization", but rather a definition of the conditions under which a process is reproducible. Optimization refers to both the
small-scale and process-scale experiments in which the methods and boundaries of operation are defined (e.g., packing, mobile phases, capacity, flow,
cleaning procedures, sample clearance). Even at process-scale, optimization
work may involve significant changes in each run. Validation, on the other
hand, is the end result of a good job at optimization. In chromatography,
validation runs are process-scale runs in which no deliberate changes are
introduced. Uncontrolled variation due to feedstocks and process methodology are often uncovered. These runs provide the benchmark by which a
manufacturer can ensure itself and the FDA that the chromatographic process is fully under control. In drug manufacturing, changes from the validation

conditions may mean the process is no longer in compliance; revalidation of one or more parts of the process will be required. From both a regulatory and business viewpoint, it is necessary to get validation underway as soon as possible. In a recent review, Bala[66] outlined some of the steps required for successful process validation. For a successful product launch, validation of the purification process should be only part of a master validation plan. This plan must be designed and implemented very early in the project to avoid delays in time and keep the entire project on track.

Validation activities have occupied an increasingly large place in the biopharmaceutical industry over the last 20 years. The history of validation tracks the history of cGMP regulations. While the concept of validation had been applied to analytical test methods prior to 1976, when the FDA introduced its proposed new cGMPs in 1976, validation became a prime focus in manufacturing as well. Chapman[67] reviewed the history of validation in the U.S. and traced the development of validation since 1976 under cGMPs to FDA concerns in five areas: sterilization, aseptic processing, water treatment processing, nonaseptic processing, and computer-related systems. Problems in the pharmaceutical industry were recognized by the FDA and triggered issuance of regulatory guidelines by the FDA. These guidelines had wide-ranging implications not only for the areas of original concern, but also for all other phases of pharmaceutical manufacturing.

Validations fall into two types: prospective and retrospective. In prospective validation (see flow chart in Figure 2) the validation is done in a sequential manner, involving installation qualification and operational qualification (IQ/OQ) of equipment (e.g., chromatography instrumentation or column hardware). Appropriate calibrations accompany the IQ/OQ. Process qualification, or PQ, involves formal review and approval of a PQ protocol, execution of this protocol, and issuance of a formal PQ report which includes data analysis and recommendations (i.e., approval/certification of the process). If the process is not approved, the report may recommend a redesign or redoing of the validation protocol and, in some cases, a return of the process to process development for further optimization.

In 1992 the Parenteral Drug Association outlined guidelines for validation of column-based separations in which recommendations for the content of the IQ, OQ, and PQ were given.[10] The IQ provides documented evidence that the equipment used was installed correctly and adheres to specification, and it also includes a statement of system application, an equipment summary, description of utilities, list of standard operating procedures (SOPs) and manuals, spare parts, operating logs, process instrumentation, and materials of construction. The OQ documents that the equipment will perform its function and includes information on system integrity, flows and pressures, gradient formation, detectors, and computer control. The PQ includes extensive testing to document that the process will perform reproducibly and includes detailed data on column performance, purity of feed and fractions, cleaning, operation under SOPs and batch records. PQs are sometimes done concurrently, with data from actual production batches being used.

Figure 2   Flow chart for prospective validation.

The FDA generally insists on no less than triplicate runs to prove consistency. Each critical parameter must have been defined previously as to an acceptable range (i.e., upper and lower limits) and validation runs must be done within these limits.[67] Of particular importance is documentation of the clearance from each step of potential contaminants of the product of interest. These can include heterologous contaminants of the process (e.g., DNA and host cell proteins from a fermentation process, endotoxins, viruses, tritanol from a synthetic oligonucleotide synthesis) as well as preservatives used for column storage by the manufacturer (e.g., sodium azide, ethanol). One approach to demonstrating clearance is by the use of spiking experiments.[68,10] In this method, each chromatographic step is challenged by a spike of the contaminant of interest, given either as an addition at high concentration to the feedstock or, if possible, as a radiolabeled tracer. With proteins expressed in mammalian cultures, collections of model viruses are often employed in clearance studies. Each step in the purification is challenged separately. These experiments are best done at smaller scale to avoid contaminating expensive process equipment and columns.[10]

Retrospective validation uses historical information gathered in actual process runs to evaluate the process. For example, batch records can provide extensive data on column performance and analytical data of fractions and final product can provide valuable information on the efficiency of the chromatographic steps in removing contaminants. Chapman[67] cautions that while retrospective validation is a valid and valuable approach, it is not meant to be retroactive — validation must be done before product is released to market.

The most recent area of validation development concerns computer systems. This area is in a state of flux. Process-scale chromatography instrumentation is largely automated today, and this automation relies extensively on vendor-generated and vendor-validated software packages for process control. Tetzlaff summarized FDA thinking on GMP requirements for automated systems in a series of articles in 1992.[69,70] He pointed out that GMP regulations place ultimate responsibility for verifying the validation of a vendor's software with the user, even though the user firm is not involved in software development and testing. Proper validation documentation of software by the user is an area that is receiving increased attention from the FDA. New guidelines relating specifically to computer concerns are expected to be issued in the near future.

Process validation is, at least to some extent, a moving target due to the rapid pace of change seen in the pharmaceutical industry over the last 20 years. Up-to-date knowledge of industry trends and FDA guidelines must be taken into consideration as early as possible when designing and validating new processes. Indeed, the many changes seen in the pharmaceutical industry in this time period resulting from the use of recombinant DNA technology, barrier technology, computer technology, and improved facility designs recently prompted the PhRMA (Pharmaceutical Regulatory Manufacturing Association) Quality Control Bulk Pharmaceuticals Work Group to publish guidelines that reflect the industry's understanding of *current* GMPs in the manufacturing of drug substances.[71,72]

3.4 Case history

The methods in the literature for purification of oligonucleotide phosphorothioates are largely unsuitable for commercial scale use. Process-scale industrial production of this class of biopharmaceuticals is in its infancy as compared to, for example, recombinant proteins, with the first DNA therapeutics just beginning to enter clinical trials. It is very likely, however, that oligonucleotide drug development and use are poised for explosive growth in the next 5 to 10 years. There is already a pressing need, even at the relatively small production scales currently used, for chromatographic purification techniques that are commercially scalable and suitable for use in GMP processes.

Puma et al.[73] have recently developed a method for large-scale purification of oligonucleotides which employs many of the design characteristics reviewed in the sections above. The purification process was designed from the very start for use in large-scale processing of oligonucleotides, as opposed to being adapted from a previously used analytical methodology. Older purification schemes, which were adapted from successfully used analytical and semi-preparative procedures, separated DMT-on sequences from DMT-off failure sequences by use of reverse phase liquid chromatography (RPLC). We have replaced RPLC with hydrophobic interaction chromatography (HIC). HIC was originally investigated as a relatively simple, scalable technology for preparation of the crude ammoniacal solutions of phosphorothioates for subsequent RPLC purification. This preparation step was previously done using evaporation under reduced pressure to remove ammonium hydroxide and concentrate the crude feedstock. It was desirable from an economic and practical scale-up perspective to eliminate this technique, as it is relatively labor intensive and not easily scaled. While initial results indicated HIC serves this purpose, we later found that proper manipulation of load and elution conditions eliminates all need for subsequent RPLC. This offers two very significant advantages to the process: (1) use of packing materials which are stable at alkaline pH, and (2) elimination of organic solvents from mobile phases in the downstream purification stream. The HIC elution pool is detritylated and the oligonucleotide can be further purified by passage over an anion ion-exchange chromatography column. As mentioned previously, there are few published reports of successful and scalable purification of phosphorothioates on ion-exchange supports. The method developed here can give PS purities of 98 to 100%, with efficient removal of shorter oligonucleotides and very good yields.

For the studies discussed below, a 25-mer phosphorothioate with the sequence CTCTCGCACCCATCTCTCTCCTTCT was used. The HIC packing material used was Phenyl Sepharose™ fast flow, high substitution (Pharmacia). The anion IEC packing material was DEAE 5PW™ (TosoHaas; Philadelphia, PA). The DEAE elution pool was desalted using ultrafiltration on tangential flow filtration membrane cassettes (Pall Filtron; Northborough, MA). The entire process took 2 days, as opposed to 4 days for a previously used RPLC procedure.

The HIC column was equilibrated in ammonium acetate, pH 8.5 to 11.0, and elution was achieved by washing in a solution of lowered salt concentration. Preliminary experiments indicated that the salt concentration of the crude ammoniacal solutions had to be increased by addition of ammonium acetate to ensure binding of DMT-on product to the column. While elution with low concentrations of ammonium acetate was tried, the most successful procedures involved use of plain water (in this case, water for injection). Flow-through and wash fractions contain the DMT-off failure sequences. The water wash contains the DMT-on sequences. Table 2 contains a summary of some of the key experiments used to develop the final HIC protocol.

Table 2 Development of HIC Protocol

pH	Configuration (height:diameter)	Load (mg oligonucleotide loaded per ml packing)	Elution purity (% DMT-on)	Recovery in elution pool DMT-on oligo (%)	Total oligo (%)
Low	2.5:1	8.7	77	63.9	79.7
Low	2.2:1	16.7	98	42.4	66.7
Low	2.5:1	8.7	91	57.9	76.6
High	1:1	13.0	99	48.5	73.5
High	1:1	8.7	96	64.0	89.0
High	1:1	8.7	96	56.6	83.5

Load and elution conditions for experiments in Table 2 were as follows:

Experiment #1:   The load and column equilibration buffer were adjusted to 2 M ammonium acetate. The DMT-on product was eluted in 0.5 M ammonium acetate followed by a linear gradient of 0 to 0.5 M ammonium acetate.

Experiment #2:   The load and the column equilibration buffer were 1 M ammonium acetate. The DMT-on product was eluted with water.

Experiment #3:   The load and the column equilibration buffer were 0.75 M ammonium acetate. The DMT-on product was eluted in 0.01 N NaOH.

Experiments #4, 5, and 6:   The load and column equilibration buffer were 0.75 M ammonium acetate. The DMT-on product was eluted in water. The load for Experiment #4 was 69% DMT-on, 9.6% PO, 57.6% n, and 6.5% n − 1; the load for Experiment. #6 was 63% DMT-on, 9.6% PO, 56.7% n, and 10.3% n − 1.

For linear velocities, the crude product in Experiment #2 was loaded at 75 cm/hr; wash and elution were done at 315 cm/hr. For all other experiments the crude was loaded at 75 cm/hr; wash was done at 315 cm/hr; elution was done at 158 cm/hr.

Lower pHs were used in equilibration and wash buffers in the first three runs, and this is most likely responsible for the lowered DMT-on purities of the elution pool in the first two and the lowered overall yield of DMT-on product in the third. Equilibration and load conditions were adjusted in subsequent experiments to effectively use a relatively short, wide column with a height-to-diameter ratio of 1:1. This is important for a primary recovery step in terms of throughput. Processing can proceed at relatively high flow rates, reducing cost in terms of time and minimizing loss of the DMT groups from the product during recovery. In the second experiment, we tried to further diminish loss of DMT-on product by increasing the linear velocity at which the column was eluted. While this improved purity of the elution pool, yield suffered. The increased load may also have played a role in this;

Table 3 IEC of Sodium Salt of Oligonucleotide

| Configuration (height:diameter) | Buffer (NaCl) | Elution pool | | Purity by IEC-HPLC |
		Product recovery (%)	Total recovery (%)	
5:1	1.0 *M*	93.4	83.5	98.4
1.2:1	1.0 *M*	73.0	67.0	99.0
1.2:1	0.85 *M*	80.0	86.0	97.0

even though there was no loss of DMT-on product in the breakthrough fractions, the functional capacity of the packing to resolve product may be compromised at the higher loading level.

Differences in the quality of the crude ammoniacal solutions also play a significant role in yield from HIC. The effect of the slightly lowered DMT-on purity and significantly increased level of n – 1 failure sequences in the starting material of the fifth experiment as compared to the sixth are readily apparent in the differing yields. While the percentages of DMT-on purities in Experiments #5 and 6 were somewhat less than the target of 98% DMT-on purity we originally wanted, subsequent experiments using fresher, higher-quality crude materials more indicative of materials used in production have given consistent purities in the elution pools of ≥98% DMT-on.

The IEC column used to obtain more highly purified 25-mer is equilibrated in Tris-HCl (pH 7.2 to 8.0), containing 0 to 0.85 *M* NaCl, and elution is achieved by use of Tris buffer containing 0.85 to 2 *M* NaCl. Initial experiments to optimize run conditions on the DEAE 5PW™ column were done using the sodium salt of the oligonucleotide and were extremely interesting as an illustration of the sometimes drastic effect a change in column configuration can have during scale-up. There appeared to be a very critical relationship between configuration and concentration of NaCl in the equilibration buffer. The longer the column in relation to the diameter, the greater the concentration of NaCl that was required to elute the product. Table 3 highlights some key experiments done to investigate this phenomenon. The use of the ammonium salt of the oligonucleotide, which was explored subsequently, indicated a reduced sensitivity to column configuration.

Contaminating PO and short oligonucleotides of n ≤ 23 are separated in the breakthrough and initial wash fractions. PS oligonucleotide eluted in the latter wash fractions and with application of 2 *M* NaCl. The load level was absolutely critical for this separation to succeed, with the optimum load level at 6.5 mg/ml DEAE 5PW™. An increase in load caused poor product recoveries of <60%, although purity of the elution is very good; a decrease in load significantly below 6.5 mg/ml packing produced even worse results — no purification occurred, and in some experiments, purity in terms of IEC-HPLC analysis actually decreased. We are currently exploring the mechanism behind this phenomenon. Preliminary results suggested that a mixed-mode displacement effect was responsible. At 6.5 mg/ml packing, the column was sufficiently overloaded that the PO front was pushed forward or displaced by the PS front in a very sharp band. At higher levels, the bands were not as sharp and mixing occurred. At lower levels, however, the column was functionally underloaded and did not work in the displacement mode; separation of PO and PS species, with their essentially identical charges, must rely at this load level on a purely ion-exchange mechanism, and the chromatographic packing was inadequate to accomplish this.

Preliminary cost analysis indicated that even though the packing material and column hardware initially were more costly than those used in the original RPLC purification process, the savings in cycle time and improved purity of the product will make the HIC/IEC approach an economically favorable one. Other phosphorothioates of differing size and sequence have been successfully purified using this approach. Results with regard to both yields and purities are equivalent to work with the 25-mer and suggest that the combination of hydrophobic interaction chromatography/anion ion-exchange chromatography/tangential flow filtration will provide a generalized scheme for economically purifying oligonucleotides at process scale. While base-stable polymer RPLC packing materials such as 20-μ nonfunctionalized poly(styrene-divinyl benzene) are commercially available and should also work very well with regard to streamlining purification and providing good cycle times, these polymers are significantly more costly than the HIC support while offering degrees of purity and yield similar to what we can achieve on HIC, as well as possessing the disadvantage of having to use organic solvents in downstream purification.

3.5   Summary

The principal considerations involved in design of a process-scale chromatographic purification include scalability, reproducibility, safety, and validatability. Cost factors, however, must by necessity enter into all industrial decisions. Due to the high value-added nature of most biopharmaceuticals, this cost factor is driven by throughput, rather than by capital investment cost.

References

1. Ladisch, M. R. and Kohlmann, K. L., Recombinant human insulin, *Biotechnol. Prog.*, 8, 469, 1992.
2. Colin, H., Large-scale high-performance preparative liquid chromatography, in *Preparative and Production Scale Chromatography*, Ganetsos, G. and Barker, P. E. Eds., Marcel Dekker, New York, 1988, 11.
3. Leser, E. W. and Asenjo, J. A., Rational design of purification processes for recombinant proteins, *J. Chromatogr.*, 584, 43, 1992.
4. Wheelwright, S. M., Designing downstream processes for large-scale protein purification, *Bio/Technology*, 5, 89, 1987.

5. Schein, C. H., Production of soluble recombinant proteins in bacteria, *Bio/Technology*, 7, 1141, 1989.

6. Titchener-Hooker, N. J., Gritsis, D., Mannweiler, K., Olbrich, R., Gardiner, S. A. M., Fish, N. M., and Hoare, M., Integrated process design for producing and recovering proteins from inclusion bodies, *BioPharm*, July/Aug., 34, 1991.

7. Lowe, P. A. and Rhind, S. K., Solubilization, refolding, and purification of eukaryotic proteins expressed in *E. coli*, in *Protein Purification — Micro to Macro*, Burgess, R., Ed., Alan R. Liss, New York, 1988, 429.

8. Knuth, M. W. and Burgess, R. R., purification of proteins in the denatured state, in *Protein Purification — Micro to Macro*, Burgess, R., Ed., Alan R. Liss, New York, 1988, 279.

9. Grandics, P., Szathmary, S., Szathmary, Z., and O'Neill, T., Integration of cell culture with continuous, on-line sterile downstream processing, *Ann. N. Y. Acad. Sci.*, 646, 322, 1991.

10. Levine, H. L., Tarnowski, S. J., Dosmar, M., Fenton, D. M., Gardner, J. N., Hageman, T. C., Lu, P., Sofer, G., and Steininger, B., Industry perspectives on the validation of column-based separation processes for the purification of proteins, *J. Parenteral Sci. Technol.*, 46, 87, 1992.

11. Muller, F., Bruhl, K., Freidel, K., Kowallik, K. V., and Ciriacy, M., Processing of TY1 proteins and formation of Ty1 virus-like particles in *Saccharomyces cerevisiae*, *Mol. Gen. Genet.*, 207, 421, 1989.

12. Righetti, P. G., Isoelectric focusing of proteins in conventional and immobilized pH gradients, in *Protein Structure: A Practical Approach*, Creighton, T. E., Ed., IRL Press, New York, 1989, 23.

13. Asenjo, J. A., Parrado, J., and Andrews, B. A., Rational design for purification processes for recombinant proteins, *Ann. N. Y. Acad. Sci.*, 646, 334, 1991.

14. Bonnerjea, J., Oh, S., Hoare, M., and Dunnill, P., Protein purification: the right step at the right time, *Bio/Technology*, 4, 954, 1986.

15. Kroeff, E. P., Owens, R. A., Campbell, E. L., Johnson, R. D., and Marks, R. I., Production scale purification of synthetic human insulin by reversed-phase high-performance liquid chromatography, *J. Chromatogr.*, 461, 45, 1989.

16. Downham, M., Busby, S., Jefferis, R., and Lyddiatt, A., Immunoaffinity chromatography in biorecovery: an application of recombinant DNA technology to generic adsorption processes, *J. Chromatogr.*, 584, 59, 1992.

17. Porath, J., Carlsson, J., Olsson, I., and Belfrage, G., Metal chelate affinity chromatography, a new approach to protein fractionation, *Nature*, 258, 598, 1975.

18. Smith, M. C., Engineering metal binding sites into recombinant proteins for facile purification, *Ann. N. Y. Acad. Sci.*, 646, 315, 1991.

19. Takacs, B. J. and Girard, M.-F., Preparation of clinical grade proteins produced by recombinant DNA technologies, *J. Immunol. Meth.*, 143, 231, 1991.

20. Gagnon, P., Grund, E., and Lindback, T., Large-scale process development for hydrophobic interaction chromatography, part I: gel selection and development of binding conditions, *BioPharm*, 8, 21, 1995.

21. Gagnon, P. and Grund, E., Large-scale process development for hydrophobic interaction chromatography, part III: factors affecting capacity determination, *BioPharm*, 9, 34, 1996.

22. Gagnon, P., Grund, E., and Lindback, T., Large-scale process development for hydrophobic interaction chromatography, part II: controlling process variation, *BioPharm*, 8, 36, 1995.

23. Caruthers, M. H., Synthesis of oligonucleotides and oligonucleotide analogues, in *Oligodeoxynucleotides — Antisense Inhibitors of Gene Expression*, Cohen, J. S., Ed., CRC Press, Boca Raton, FL, 1989, 7.
24. Stein, C. A. and Cohen, J. S., Phosphorothioate oligodeoxynucleotide analogues, in *Oligodeoxynucleotides — Antisense Inhibitors of Gene Expression*, Cohen, J. S., Ed., CRC Press, Boca Raton, FL, 1989, 97.
25. Zieske, L. R., Novel approaches in synthetic polynucleotide purification, *Biochromatography*, 3, 112, 1988.
26. Kasai, K., Size-dependent chromatographic separation of nucleic acids, *J. Chromatogr.*, 618, 203, 1993.
27. Agrawal, S. A., Tang, J. Y., and Brown, D. M., Analytical study of phosphorothioate analgues of oligodeoxynucleoides using high-performance liquid chromatography, *J. Chromatogr.*, 509, 396, 1990.
28. Warren, W. and Vella, G., Analysis and purification of synthetic oligonucleotides by high-performance liquid chromatography, in *Oligonucleotide Synthesis Protocols*, Agrawal, S., Ed., Humana Press, Totowa, NJ, 1993, 235.
29. V. Metelev, K. Misiura, and S. Agrawal, Ion-exchange HPLC analysis of oligoribonucleotides and chimeric oligodeoxyribonucleotides, *Ann. N. Y. Acad. Sci.*, 660, 321, 1992.
30. Gerstner, J. A., Economics of displacement chromatography — a case study: purification of oligonucleotides, *BioPharm*, 9, 30, 1996.
31. Snyder, L. R. and Kirkland, J. J., *Introduction to Modern Liquid Chromatography*, John Wiley & Sons, New York, 1979, 168.
32. Golshan-Shirazi, S. and Guiochon, G., Theory of optimization of the experimental conditions of preparative liquid chromatography: optimization of column efficiency, *Anal. Chem.*, 61, 1368, 1989.
33. Verzele, M., Preparative LC, *Anal. Chem.*, 62, 265A, 1990.
34. Ghodbane, S. and Guiochon, G., Effect of mobile phase flow-rate on the recoveries and production rates in overloadde elution chromatography: a theoretical study, *J. Chromatogr.*, 452, 209, 1988.
35. Colin, H., Hilaireau, P., and de Tournemire, J., Dynamic axial compression columns for preparative high performance liquid chromatography, *LC-GC*, 8, 302, 1990.
36. Mazsaroff, I. and Regnier, F. E., An economic analysis of performance in preparative chromatography of proteins, *J. Liq. Chromatogr.*, 9, 2563, 1986.
37. Snyder, L. R. and Kirkland, J. J., *Introduction to Modern Liquid Chromatography*, John Wiley & Sons, New York, 1979, 615.
38. Snyder, L. R., Dolan, J. W., and Cox, G. B., Preparative high-performance liquid chromatography under gradient conditions. III. Craig simulations for heavily overloaded separations, *J. Chromatogr.*, 484, 437, 1989.
39. Eble, J. E., Grob, R. L., Antle, P. E., and Snyder, L. R., Simplified description of high-performance liquid chromatographic separation under overload conditions, based on the Craig distribution model. I. Computer simulations for a single elution band assuming Langmuir isotherm, *J. Chromatogr.*, 384, 25, 1987.
40. Golshan-Shirazi, S. and Guiochon, G., Analytical solution for the ideal model of chromatography in the case of a Langmuir isotherm, *Anal. Chem.*, 60:, 2364, 1988.
41. Golshan-Shirazi, S. and Guiochon, G., Comparison between experimental and theoretical band profiles in nonlinear liquid chromatography with a binary mobile phase, *Anal. Chem.*, 61, 1276, 1989.

42. Zhu, J., Ma, Z., and Guiochon, G., The thickness of shock layers in liquid chromatography, *Biotechnol. Prog.*, 9, 421, 1993.

43. Horváth, Cs., Displacement chromatography: yesterday, today, and tomorrow, in *Science of Chromatography*, Brunner, S., Ed., Elsevier Press, Amsterdam, 1985, 179.

44. Cramer, S. M. and Subramanian, G., Recent advances in the theory and practice of displacement chromatography, *Sep. Purif. Meth.*, 19, 31, 1990.

45. Cramer, S. M., El Rassi, S., and Horváth, Cs., Tandem use of carboxypeptidase Y reaction and displacement chromatograph for peptide synthesis, *J. Chromatogr.*, 394, 305, 1987.

46. Torres, A. R., Krueger, G. G., and Peterson, E. A., Purification of Gc-2 globulin from human serum by displacement chromatography: a model for the isolation of marker proteins identified by two dimensional electrophoresis, *Anal. Biochem.*, 144, 469, 1985.

47. Torres, A. R., Edberg, S.C., and Peterson, E. A., Preparative high-performance liquid chromatography of proteins on an anion exchanger using unfractionated carboxymethyl displacers, *J. Chromatogr.*, 389, 177, 1987.

48. Lee, A. L., Liao, A. W., and Horváth, Cs., Tandem separation schemes for preparative high-performance liquid chromatography of proteins, *J. Chromatogr.*, 443, 31, 1988.

49. Liao, A. W. and Horváth, Cs., Purification of β-galactosidase by combined frontal and displacement chromatography, *Ann. N. Y. Acad. Sci.*, 589, 182, 1990.

50. Jen, S. C. D. and Pinto, N. G., Dextran sulfate as a displacer for the displacement chromatography of pharmaceutical proteins, *J. Chromatogr. Sci.*, 29, 478, 1991.

51. Ghose, S. and Mattiasson, B., Evaluation of displacement chromatography for the recovery of lactate dehydrogenase from beef heart under scale-up conditions, *J. Chromatogr.*, 547, 145, 1991.

52. Gerstner, J. A. and Cramer, S. M., Cation-exchange displacement chromatography of proteins with protamine displacers: effect of salt-induced gradients, *Biotechnol. Prog.*, 8, 540, 1992.

53. Suresh, V., Gallant, S., and Cramer, S., Immobilized metal affinity chromatography: displacer characteristics of traditional mobile phase modifiers, *Biotechnol. Prog.*, 12, 84, 1996.

54. Zhu, J., Katti, A. M., and Guiochon, G., Effect of displacer impurities on chromatographic profiles obtained in displacement chromatoraphy, *Anal. Chem.*, 63, 2183, 1991.

55. Sütfeld, R., Preparative liquid chromatography with analytical separation quality: interval injection/displacement reversed-phase high-performance liquid chromatography, *J. Chromatogr.*, 464, 103, 1989.

56. Ramsey, R., Katti, A. M., and Guiochon, G., Displacement chromatography applied to trace component analysis, *Anal. Chem.*, 62, 2557, 1990.

57. Golshan-Shirazi, S. and Guiochon, G., Optimization in preparative liquid chromatography, *Am. Biotechnol. Lab.*, 8, 26, 1990.

58. PDA, supplier certification — a model program, *J. Parenteral Sci. Technol.*, 43, 151, 1989.

59. Seely, R. J., Wight, H. D., Fry, H. H., Rudge, S. R., and Slaff, G. F., Biotechnology product validation. Part 7. Validation of chromatography resin useful life, *Biopharmacology*, 41, 48, 1994.

60. Colin, H., Hilaireau, P., and Martin, M., Flip-flop elution concept in preparative liquid chromatography, *J. Chromatogr.*, 557, 137, 1991.

61. Little, J. N., Cotter, R. L., Prendergast, J. A., and McDonald, P. D., Preparative liquid chromatography using radially compressed columns, *J. Chromatogr.*, 126, 439, 1976.

62. Saxena, V., Subramanian, K., Saxena, S., and Dunn, M., Production-scale radial flow chromatography, *Biopharmacology*, 36, 46, 1989.

63. Weaver, K., Chen, D., Walton, L., Elwell, L., and Ray, P., Uridine phosphorylase purified from total crude extracts of *E. coli* using Q Sepharose™ and radial-flow chromatography, *Biopharmacology*, 37, 25, 1990.

64. McCartney, J. E., Rapid purification of a recombinant protein using tandem radial flow ion-exchange column chromatography, *BioTechniques*, 11, 648, 1991.

65. Code of Federal Regulations, 21, Parts 170 to 199, revised as of April 1, 1993.

66. Bala, G., An integrated approach to process validation, *Pharm. Eng.*, May/June, 57, 1994.

67. Chapman, K. G., A history of validation in the United States, Part I, *Pharm. Technol.*, Oct., 82, 1991.

68. Hageman, T. C., An analysis of clearance factor measurements performed by spiking experiments, *Biopharmacology*, July/Aug., 39, 1991.

69. Tetzlaff, R. F., GMP documentation requirements for automated systems, Part I, *Pharm. Technol.*, 16, 112, 1992.

70. Tetzlaff, R. F., GMP documentation requirements for automated systems, Part II, *Pharm. Technol.*, 16, 60, 1992.

71. Lazar, M. S., PhRMA guidelines for the production, packing, repacking, or holding of drug substances, Part I, *Pharm. Technol.*, 19, 22, 1995.

72. Lazar, M. S., PhRMA guidelines for the production, packing, repacking, or holding of drug substances, Part II, *Pharm. Technol.*, 20, 50, 1996.

73. Puma, P., Duffey, D., and Dawidczyk, P., U.S. Patent appl. 94,944, Purification of oligodeoxyribonucleotide phosphorothioates using DEAE-5PW anion ion-exchange chromatography and hydrophobic interaction chromatography, 1994.

chapter four

Reversed phase HPLC

Rekha D. Shah and Cynthia A. Maryanoff

0-8493-2682-6/97/$0.00+$.50
© 1997 by CRC Press, Inc.

4.1   Introduction

4.1.1   Chromatographic modes

Of the various modes of high performance liquid chromatography (HPLC), reversed phase (RPLC) is the most commonly used, with normal phase, ion exchange, gel permeation, chromatofocusing, metal interaction, and affinity chromatography being the principal alternative modes. Normal phase is a phrase used to indicate that the stationary phase is more polar than the solvent, and reversed phase indicates that the stationary phase is less polar than the solvent. The development of aqueous-organic gradient generation, which allows the polarity to be varied over a wide range during a single run, has made the distinction between reversed and normal phase somewhat artificial, although the distinction is still useful. In general, adsorption of a solute to a reversed phase is driven by hydrophobic interactions, while adsorption to normal phase is often driven by hydrogen bonding between the solute and stationary phase. Normal phase and reversed phase HPLC predominate in the analysis of small organic molecules. Reversed phase was extended long ago to the analysis of macromolecules, such as proteins, while the extension of normal phase to macromolecules has been slower. A considerable variety of very efficient reversed phase columns is available. The present chapter will examine the theory and practice of reversed phase chromatography, beginning with an examination of the separation of small molecules.

4.1.2   Isocratic column performance

The mechanism of reversed phase chromatography can be understood by contrast with normal phase chromatography. Normal phase liquid chromatography (NPLC) is usually performed on a polar silica stationary phase with a nonpolar mobile phase, while reversed phase chromatography is performed on a nonpolar stationary phase with a polar mobile phase. In RPLC, solute retention is mainly due to hydrophobic interactions between the solutes and the nonpolar hydrocarbon stationary surface. The nonpolar components of a sample interact more with the relatively nonpolar hydro-carbon column packing and thus elute later than polar components. In NPLC, many types of interaction have been described, but hydrogen bonding often predominates. The elution order of solutes in RPLC is in the order of decreasing polarity, i.e., increasing hydrophobicity, while in NPLC the least polar compound elutes first. Thus, NPLC and RPLC are complementary in elution order. The general theory to describe separation is, however, similar. The analyte is capable of adsorbing to and desorbing from the stationary phase, so it migrates through the column more slowly than the solvent. The width of the analyte band is limited at a theoretical minimum by the volume in which the analyte is injected and is increased by processes that tend to disperse the band, such as diffusion of the analyte in the mobile phase. All things being equal, the later that an analyte elutes, the broader the band will be. Because on-line detectors are in such common use, chromatographic bands are generally referred to as peaks.

The isocratic reversed phase solvent system consists of water (polarity, $p' = 10.2$), the most polar solvent in RPLC, as a primary solvent to which water-miscible organic solvents such as methanol ($p' = 5.1$), acetonitrile ($p' = 5.8$), or tetrahydrofuran ($p' = 4.0$) are added. In order to optimize the speed of separation for an analyte pair, the proportions of water to nonpolar solvent are chosen such that the capacity factor of the last-eluting analyte of interest has a value of about 2.[1-3]

The capacity factor, k', is defined as

$$k' = (t_r - t_o)/t_o,$$

where t_r is the retention time and t_o is the void time, i.e., the elution time of a nonretained component. On a column with a void time of 2 minutes, the retention time of an analyte would have to be three times the void volume (6 minutes) to achieve a k' of 2. If the mobile phase is made more polar, the retention time, and therefore k', would increase. The void time t_o should be measured with care, as the value of k' is dependent on proper measurement. It has been shown that the first peak seen may not represent the actual void volume.[4] The first peak may elute later than the void volume, due to adsorption, or may elute earlier than the void volume, perhaps due to an electrophoretic effect. Injection of a nanomolar solution of a simple anion has been recommended for measurement of the void volume.[4] As t_r and k' increase, peak width increases at the expense of peak height, an effect that is termed the general elution problem.[1] The general elution problem is that early-eluting peaks are poorly resolved, while late-eluting peaks are broad and difficult to detect, with excessive elution times.

For a complex sample, it may not be possible to optimize the k' values only for maximum speed, since resolution is also a consideration. The resolution of an analyte pair is defined as:[1,2]

$$R = 2(t_2 - t_1)/(w_2 + w_1)$$

where t_2 and t_1 are the retention times of the two analytes, and w_2 and w_1 are the baseline widths, set by taking the tangent to the inflection points of each peak. If the peaks are Gaussian, the baseline width is four times the peak variance σ, which is measured as the peak width at half-height in units of time.

If the peak widths are similar, the resolution can be written as

$$R = (t_2 - t_1)/4\sigma$$

Therefore, a 4σ separation ($R = 1$), in which peak retention times differ by four times the width at half-height, corresponds to a 2% area overlap between peaks.[1] The maximum number of peaks that could be separated in a given time period assuming a given value of R, is defined as the peak capacity.[1] The peak capacity must be greater — usually *much* greater — than the number of components in the mixture for a separation to succeed. The resolution of two compounds can also be written in terms of the number of plates of a column, N, the selectivity, α, and the capacity factors, k'_1 and k'_2, as[1,2]

$$R = (1/4)\ (\alpha - 1)\ N^{1/2}\ [k'_2/(1+k'_2)].$$

The selectivity is calculated as the ratio $\alpha = k'_2/k'_1$.

The efficiency, or plate count of a column N is often calculated as 5.54 $(t_r/\sigma)^2$, where t_r is the retention time of a standard and σ is the peak width in time units at half-height.[1,2,5] This approach assumes that peaks are Gaussian; a number of other methods of plate calculation are in common use. Values measured for column efficiency depend on the standard used for measurement, the method of calculation, and the sources of extra-column band broadening in the test instrument. Therefore, efficiency measurements are used principally to compare the performance of a column over time or to compare the performance of different columns mounted on the same HPLC system.

Because column efficiency is so central to obtaining resolution, a great deal of study has gone into defining the dependence of peak broadening on column parameters, such as particle size, flow rate, temperature, and other fundamental operating parameters. To eliminate the column length as a parameter, the efficiency of the column is generally expressed, not in plates, but in plates per unit length. The inverse of plates per unit length is the length of column required to generate one plate, and this is called the plate height, H. The smaller the plate height, the better the column. The dependence of plate height on operating parameters can be expressed as

$$H = c_1 + c_2\ f(T)/v + c_3\ v/f(T) + c_4[v/f(T)]^{2/3} + V/g(T)$$

where c_1 through c_4 are positive constants dependent principally on the particle size, v is the flow velocity, $V = vk'/(1 + k')^2$, and $f(T)$ and $g(T)$ are the temperature-dependent solute diffusivity and desorption rate constant, respectively.[6] The interesting point about the temperature-dependent terms is that the plate height can be reduced at higher temperatures by better transfer from the adsorbed state to the stationary phase through g(T), even though higher temperatures promote band-broadening by increased solute diffusion through f(T). Often, increasing the temperature does not improve column performance,[7] but (particularly with macromolecules) higher temperatures may substantially improve column performance.

In simpler formulations,[1,8-11] the fourth and fifth terms are neglected, and the temperature is taken to be constant, so the plate height equation is written as

$$H = c_1 + c_2'/v + c_3'v,$$

which is known as the van Deemter equation. The constants are related to eddy diffusion (c_1), longitudinal diffusion (c_2'), and the effects of mass transfer (c_3'). The van Deemter equation emphasizes the dependence of the plate height on flow. At low flow rates, the second term in the van Deemter equation becomes large, so performance deteriorates. At high flow rates, the third term in the van Deemter equation becomes large, so performance deteriorates. There is a an optimum flow rate at intermediate flow rates. Sometimes, longitudinal diffusion is not important, so the second term may be neglected.[1] In such cases, the slower the flow, the better; however, often column performance improves with increasing flow.[7] Therefore, in optimizing separations, slower is not necessarily better.

Peak dispersion does not come only from broadening on the column. Among the factors that have been identified as important in the performance of a chromatographic system are variances arising from the injector, connecting tubes, and the detector.[12] In evaluating a chromatographic system, it is useful to compare the peak volume (i.e., the volume of mobile phase in the eluted peak) with the injection volume, the internal volume of tubing, and the internal volume of the detector cell.

Peak shape is also an important issue in column performance. Much of chromatographic theory idealizes peaks as having a Gaussian shape, although invariably there is some degree of asymmetry. The simplest method of quantitating asymmetry is to draw a vertical perpendicular to the peak maximum. At an arbitrary fraction of the peak height, the width from the leading and trailing end of the peak to that perpendicular is measured. The ratio of those two widths, taken such that it is greater than one, is called the "asymmetry".

A more sophisticated means of quantitating peak asymmetry is through the theory of peak moments.[13,14] Briefly, the zeroth moment is the peak area, the first moment is the average retention time

$$\int_0^\infty t\, h(t)\, dt / M_o$$

and the second moment is the peak variance. The third moment is the skew, and the fourth moment is the excess. The higher moments are defined as

$$\int_0^\infty (t - M_1)^n h(t)\, dt / M_o$$

where t is the time, M_1 is the average retention time, h(t) is the chromatographic peak height at time t, and M_o is the peak area. Peak distortion may be studied by assuming that the peak shape is an exponentially modified Gaussian form, in which model the peak moments may be solved to generate two parameters, σ and τ, which express the standard deviation of the Gaussian components and the time constant of the exponential modifier, respectively.[13,14] The co-elution of a second component may be another cause of peak asymmetry. Multiwavelength detection has, as discussed in Chapter 1, proved useful for differentiating peak distortion from the coelution of another component.[15]

Other parameters that have been used to characterize column performance include the column pressure drop and the column flow resistance.[16,17] The column pressure drop is simply the difference in pressure observed when the column is or is not in-line. The column flow resistance normalizes for particle diameter, solvent viscosity, and column length. One may also wish to compare issues of cost per analysis and column lifetime in evaluating a column.[18]

4.1.3   Isotherms

So far, it has been assumed that elution is independent of analyte load or the presence of multiple components in a mixture. If this condition holds, then the analyte concentration in the mobile phase is directly proportional to the concentration in the stationary phase, no matter what the concentration is. Experimentally, this could be determined by incubating various concentrations of an analyte with a fixed amount of stationary phase and measuring the amount adsorbed. A plot of the concentration of analyte in the mobile phase on the x-axis vs. that in the stationary phase on the y-axis would be linear, and such a plot is called a "linear isotherm". A convex isotherm implies that tailing would be expected, and a concave isotherm implies that fronting is expected.

4.1.4   Isocratic separations of mixtures

For complex mixtures, the k' values should be distributed in the range of 0.5 to 20,[3] corresponding to retention times of about 3 to 45 minutes on a column with a void time of 2 minutes. If the components of a sample are neutral, i.e., nonionizable, the correct proportion of water and organic solvent gives the optimum k' value by which symmetrical, well-resolved peaks

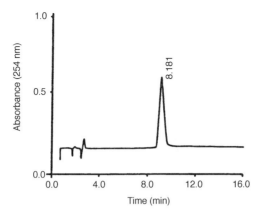

Figure 1 Chromatogram of a neutral compound (toluene) with water:acetonitrile mobile phase. Chromatographic conditions — column: 30 cm × 3.9 mm μ-Bondapak™ C$_{18}$ (10-μm particle size); mobile phase: water:acetonitrile (50:50); flow rate: 1.5 ml/min; column temperature: ambient; detector wavelength: 254 nm.

are obtained. Figure 1 is a chromatogram of the separation of a neutral molecule (toluene) using a water:acetonitrile (50:50) mobile phase system. Very polar compounds, especially those which ionize in water such as benzoic acid, may be poorly retained on the column (k′ < 0.5) and therefore elute almost coincident with the injection front. The peak symmetry of such analytes is poor, as shown in Figure 2A. To obtain a symmetrical peak of an ionic compound, modifiers such as an acid, base, or counterion may be added to the water:organic solvent system. Such modifiers can convert the ionized compounds into nonionized or ion-paired species, which are retained longer on the column and elute with a symmetrical peak shape. Figure 2B is a chromatogram of benzoic acid where ammonium acetate was added to the mobile-phase. The ammonium ion paired with the benzoate and the resultant ion pair eluted with a sharp symmetrical peak at a k′ of about 1. Another approach would be to adjust the pH. Since neutral molecules are unaffected by pH, but ionizable compounds titrate, an ionizable compound can be converted to neutral form by adjusting the pH. In general, the nonionized form will be retained longer on the column. The control of ionization by the addition of acid or base is called "ion suppression", while the addition of a counter-ion is called "ion-pairing".[19] With benzoic acid, the following equilibria are important:

Ion suppression: $RCOO^- + H_3O^+ \leftrightarrows RCOOH + H_2O$

Ion pairing: $RCOO^- + NH_4^+ \leftrightarrows RCOO^- \cdots NH_4^+$

Silica-based columns are only stable in the range of pH 2 to 8, so ion suppression is particularly useful for weak acids and bases. The advent of

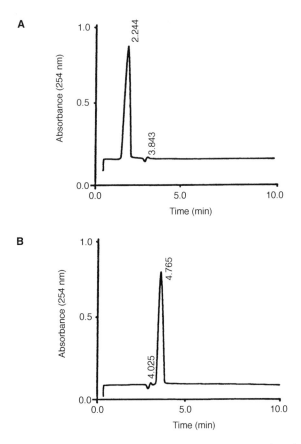

Figure 2   Chromatogram of a polar (benzoic acid) compound with water:acetonitrile mobile phase. Chromatographic conditions — column: 30 cm × 3.9 mm µ-Bondapak™ C_{18} (10-µm particle size); mobile phase: water:acetonitrile (90:10); flow rate: 1.0 ml/min; column temperature: ambient; detector wavelength: 254 nm. **(A)** No ion pairing agent. **(B)** With ammonium acetate as an ion pairing agent.

alumina-based and polymer-based supports has greatly extended the pH range (to pH 2 to 13) in which ion suppression may be performed. Retention of compounds can be greatly increased by using a hydrophobic counterion to form the ion pair. Since ion pairing is dependent on the pairing equilibrium, the concentration of the counterion can affect the separation.

Most small organic molecules are soluble in mixed organic-aqueous solvents and can be easily analyzed using RPLC. However, there are some polar compounds which are not soluble in typical RPLC solvent systems or are unstable in an aqueous mobile phase system. These compounds can be analyzed on an RPLC column with a nonaqueous solvent system. This technique is called "nonaqueous reversed phase chromatography" (NARP).[20,21] The NARP technique is primarily used for the separation of lipophilic compounds having low to medium polarity and a molecular weight larger than

250 to 300 (e.g., polystyrene, olefins, and fatty acids).[8] This technique is also useful if a sample is unstable in aqueous media. In general, the NARP system consists of a nonpolar stationary phase (C_8 or C_{18}) and a polar mobile phase. The mobile phase employed in NARP consists of solvents such as methanol, acetonitrile, tetrahydrofuran, methylene chloride, ethyl acetate, and hexane. This technique is rarely used in the pharmaceutical industry since most drug substances are soluble and stable in common reversed phase solvent systems.

4.1.5   Gradient chromatography

In an isocratic system, the separation of early-eluting compounds can be increased by reducing the organic content of the mobile phase, but only at the expense of greatly increasing the retention times of later-eluting components. This is referred to as the general elution problem. By increasing the organic component over the course of a separation, the late-eluting peaks can be eluted more rapidly without affecting the separation of the early-eluting peaks.

The conceptual basis for understanding the connection between isocratic and gradient elution is well established and is called "linear solvent strength theory".[22-27] Linear solvent strength theory proposes that, for a given solute, mobile phase, and column, if one measures the retention time of an analyte at two organic component concentrations, it will be possible to predict the retention time with any other mobile phase composition. The k′ value that would be observed in pure water, k_w, is related to the actual k′ by the relationship

$$\log k' = \log k_w - S\varphi,$$

where S is a constant, and φ is the volume fraction of organic solvent. The parameter S is dependent on the identity of the solute, increasing with hydrophobicity. S also depends on the column at mobile phase. By plotting log k′ vs. the composition at elution, one may measure the slope, S, and the intercept, $\log k'_w$.

For gradient chromatography, the mobile phase composition varies from the starting to the final composition, covering a range $\Delta\varphi$ over a time period t_G. Assuming that the rate of change $\Delta\varphi/t_G$ is constant, a parameter b can be defined as

$$b = S \, \Delta\varphi t_o / t_G.$$

By varying the gradient rate or the flow rate, F, one may measure S using the relationship

$$\bar{k} = t_G F / 1.15 V_m \, \Delta\varphi S$$

where V_m is the void volume. The slope of a plot of log $\bar{k}$ vs. $\bar{\varphi}$ is b. There is one very important experimental implication from the discussion above. The elution order of two components may invert as the organic composition of an isocratic separation or the gradient rate or the flow rate of a gradient separation are changed. Therefore, if the components of a mixture are identified with standards at one composition, gradient rate, or flow rate, the identification will need to be repeated if the conditions are changed.

There may be deviations from linear solvent strength theory due to instrumental factors such as gradient dispersion between the mixer and the column head or gradient misproportioning by the mixing system.[28] Also, a correction may have to be made if the analyte is large enough to be excluded or partially excluded from the pores.[25] Macromolecules may show significant deviations from the retention predicted by linear solvent strength theory due to conformational effects. However, in the absence of conformational effects, macromolecular RPLC can be described according to linear solvent strength theory, with the caveat that desorption tends to be more cooperative than small molecules, exhibiting a steep slope on a graph of log(k') vs. mobile phase composition.[29] Another source of deviation from linear solvent strength theory may be the existence of nonlinear or multicomponent competitive isotherms.[30,31]

4.1.6   Prediction of retention time

It is possible to predict retention times with a reasonable degree of accuracy based on a knowledge of the analyte structure. A number of schemes have been proposed.[32-38] Usually, approaches to prediction of retention time assume that the hydrophobicity of a complex molecule can be divided into contributions of hydrophobicity or hydrophilicity of each functional group comprising it. For example, the retention of a peptide may be estimated by adding up the retention times of the amino acids comprising it and corrected for the number of free polar groups. Although there are frequent and significant deviations from prediction, the values obtained from theory can be extremely useful in selecting initial conditions for separation. Principal component factor analysis[39] has demonstrated that, given the retention values of a few test compounds, the retention times of other compounds, even ionizable ones, can be predicted within 3%; only four eigenvectors, representing mutually independent factors, are required.[40] Temperature, of course, influences retention. Inclusion of temperature in the factor analysis can allow one to calculate the degree of polymerization of a polymer such as polyethylene glycol using appropriate standards.[41]

4.1.7   Column packings

Although open tubular columns have been used for RPLC,[42] most applications use packed columns. The most widely used column packings are formed by chemically bonding butyl (C_4), octyl (C_8), or octadecyl (C_{18}) chains to a silica surface. Phenyl (Ph), cyano (CN), and amino (NH_2) functionalities are also used. Polymeric packings which are finding broad acceptance include alkyl-grafted poly(methylmethacrylate) and alkyl-grafted or unmodified

Table 1 A Partial List of Types of RPLC Columns

Manufacturer	Stationary phase
Alltech	Absorphere® C_8, C_{18}, TMS, CN, NH_2
Beckman	Ultrasphere® C_8, C_{18}
Brownlee Labs	Spheri® RP-8, RP-18, phenyl, CN, NH_2
E. Merck	Lichrosorb® RP-8, RP-18, diol, NH_2, CN
Macherey-Nagel	Nucleosil™ C_8, C_{18}, phenyl, CN, NH_2, nitrophenyl, diol
Phase Separation	Spherisorb® ODS1, ODS2, C_8, C_6, phenyl, methyl, CN, NH_2
Supelco	Supelcosil™ C_1, C_4, C_8, C_{18}
Waters™	Nova-Pak® C18, CN, phenyl, μ-Bondapak™ C_{18}, CN, phenyl Resolve™ C_8, C_{18}, CN
Whatman	Partisil® C_8, ODS-2, ODS-3, ODS, PAC

poly(styrene-divinyl benzene) polymeric columns. The polymer may be coated onto the surface of other particles such as silica, combining the solvent resistance of the polymer to the incompressibility and good mechanical properties of the rigid support.[43] Derivatization, whether to silica or to a polymeric surface, does not generally go to completion, leaving patches of column material with adsorptive properties different than neighboring patches. For this reason, grafted materials often exhibit some degree of tailing and irreversible adsorption. A common example is the tailing exhibited by amines on silica-based materials. In recent years, endcapping of silica-based materials has improved column performance. The causes of poor performance of silica-based materials in separation of basic compounds may be the incorporation of heavy metal impurities into the surface through heat treatment, thus increasing the acidity of surface silanols.[44]

Particle sizes of analytical columns typically are described as 3-, 5-, and 10-μ packings, although the actual mean particle size often differs from the label claim. In isocratic chromatography, resolution generally improves as particle size decreases down to a limit of about 2 μ, i.e., smaller particles resolve better. For a given column length, however, columns packed with small particles have greater back pressures and are more prone to clogging than columns with larger particle sizes. Examples are given in Table 1. A review of characterization of column packings by reflectance infrared spectroscopy, cross-polarization, magic angle nuclear magnetic resonance, and other spectroscopic techniques is available.[45] A study of column degradation showed that small particles and debris accumulate principally at the outlet of silica-based columns, while a poly(styrene-divinyl benzene) column showed stress wrinkles but no apparent particle fracture.[46]

4.1.8   Detectors

Many different detectors are used in RPLC, including ultraviolet-visible spectrophotometers (UV-VIS), refractive index (RI) detectors, electrochemical (EC) detectors, evaporative light-scattering detectors, fluorimeters, and others. These are described elsewhere in this book. By far the most popular is the UV-VIS, so a brief review of the points relevant to this detector are reviewed here. The UV-VIS detects the difference in the amount of light absorbed by an analyte transiting the detector relative to the amount of light absorbed by the solvent. If the solvent absorbs light strongly at the wavelength of detection, the spectrophotomer is unable to differentiate the small additional amount of absorbance when the analyte transits the detector. Anomalous results may be obtained. For RPLC, the most commonly used organic solvents are methanol and acetonitrile, since these do not absorb appreciably above 200 nm. Ethanol and propanol are also relatively transparent to 200 nm. Unstabilized tetrahydrofuran may be useful above about 240 nm. Of the common ion suppression agents, only HCl is transparent to 200 nm. Phosphoric acid and alkali phosphates may be used in high dilution. Very dilute (0.1%) trifluoroacetic acid may also be used. Many common buffers absorb strongly below 220 nm and must therefore be used in very high dilution. Acetates, carbonates, borates, and other buffers are difficult to use below 240 nm, unless the dilution is sufficiently high.

4.2   Chromatographic practice in RPLC

4.2.1   Mobile phase preparation

The mobile phase should be free of particles and, especially in gradient chromatography, free of UV-absorbing impurities. HPLC grade solvents and buffers are required, and these are generally filtered through 0.45- or 0.2-μ filters. Since filtration may result in the evaporation of volatile components, the organic phase is usually filtered separate from the aqueous phase prior to blending. Also, volatile acids such as HCl and TFA may be added to the mobile phase after the mobile phase has been filtered. For gradient separations, it is common to add equal amounts of any UV absorbing buffer to both aqueous and organic components. In this way, the change in absorbance during a run is not very much altered. In the name of cost savings, some laboratories keep mobile phases on the shelf for months. This is often a false economy, since aqueous mobile phases can grow bacteria, and both aqueous and organic mobile phases can develop UV-positive components with time. The latter is particularly troublesome in gradient chromatography. Even in isocratic chromatography, in which a UV-positive component cannot form a peak since it is being continuously eluted from the column, mobile phase impurities can lead to shifts in retention times. Water is best generated on-site, using an ion exchange system and a carbon filter. High resistivity water is less prone to bacterial growth, and the carbon cartridge removes most UV positive components. A simple way to identify whether a peak in gradient chromatography is due to a mobile phase component is to run the mobile through the column for a variable period of time before running the gradient. If a peak grows with pre-gradient time, it is being concentrated from the mobile phase onto the column and displaced by the gradient.

4.2.2   Sample preparation

The solvent in which the sample is dissolved plays an important role in terms of peak band broadening and retention time of the solute. A mismatch between the injection solvent and the mobile phase can lead to artifacts in the chromatogram or to variable retention times. If the sample solvent is different than the mobile phase solvent, one or more distinct system peaks, either positive or negative, may be obtained. The best chromatographic result is obtained if the mobile phase is used to prepare the sample. If a mismatch between mobile phase and sample is suspected, varying the injection volume can help in troubleshooting. In some cases, because of the lack of solubility, it is difficult to dissolve a sample in the mobile phase solvents. In such situations, a stronger, i.e., less polar solvent may be needed to dissolve the sample. The use of a less polar solvent can cause peak band broadening or distortion and reduction in retention time. It is important to inject a blank of the sample solvent alone to identify system peaks.

4.2.3   Sample concentration effects

The peak symmetry, resolution, and detector response are directly dependent on the concentration of the sample. As the concentration of a sample increases, the retention time, separation, and peak symmetry generally decrease. These phenomena are due to isotherm nonlinearity. The detector response may also be nonlinear above or below certain concentrations. In some cases, small amounts of a dilute component are irreversibly adsorbed to the column, leading to reduced recovery. Above some concentration, the response of any detector will cease to be linear. The UV-VIS is one of the most linear detectors, generally exhibiting at least three decades of linearity, while RI, electrochemical, and fluorimetric detectors have a markedly narrower range of linearity.

4.2.4   Anomalies

A very good series of reports on anomalies in HPLC has been published.[47-49] Among the phenomena reported are analyte decomposition in the mobile phase, concentration-dependent analyte dimerization, and analyte isomerization. These phenomena are very frequently seen in protein HPLC. An understanding of how to identify each of these phenomena is, however, important in chromatography of any compound. Analyte decomposition in the mobile phase is usually established by incubating the sample with mobile phase for various times before injection. If the decomposition is rapid, it may be necessary to modify the solution in which the sample is dissolved. Dimerization and isomerization generally cause multiple peaks. Dimerization and isomerization can sometimes be established by collecting fractions and reinjecting. In either case, each peak should convert into the other. If not, decomposition may be the cause. If concentration dependence is seen in the formation of multiple peaks, dimerization or column overloading may be the cause. A study of the wetting of C_{18} column in methanol-water demonstrated that at room temperature and low pressure, full solvation of a water-equilibrated phase does not occur below 65% methanol content, leading to very long equilibration times and retention artifacts.[50] Below a 10% methanol content, the ODS particles are not wetted.

4.2.5   Development of an RPLC method

Method development starts from an understanding of the physical and chemical properties of the sample molecule and/or the impurities present (i.e., the properties such as solubility, stability, polarity, dissociation constant, molecular weight, electroactivity, and the UV absorption spectrum). If the sample is of a low-molecular-weight (< 2000) and is soluble in the reversed phase (water/organic) solvents, then RPLC is one of the methods of choice. Preliminary estimates of retention may be made on structural bases alone.[22,51] If the sample is ionizable, with a pK_a in the range of 3 to 8, ion-suppression may be useful in reversed phase. If the sample components are of a very polar nature (strong acid or strong base) ion-pairing may be helpful. Having used molecular structure to determine the general outline of the method, it is necessary to select an appropriate sample solvent, detector (e.g., UV, RI, fluorescent, or EC), and sample concentration. If the sample is sufficiently complex, the use of a gradient may be necessary. Gradient RPLC is far more prone to artifacts than isocratic RPLC but is also far more versatile.

Assuming the components of a sample are unknown, a gradient run should be employed to determine the polarity of the sample. By gradually adjusting the starting composition, one can control the elution times of the early components such that they are well resolved in a minimal time, and by reducing the gradient rate, one may arrive at an isocratic solvent composition at which all components elute within reasonable time. One may estimate the effect of solvent composition on retention by the approximate rule that an increase of the aqueous component by 10% doubles k'.[52] Further adjustments in additives to the mobile phase, changes in the column type, and changes in temperature may help to optimize the separation. The resolving characteristics of columns have been characterized.[52] One surprising observation is that the batch-to-batch differences in selectivity between one brand of stationary phase may actually be greater than the differences between C_8 and C_{18}.

A number of recommendations have been made in the development of quantitative chromatographic methods. The American Society for Testing Materials — using as a benchmark the reversed phase separation of benzyl alcohol, acetophenone, benzaldehyde, benzene, and dimethylterephthalate — discovered substantial laboratory-to-laboratory differences in quantitative analysis.[53] These compounds are routinely used to test column performance or for system suitability testing. A followup study, using benzyl alcohol, acetophenone, p-tolualdehyde, and anisole, showed that measurement of peak height was more accurate than peak area for poorly resolved peaks.[54] Because some columns exhibit interactions with amines, aniline has been proposed as a test substance sensitive to column degradation.[52] Uric acid, tyrosine, hypoxanthine, xanthine, thymine, inosine, guanosine, adenosine, theophylline, and caffeine also have been used as test substances.[18] Flow rate, temperature, and signal filtering have been shown to be significant factors in accuracy of peak measurement,[55] as has the method of peak cutting.[56] An important recent review described criteria required for validation of bioanalytical methods.[57]

The modern HPLC system is a very powerful analytical tool that can provide very accurate and precise analytical results. The sample injection volume tends to be a minor source of variation, although fixed-loop detectors must be flushed with many times their volume in sample to attain high precision. Assuming adequate peak resolution, fluorimetric, electrochemical, and UV detectors make it possible to detect impurities to parts per billion and to quantitate impurities to parts per thousand or, in favorable cases, to parts per million. The major sources of error in quantitation are sample collection and preparation. Detector response and details of the choice of chromatographic method may also be sources of error.

Error in sampling and sample handling occur in several ways. First, solid samples may be inhomogenous, so special precautions need to be taken in sampling. In liquid samples, there may be phase separation or adsorption of a component to glassware. Adsorption is often observed in analyzing macromolecules such as proteins. Solid samples, particularly macromolecules, may be slow to dissolve to a monomeric form. Even though a clear solution is obtained, the solute may be in an aggregated form that chromatographs differently than the monomer. Samples may also undergo chemical changes on standing, so analyzing a fresh and an aged sample may help to expose errors due to stability. If sample preparation includes prefiltration or chemical derivatization, additional errors may occur due to physical losses or selective derivatization. There may also be selective losses onto column frits or the stationary phase that may introduce error. Spiking and recovery experiments may be helpful in establishing the extent of error introduced by physical losses. An internal standard should not be used to estimate the effects of chemical losses.[57]

A second source of error may be in the detector. Detector linearity is an idealization useful over a certain concentration range. While UV detectors are usually linear from a few milliabsorbance units (MAU) to 1 or 2 absorbance units (AU), permitting quantitation in the parts per thousand level, many detectors are linear over only one or two decades of operation. One approach in extending the effective linear range of a detector is high-low injection.[58] In this approach, an accurate dilution of a stock sample solution is prepared. The area of the major peak is estimated with the dilution, and the area of the minor peak is estimated with the concentrated stock. This method, of course, relies on linear recovery from the column. Another detector-related source of error that is a particular source of frustration in communicating analytical results to nonanalysts is that of relative detector response. Detectors usually are not equally sensitive to all components of a mixture, so a small peak may not represent a minor component. Measurement of response factors is an important part of method development and requires pure samples of impurities as well as the principal component.

The importance of verifying the linearity of a method cannot be overstated. Aside from detector nonlinearity, other factors can cause a method to be nonlinear. Trace components can adsorb to the surfaces to which they are exposed in sample handling, transfer, and chromatography, leading to underestimation of the amount of that component. Also, self-association of an analyte can lead to concentration-dependent chromatography. In this context, it should be noted that observations of nonlinearity of a method may be erroneously ascribed to the detector. Accurate dilution is essential in studies of linearity. Since the precision of modern analytical balances is far greater than that of many volumetric devices, such as pipettes, one approach to ensuring accuracy of dilution is to use an analytical balance to monitor solution preparation. To check linearity of the method, a dilution series of standards of known concentration is prepared and analyzed. A graph of concentration vs. detector response (as either area or height) is prepared. Deviation from the least squares fit, as indicated by a correlation coefficient $r^2 < 0.99$, is often a signal of nonlinearity of the method.

4.2.6   *Method calibration*

The determination of the relationship between detector response and the sample concentration is termed the calibration of the method. There are two types of methods in use for the quantitative analysis of a sample, i.e., the external standard and the internal standard method. An external standard method is a direct comparison of the detector response of a pure compound (standard) to a sample.[2] The calibration of the method is performed by preparing standards of varying concentration and analyzing them by a developed method. Method 1 (below) was developed for toluene, and standards of varying concentration were prepared and analyzed. The results obtained are summarized in Table 2; see Figure 3.

Table 2   Calibration Data for
External Standard Method

Standard no.	Toluene (mg/ml)	Area ($\times 10^4$)
1	1.400	30.08
2	0.700	15.44
3	0.350	6.38
4	0.175	3.28

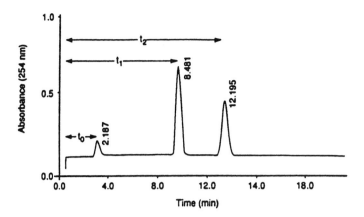

Figure 3 A two-component (toluene/2,4-dichlorophenol) separation. Chromatographic conditions — column: 30 cm × 3.9 mm µBondapak™ C$_{18}$ (10-µm particle size); mobile phase: 0.1 M ammonium acetate:acetonitrile (50:50); flow rate: 1.0 ml/min; column temperature: ambient; detector wavelength: 254 nm.

Method 1: External Standard Method for Toluene

Instrument:	Waters™ HPLC System (Waters Assoc.; Milford, MA)
Column:	30 cm × 3.9 mm µ-Bondapak™ C$_{18}$, 10-µ particle size (Waters Assoc.; Milford, MA)
Mobile phase:	50% acetonitrile:50% water
Flow:	1.5 ml/min
Column temperature:	Ambient
Detector wavelength:	254 nm
Detector sensitivity:	0.5 AUFS (absorbance unit full scale)
Sample concentration:	1.400 mg/ml to 0.175 mg/ml
Sample size:	20 µl
Sample:	Toluene
Retention time:	8.1 min

To determine the linearity of the method, a graph (standard curve) of concentration vs. detector response (area/or height) is plotted (Figure 4A). A straight line is obtained if the method is linear. Using linear regression analysis, the standard curve is constructed and correlation coefficient (r^2) of the regression line is determined. A method is acceptable if the r^2 value obtained is 0.99 or better. The best result (weight percent, or wt%) is obtained if the concentration of the sample is within the concentration range evaluated (i.e., standard curve).

The weight percent purity of a sample can be determined, using c_y as the actual concentration of an analyte in mg/ml of the unknown, c_x as the concentration in mg/ml in the standard, c_o as the nominal concentration of

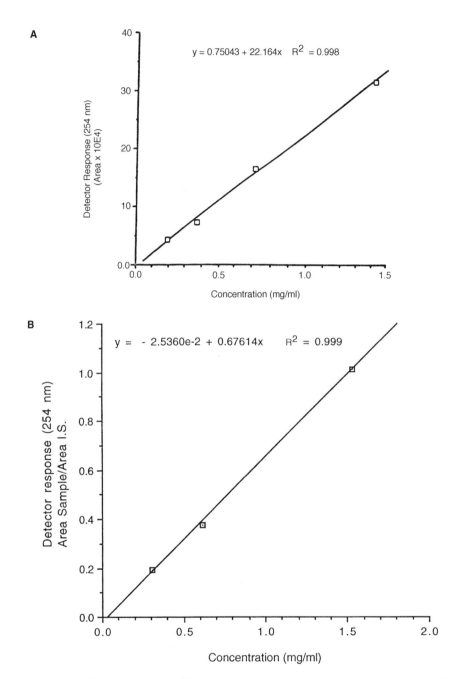

Figure 4 Calibration of external and internal standard method. Chromatographic conditions — column: 30 cm × 3.9 mm μ-Bondapak™ C$_{18}$ (10-μm particle size); mobile phase: water:acetonitrile (50:50); flow rate: 1.5 ml/min; column temperature: ambient; detector wavelength: 254 nm. **(A)** External standard method, **(B)** internal standard method.

the analyte in the sample in mg/ml, and A_x and A_y as the area in the sample and standard, respectively, as

$$c_y = c_x \, A_y \, / \, A_x$$

$$Wt\% = 100 \, c_y \, / \, c_o$$

Because this method is a direct comparison of detector response (i.e., the peak area and/or peak height) of standard to sample, high accuracy in preparing the sample and then injecting the precise volume is very critical to obtaining an accurate and reliable result. The use of an autoinjector or washing through the sample loop with a tenfold excess of sample eliminates or minimizes an error due to injection. As sample loops can leak, it is useful to inject the sample onto the column while the injection syringe is still moving to minimize volume error.

An internal standard method gives more reliable results when elaborate sample preparation is required, as in extraction of a drug substance from biological fluids, or extraction of pesticides and herbicides from soil and plant matter. The addition of internal standard (IS) to the sample and standard acts as a marker to give accurate values of the recovery of the desired compound(s). Since the determination of wt% involves the ratio of the detector responses in the two chromatograms, the injection volume is not critical as in an external standard method.

Calibration of an internal standard method is done by preparing standard samples of varying concentration. The same amount of IS is added to each, and the standard samples are analyzed using a developed method. The detector response, area or height, of each standard is determined, and a ratio is calculated. The graph of concentration vs. area ratio is plotted. The method is considered linear if the correlation coefficient is 0.99 or better. The response factor RF is calculated as

$$RF = (A_{IS\text{-}St} / A_{An\text{-}St}) \, (c_{An\text{-}St} / c_{IS\text{-}St})$$

where $A_{IS\text{-}St}$ is the area of the internal standard in a standard sample, $A_{An\text{-}St}$ is the area of the analyte in the standard sample, $c_{An\text{-}St}$ is the concentration of the analyte in the standard, and $c_{IS\text{-}St}$ is the concentration of the internal standard in the standard. Then, analyzing for the concentration c_y of an unknown,

$$c_y = RF \times c_{IS\text{-}Sa} \times A_{An\text{-}Sa} / A_{IS\text{-}Sa}$$

where $c_{IS\text{-}Sa}$ is the concentration of the internal standard in the sample, $A_{An\text{-}Sa}$ is the area of the analyte peak in the sample, and $A_{IS\text{-}Sa}$ is the area of the internal standard in the sample. Then, the wt% of the analyte in the sample is simply

$$100 \, c_y \, / \, c_o$$

where c_o is the nominal concentration.

For the example of toluene given above, the external standard method can be converted into an internal standard method by adding anisole (an appropriate internal standard) to both standard and sample. The retention time of anisole is 4.5 minutes if analyzed by the method above. To calibrate the internal standard method for toluene, toluene standards of concentration 0.3 to 1.5 mg/ml containing 0.5 mg/ml anisole were prepared. The detector response as a function of the amount of sample injected is shown in Figure 4B.

The key step in the internal standard method is to choose an appropriate internal standard, which has polarity similar to the analyte, is inert to the conditions of extraction and processing, and elutes before or well after the peak of interest. An internal standard method is useful only for correcting for losses due to transfer or variability in dilution or injection, and it is inappropriate to use an internal standard to correct for losses due to degradation.[57] This technique gives reliable, accurate, and precise results. If the internal standard is truly inert, the method is useful for determining the rate of analyte conversion in a chemical reaction.

4.3   Selected applications

It would be easier to describe those classes of compounds not normally separated by RPLC than to catalogue the applications to which RPLC has been turned. Applications for reversed phase can be found in virtually every area of analysis and are reviewed regularly in the journal *Analytical Chemistry*. RPLC has not been in general use for the analysis of inorganic ions, which are readily separated by ion exchange chromatography; polysaccharides, which tend to be too hydrophilic to separate by RPLC; polynucleotides, which tend to adsorb irreversibly to the reversed phase packing; and compounds which are so hydrophobic that reversed phase offers little selectivity.

4.3.1   Homologous series of simple hydrocarbons; aromatic hydrocarbons

The central issue in separations of homologous series is column selectivity. Separations of a homologous series of compounds differing by a single methylene are a common method of studying the separating power of a column. Compilations of methylene selectivity are available in the literature.[59-61] Applications have included the separation of alkyl benzenes, dialkyl phthalate, and alkylanthroquinone;[62] alkanes;[63] mesogenic esters of biphenyl-4,4'-dicarboxylic acid;[64] carboxylic acids as the 6,7-dimethoxycoumarin derivatives in a methanol:water mobile phase;[65] amines;[66] and ketones as the ketone semicarbazone derivative.[67] Porous graphitic carbon showed a greater selectivity in the retention behavior of 36 positional isomers of ionizable substituted benzenes than an octadecylsilane phase.[68] A comparison of the separation of aromatic hydrocarbons such as benzene, fluorene, phenanthrene, anthracene, pyrene, and 1,2-benzanthracene on phenylsilicone and ODS phases in acetonitrile:water was performed.[69] A separation of biphenyl, fluorene, phenanthrene, and anthracene on a 120-cm, 5-µ interior diameter (i.d.) open tubular C_{18} column showed plate counts equivalent to 500,000 plates per meter.[42]

Detectability may be a significant problem with homologous series of unsaturated compounds, particularly *n*-alkanes. For these compounds, refractive index detection or evaporative light-scattering, both of which are described elsewhere in the book, may be of use. Indirect photometry is a useful detection scheme for compounds that do not absorb in the UV. Acetone, methylethyl ketone, methyl propyl ketone, methyl isopropyl ketone, methyl isobutyl ketone, and acetophenone are added to an acetonitrile:water mobile phase, generating a negative vacancy peak when the nonchromophoric analyte emerges and a positive peak if the ketone is adsorbed and displaced.[70] Dodecyl, tetradecyl, cetyl, and stearyl alcohols also have been derivatized with 2-(4-carboxyphenyl)-5,6-dimethylbenzimidazole and the derivatives separated on Zorbax® ODS in a mobile phase of methanol and 2-propanol.[71]

For aromatic hydrocarbons, excessive retention in reversed phase may be the principal problem. For this reason, normal phase chromatography is sometimes preferred. Phthalic anhydride in workplace air was determined on a µ-Bondapak™ using methanol:water:acetic acid as the mobile phase.[72] Polynuclear aromatic hydrocarbons including naphthalene, acenaphthylene, acenaphthene, fluorene, phenanthrene, anthracene, fluoranthrene, pyrene, benz[a]anthracene, chrysene, benzo[b]fluoranthrene, benzo[k]fluoranthrene, benzo[a]pyrene, dibenz[a,h]anthracene, benzo[ghi]-perylene], and indeno[1,2,3-cd]pyrene have been separated on a Perkin-Elmer™ HS/5 HCODS column in acetonitrile:water.[73] 1-Nitropyrene and nitroanthracenes were separated in acetonitrile:water using Kieselguhr 60 HPLC RP-18.[74] Polychlorinated biphenyls were separated on Perkin-Elmer™ HS/3 C18 in acetonitrile:water.[75] A microcolumn packed with ROSIL-C-18-D was used to separate naphthalene, biphenyl, anthracene, and pyrene.[76] The Fusica™ C18 microcolumn, eluted with acetonitrile:water, was used to separate naphthalene, fluorene, phenanthrene, anthracene, and chrysene.[77] Sensitive amperometric detection of separated 4-chlorophenol, 2,4-dichlorophenol, 2,4,6-trichlorophenol, tetrachlorophenol, and pentachlorophenol was obtained using a Lichrosorb® C-18 column eluted with acetonitrile:water containing perchlorate.[78]

4.3.2 Carboxylic acids

Carboxylic acids present no exceptional problems in reversed phase analysis, although detectability may be a limitation in the analysis of simple fatty acids. Wine acids, including succinic, acetic, citric, lactic, malic, and tartaric, have been separated on a polymeric Hamilton™ PRP®-1 column in acetonitrile:water.[79] The phenolic acids, trihydroxybenzoate, 3,5-dihydroxybenzoate, 3,4-dihydroxybenzoate, 2-hydroxybenzoate, 4,8-dihydroxyquinoline-2-carboxylate, 4-hydro-xybenzoate, 2,6-dihydroxybenzoate, 4-hydroxy-3-methoxybenzoate, and 3,4-dihydroxycinnamate, as well as ascorbate, were separated on a Zorbax® ODS or μ-Bondapak™ C_{18} column in methanol:water buffered with acetate using dual detection on UV and a coulostatic electrochemical detector.[80] Ethylene diamine tetraacetic acid (EDTA) was determined by chelation with Fe(III) and separation on a Lichrocart® RP-18 column using acetonitrile:water, buffered with formate (pH 3.3) and containing the ion pairing agent tetrabutyl ammonium bromide.[81] The mycotoxin citrinin, (3R, 4S)-4, 6-dihydro-8-hydroxy-3, 4, 5-trimethyl-6-oxo-3H-2-benzopyran-7-carboxylic acid, was separated on a Nucleosil™ ODS column in methanol:water containing tetrabutylammonium hydroxide.[82] Post-column acidification improved detection by fluorescence.

4.3.3   Simple amines

The principal analytical issue with simple amines often has to do with mixed-mode interactions between the stationary support and the analyte, leading to tailing. For this reason, polymeric and alumina supports have found some favor in preference to silica in the analysis of simple amines. The high pH values tolerated by polymeric and alumina supports permit the chromatography to be performed under conditions such that the amine is in its neutral form. With improvements in endcapping and the use of trifluoroacetic acid and other ion pairing agents in common use, chromatography of amines on silica supports has continued. Another strategy has been the pre-column derivatization of the amino functionality, an approach which may improve detectability many orders of magnitude but at the expense of experimental complexity. Derivatives of *m*-toluoyl chloride were formed with ammonia, methylamine, ethylamine, dimethylamine, allylamine, isopropylamine, *n*-propylamine, ethylenediamine, diethylamine, and *n*-butylamine, and separation was performed on Spherisorb® C_{18} in acetonitrile:water. A mixture of 1,4-phenylenediamine, 1,3-phenylenediamine, 2-phenylethylamine, 1,2-phenylenediamine, 3-phenylpropylamine and aniline were extracted from hair dyes and analyzed on Lichrosorb® RP-18 in water containing octylammonium salicylate or octylammonium phosphate.[83] It was added that the pH of the mobile phase is critical in ion interaction chromatography of amines and amino acids, with the best results being obtained at pH 8.[84] Benzotriazole bound to polystyrene and containing a 9-fluorenyl or an *o*-acetylsalicyl labeling moiety has been used for off-line derivatization of amines and polyamines.[85] Fluorenyl methoxycarbonyl derivatives of polyamines such as putrescine, cadaverine, spermidine, and diaminohexane were determined by injecting them onto a guard column, washing extensively while directing eluent to waste, then separating on a 200 × 2.1-mm Hypersil® ODS (5-μ particle size) column with an acetonitrile:water:methanol gradient.[86] Simple nucleic acid components such as cytosine, cytidine, arabinosylcytosine, and 5-azaarabinosylcytosine in methanol:water were separated on a 5-μ Spherisorb® ODS column buffered at pH 3.5 to 7 and at temperatures of 25 to 45°C.[87]

4.3.4   *Sugars and carbohydrates*

Carbohydrates are found in simple sugars, as constituents of glycoproteins and glycolipids, and in polymeric form as structural features of bacterial and fungal cell walls, among other sources. Polymeric sugars include dextran, used as a blood expander, and starch. Sugars are most commonly analyzed by ion-exchange or other chromatographies, since they tend to be so hydrophilic as to elute rapidly from reversed phase columns. N-butyldeoxynojirimycin, deoxynojirimycin, and their degradation products were separated on a Zorbax® C_8 Rx column in water:acetonitrile containing heptane sulfonate and acetic acid.[88] The amino sugars mannosamine, galactosamine, and glucosamine extracted from tobacco were derivatized with *o*-phthaldehyde and analyzed in THF (tetrahydro furan):water on C_{18}.[89] Another derivatization agent, 6-aminoquinolyl-N-hydroxysuccinimidyl carbamate, available commercially as AccQ-Tag®, has been used for amino sugars such as α- and β-galactosamine and α- and β-glucosamine in soy protein, crustacean shells, and sewage sludge.[90] Following sodium borohydride reduction to eliminate the second anomeric peak of the amino sugars, separation was performed on a 15-cm AccQ-Tag® C_{18} (4-μ particle size) column using a ternary acetonitrile:water gradient containing triethylamine acetate and phosphoric acid (pH 5.0).

4.3.5   *Lipids and surfactants*

Simple lipids, or fatty acids, are *n*-alkyl carboxylic acids typically derived from the hydrolysis of the triglycerides of fats or of the phospholipids found in the cell wall. Many naturally occurring fatty acids contain one or more sites of unsaturation. Fatty acids are found in nature conjugated to such sugars as glycolipids or lipopolysaccharides, to amino acid residues of peptides and proteins, to glycerol to form the acylglycerols of fats, and to glycerophosphocholate and glycerophosphatidylethanolamine residues of phospholipids. Because of their amphipathic character, lipids have some tendency to form micelles, which chromatograph quite differently than the unaggregated compound. A class fractionation of plant-derived extracts into the component waxes (e.g., cetiolate), hydrocarbons, fatty acids (e.g., linoleic acid), fatty acid methyl esters (e.g., methylpalmitate), sterols (e.g., stigmasterol), sterol esters (e.g., cholesteryl linoleate), glycerylethers (e.g., chimyl alcohol), fatty alcohols (e.g., behenyl alcohol), vitamin E, and mono-, di-, and triglycerides was performed with stepped gradient elution on a 7-μ, 250 × 4-mm i.d. Nucleosil™-300 C18 column.[91] Coumarin-derivatized 1-O-alkyl-2-sn-lysoglyceryl-3-phosphoryl choline, a fluorescent-labeled phospholipid platelet activating factor, was separated on a 4-μ particle size

Nova-Pak® C18 column in a methanol:water:chloroform gradient.[92] Choline chloride was added to the mobile phase. One review of techniques used in the analysis of triacylglycerols lists over 300 references on separations of the triglyceride fraction of fats using nonaqueous RPLC, aqueous RPLC, argentation chromatography, and other chromatographic methods.[93]

Surfactants, like lipids, tend to be amphipathic. One of the simplest surfactants is sodium dodecyl sulfate. Linear alkylbenzene sulfonates were separated on a Supelcosil™ C_8-DB (deactivated base) column in methanol:water containing sodium perchlorate.[94] Ethoxylated alkylamines, used as surfactants in pesticide formulations, were analyzed on Polygosil 60 D-10 CN using methanol:water:dioxane.[95] Mono- and dialanine surfactants such as alkylaminoethylcarboxylates, laurylamino(diethylcarboxylate), and alkylaminodi(ethylcarboxylates) were separated in acetonitrile:water containing sodium perchlorate.[96]

4.3.6 Drugs, pharmaceuticals, and pharmacologically active compounds

Many, if not most, of pharmacologically active compounds are amines. For this reason, issues in the RPLC of substances of pharmaceutical interest tend to be similar to those encountered in the separation of amines. Incompletely endcapped silica-based phases may exhibit tailing. The use of ion pairing or ion suppression is common in the analysis of pharmacologically active substances. Also, derivatization of the amine functionality prior to analysis may be required. The RPLC retention indices of most common pharmaceutical compounds have been compiled.[97]

The catechols 3,4-dihydroxyphenylglycol, norepinephrine, 3,4-dihydroxyphenylalanine, epinephrine, 3,4-dihydroxyphenylacetic acid, and dopamine were separated on 3-μ Hypersil® C_{18} using an ion-pairing mobile phase.[98] Fluorescent derivatives of norepinephrine, epinephrine, and dopamine were formed with 1,2-diphenylethylenediamine, and separation was accomplished on TSK-Gel™ ODS-120T.[99] The formation of 4-nitrocatechol from *p*-nitrophenol was followed by RPLC on a 5-particle size Ultrasphere® ODS column in acetonitrile:water:0.1% trifluoroacetic acid (TFA) using UV and electrochemical detection.[100] Digoxide, lanatoside C, digoxigenin, and desacetyl lanatoside C were separated on Nucleosil™ C_{18} in dioxane:water.[101] Chiral separation of derivatized β-adrenergic blockers such as adrenaline, phenylephrine, atenolol, sotalol, pindolol and propanolol, as well as oxirane-derived β-amino alcohols such as ethyloxirane, butyloxirane, hexyloxirane, octyloxirane, and decyloxirane, was performed using 2,3,4,6-tetra-o-benzoyl-β-D-glucopyranosyl isothiocyanate as a chiral derivatization reagent, Lichrosphere® RP 100 C-18 as the column, and aqueous methanol mixtures as the eluent.[102] Barbiturates including barbital, phenobarbital, aprobarbital, butabarbital, mephobarbital, amobarbital, pentobarbital, and secobarbital have been separated on a Perkin-Elmer™ HS/3 C18 column in methanol:water (pH 8).[103] Anticonvulsants, such as primadone, phenobarbital, alphenal, carbemazapine, and dilantin, have been separated on a Perkin-Elmer™ HS/5 C18 column using acetonitrile:water (pH 5).[103] A gradient of acetonitrile:water (pH 4.4) separated theophylline, acetaminophen, caffeine, ethosuximide, primidone, acetylsalicylic acid, phenobarbital, chlordiazepoxide, butabarbital, pentabarbital, amobarbital, dilantin, ethchlorvynol, gluthemide, secobarbital, amitriptyline, methaqualone, and diazepam in about 6 minutes on a Perkin-Elmer™ HS/5 C8 column.[103,104]

Sulfa drugs, including sulfanilic acid, sulfamerazine, sulfaguanidine, sulfathiazole, sulfanilamide, and sulfamethazine, were separated on a polymeric Hamilton PRP®-1 column in acetonitrile:water.[79] Sulfmethazine was derivatized with 1-fluorenylmethyl chloroformate to form a fluorescent adduct and analyzed in methanol:water at pH 3.5 on either a 3- or a 5-μ trimethyl silane (TMS)-capped MicroPak™ ODS column.[105] The anti-infectious agent tinidazole, 1-[2-(ethylsulfonyl)-ethyl]-2-methyl-5-nitroimidazole, and several hydrolysis products were separated on a 10-μ particle size HP® 79951MO-174 RP-8 column using acetonitrile:water buffered with phosphate to pH 3.[106] The ionophoric antibiotics monensin A and B were separated from fermentation broth on a 7-μ Separon™ SGX C-18 column in methanol:water.[107]

Theophylline was separated from serum on a Synchrom® C_{18} column in acetonitrile:water containing acetic acid.[108] Aflatoxins G_1, B_1, G_2, and B_2 were separated on a Supelcosil™ C_{18} column, using acetic acid:acetonitrile:water:2-propanol[109] or on Spherisorb® ODS-1 with methanol:acetonitrile:water.[110] Ergot alkaloids aci-ergotamine, aci-ergotaminine, ergotamine, ergocornine, and ergokryptine were separated on a Knauer RP-18 column using acetonitrile:water buffered with $NaHCO_3$.[111] The antihistaminics phenindamine and isophenindamine were separated on a μ-Bondapak™ C_{18} in methanol:water containing silver ion.[112] Antabuse® (disulfiram) and two of its metabolites have been determined on Hypersil® ODS in acetonitrile:water buffered with acetate.[113] Ion pairing with either sodium octan – 1-sulfonate or tetrabutylammonium perchlorate was used in the separation of PGA_2, PGE_2, and PGB_2 on Lichrosorb® RP18 or μ-Bondapack™ C_{18} with methanol:water as eluent[114] Prostaglandins PGA_1, PGB_1, PGE_1, PGA_2, PGB_2, PGD_2, PGE_2, $PGF_{1\alpha}$, and Lutalyse ($PGF_{2\alpha}$) were separated on ODS in methanol:acetonitrile:water.[115]

The steroids aldosterone, cortisone, cortisol, 11-β-hydroxyandrostenedione, corticosterone, and rostenedione, 11-desoxycorticosterone, 17-hydroxyprogesterone, and progesterone have been performed on Ultrasphere® ODS using methanol:water.[19] Ranitidine {N-2-[[[5-[(dimethylamino)methyl]-2-furanyl]-methyl]thio]ethyl]-N^1-methyl-2-nitro-1,1-ethenediamine} has been separated using a μ-Bondapak™ C_{18} column operated with acetonitrile:methanol:water buffered with triethylamine phosphate.[117] Pyridoxal-5′-phosphate and other B_6 vitamers, including pyridoxamine phosphate, pyridoxal, pyridoxine, and 4-pyridoxic acid, were separated as bisulfite adducts from plasma using a YMC AQ-302 ODS column and a purely aqueous phase buffered at pH 3.[116] Detection was by fluorescence. Compilations of separations of derivatized pharmaceuticals are available.[118-120]

4.3.7   Amino acids and derivatives

The analysis of amino acids involves chromatographic issues similar to those encountered in analysis of simple amines. Underivatized amino acids have, with a few exceptions, weak UV absorbance and a strong tendency to interact with stationary phases in undesirable ways. Underivatized amino acids are normally separated with ion exchange chromatography, then visualized post-column by reaction with ninhydrin, *o*-phthaladehyde (OPA), or other agents. Underivatized tryptophan and the metabolites kynurenine, 3-hydroxykynurenine, kynurenic acid, and 3-hydroxyanthranilic acid, were separated on a Partisphere® 5-μ ODS column with fluorescent detection.[121]

Pre-column derivatization has found increasing favor. Phenyl isothiocyanate (PITC), OPA, dansyl chloride, dimethylaminoazobenzenesuplhonyl (dabsyl) chloride, and fluorenyl methoxycarbonyl (FMOC) are commonly used as derivatization reagents. PITC has been used to separate the common amino acids with an ammonium acetate-buffered acetonitrile gradient on 5-μ C_8 or C_{18} columns from various suppliers.[122] Tryptophan in infant formula was determined as its PITC derivative following barium hydroxide hydrolysis using an ODS-2 Spherisorb® column and an isocratic acetonitrile:water:triethylamine acetate (pH 6.8) mobile phase.[123] A Zorbax® ODS column eluted with acetonitrile: sodium acetate buffer has also been used for the OPA-derivatized amino acids.[124] Good on-column stability of OPA-mercaptoethanol derivatives has been observed.[125] An automated two-step derivatization with OPA/3-mercaptoproprionic acid to label the primary amino acids and then FMOC to label proline and hydroxyproline was used to analyze potato hydrolyzates in 18 minutes on a 100 × 4-mm Hypersil® ODS column packed with 3-μ particles.[126] Precolumn derivatization of amino acids has been accomplished with phenyl isothiocyanate followed by separation on Ultrasphere® ODS with THF-acetonitrile:water buffered with acetate.[19] Amino acids subjected to pre-column derivatization with dabsyl chloride were separated in acetonitrile buffered with acetate or phosphate on a Zorbax® ODS column.[124] A 25-cm × 2-mm narrow-bore column packed with 3 μ Supelcosil™ LC-18 and operated with acetonitrile:methanol:water buffered at pH 6.8 was used for amino acids derivatized pre-column with dabsyl chloride.[127] Pre-column derivatization with phenylisothiocyanate followed by separation on C_8 or C_{18} columns supplied by IBM, DuPont, or Altex in acetonitrile:water or methanol:water buffered by ammonium acetate-phosphoric acid was used for amino acid analysis.[122] The chiral isothiocyanate mentioned above (2,3,4,6-tetra-o-benzoyl-β-D-glucopyranosyl isothiocyanate) was useful as a derivatization agent to separate enantiomers of a number of amino acids.[102] The 6-aminoquinolyl-N-hydroxysuccinimidyl carbamate

(AccQ-Tag®) system mentioned in the section on amines has also been used for amino acids.[90]

4.3.8   Peptides

Peptide separations are performed in several distinct fields and with several distinct aims. Enzymatic digestion of a protein followed by RPLC finger-printing of the resulting peptide fragments is a powerful proof of identity. The fragments may also be isolated and sequenced by Edman degradation to provide an absolute proof of structure. In synthetic peptide chemistry, RPLC is widely used as a quantitative method. Racemization at an optically active site in a polypeptide often can be detected, since racemization typically generates a diastereomer of the principal product, and diastereomers are often (but not always) separable by RPLC. Analysis of naturally occurring peptides in biological organisms is another important application of RPLC.

Peptides, as amine-containing compounds, are susceptible to tailing in RPLC unless measures are taken to prevent it. End-capping of silica bonded-phases is particularly critical in peptide separations. The mobile phase in peptide separations often is formulated to contain trifluoracetic acid as a fairly UV-transparent ion suppression and ion pairing agent. Heptafluorobu-tyric acid and hydrochloric acid also have been used for ion suppression. Polymeric phases, while having found use in peptide separations, particu-larly for acid-labile peptides, tend to exhibit somewhat less resolving power than silica-based phases.[128] At elevated temperatures and flow, however, the fraction capacity increases substantially, permitting high-speed peptide sep-arations.[129] The elution order of peptides from polymeric phases is equivalent to that of alkyl-bonded phases.[130]

A significant difficulty in peptide separations is that peptides can exhibit multiple, reversibly interconverting conformers. An example of conformers of the peptide TTLTEPEPDL, found in the N-terminus of the protein hemen-tin, is given in Figure 5.[131] The conformations were derived as predictions from molecular modeling calculations. In panels A and B of Figure 5, *cis-trans* isomerization about a proline bond drastically alters the conformation. In typical peptides, the rate of interconversion is far more rapid than the chromatographic time scale, so that the statistical average of all conformers is observed. In some peptides, notably proline-containing peptides, the energy barrier to interconversion may be so high as to permit chromato-graphic separation.[132,133] Fortunately, interconversion is generally reversible, so irreversible loss of biological activity from chromatographed peptides is rarely an issue. The relationship among conformation, molecular surface area, and retention was explored for a series of cyclic dipeptides.[134] The capacity factor was greater for cyclic dipeptides containing L-Phe than those containing D-Phe and larger for cyclo-(D-Ala-L-Trp) than for cyclo(-L-Ala-L-Trp), an effect that was ascribed to the difference of surface area of the preferred conformer of a given diastereomer.

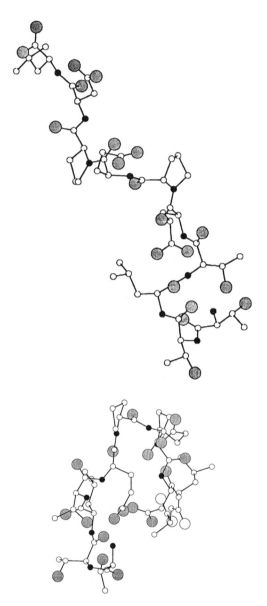

Figure 5 Conformers of the N-terminal decapeptide TTLTEPEPDL from the leech protease hementin. Molecular modeling predictions using Quanta/CharmM of two low-energy conformers.[92]

A high-speed separation of tryptic peptides of horse cytochrome c is shown in Figure 6. As the temperature and flow rate were increased, the separation substantially improved, an observation consistent with data demonstrating that conformational issues may be increasingly important factors in peak broadening as the time scale of the chromatography approaches that

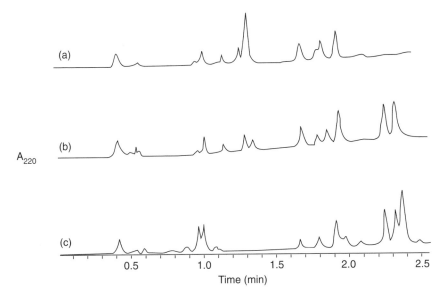

Figure 6 High-speed tryptic fingerprint. Horse cytochrome c was digested with trypsin and the peptide chromatographed in acetonitrile:water:0.1% TFA at various temperatures and flow rates on a 15×0.2-cm PS-DVB column packed with 3-μ, 300-Å particles. **(a)** 26°C and 0.5 ml/min. **(b)** 42°C and 0.7 ml/min. **(c)** 70°C and 1.1 ml/min. Detection at 220 nm. Note that the resolution rises with the speed of separation.[89] (From Swadesh, J., *BioTechniques*, 9, 626, 1990. With permission.)

of conformational interconversions. The importance of conformational effects in retention increases with peptide molecular weight.[120,135]

Despite all of these considerations, most peptide separations are uncomplicated. The decapeptide Nafarelin, an analog of luteinizing hormone-releasing hormone, was separated on a Vydac® C_{18} column in acetonitrile:water containing dibutyl ammonium phosphate as an ion pairing agent.[136] Peptide polymers with the sequence Ac-(Lys-Leu-Glu-Ala-Leu-Glu-Gly)$_n$-Lys-CONH$_2$, with $1 \leq n \leq 3$ were found to be monomers in acetonitrile:0.1% TFA, although for $4 \leq n \leq 5$, the peptides are dimers in 0.1% TFA; under reversed phase conditions, all eluted as monomers.[130] Large-pore phases were found to be superior to small-pore phases in the separation of peptides and proteins, presumably because of exclusion effects.[137-139] Twenty-three epimeric pairs of benzyloxy-(D/L)Ala-X-Val-O-methyl ester, where X is any of the naturally occurring amino acids, were prepared and separated on a Lichrosphere® 100 RP-18 column (5-μ particles) using a methanol:water eluent.[140] Bovine pituitary peptides were fractionated on μ-Bondapak™ C_{18} in acetonitrile:water:0.1% TFA.[141] Hemorphins were separated on a Delta Pak™ C_{18} column in acetonitrile:water buffered with ammonium acetate (pH 6). Their composition of aromatic amino acids was estimated by second-order derivative spectroscopy.[142] Defensins HNP-1, -2, and -3 (cationic antibiotic peptides that permeabilize cell membranes) were separated from dialyzates of lipid vesicles using a 5-cm guard column packed with 15 to 20 μ C4 resin and operated in acetonitrile:water:0.1% TFA.[143] Tryptic glycopeptides of reduced and alkylated fetuin were separated without significant neuraminic acid loss on a Zorbax® C_8 RP-300 column in acetonitrile:water:0.1% TFA.[144] A fluorocarbon phase was used to separate peptides, including enkephalin and Met-enkephalin, in acetonitrile:water containing trifluoroacetic acid, sulfuric acid, or ammonium hydroxide.[145] Hydrophilic peptides derived from tryptic digestion of peanut stripe potyvirus protein required derivatization with phenyl isothicyanate prior to separation on a 5-μ Vydac® C_{18} column in acetonitrile:water:0.1% TFA.[146] Issues in peptide characterization have been reviewed.[147]

4.3.9   Proteins

Proteins, like peptides, exhibit interactions with silica and multiple, interconverting conformations. Unlike peptides, protein conformational changes are often very slow relative to chromatographic time scales and may be effectively irreversible, leading to a loss of biological activity. When chymotrypsin was successfully chromatographed on RPLC, it was recognized that biological activity was not necessarily lost due to exposure to organic solvents.[148] The presence or absence of metals such as Ca^{+2} was recognized to alter the retention of parvalbumins due to denaturation at the metal binding site.[149]

Another effect of the slow interconversion rate is that multiple or very broad peaks may be observed in RPLC. Since the denatured conformation of a typical globular protein has more exposed hydrophobic residues than the native conformation, the denatured conformer elutes later in RPLC.[150] Raising the temperature or lowering the pH below the titration point of aspartic and glutamic acids typically drives the conformation toward the denatured form. On the other hand, some solvents may induce an ordered, non-native conformation.[151,152] At very low or especially at very high pH, chemical changes may occur, notably the deamidation of asparagine and glutamine and disulfide bond reshuffling.[153] Due to their high molecular weight, proteins are difficult to dissolve when aggregated, and aggregation occurs readily on concentration. Exposure to gas-aqueous interfaces can cause denaturation. Finally, proteins are readily lost onto surfaces, including frits.

Species variants of cytochrome c, differing only in a few residues, were successfully separated on Nucleosil™ $7C_{18}$, Nucleosil™ $7C_8$, and Nucleosil™ 5CN by RPLC in acetonitrile:water:sodium sulfate.[154] Five standard proteins were separated in as little as 90 seconds on Exsil A300.[155] RPLC on Vydac® C_4 with 2-propanol:water:0.1% TFA was a critical, activity-preserving step in the isolation of the leech protease hementin from a low-molecular-weight contaminant that interfered with sequencing and an inactive impurity of identical molecular weight.[156] By varying the polymer composition of silica supports coated with poly(2-hydroxyethylmethacrylate), co-poly(ethyl methacrylate-2-hydroxyethyl methacrylate), poly(ethylmethacrylate), and co-poly(octadecylmethacrylate-methylmethacrylate), the degree of unfolding could be controlled.[157] Hydroxyethyl methacrylate-based sorbents, available as TESSEK Separon™ HEMA (Spheron) BIO 1000, has been used to separate phospholipase in bee venom.[158] Protein RPLC has been reviewed.[159,160]

4.3.10   Herbicides and pesticides

Herbicide and pesticide samples are often derived from complex matrices and are present in trace quantities; therefore, column resolution is often a principal concern in the analysis of these compounds. Gas chromatography has been used much more widely than liquid chromatography for herbicide and pesticide analysis. With the advent of high resolution liquid chromatographic columns, there has been a resurgence of interest in the use of RPLC in this area of analysis. Limits of detection are still a restraining factor. Some of the major classes of pesticides are the organophosphorus, carbamate, and organochlorine groups. Herbicides include aryloxyproprionate, triazine, phenylurea, chlorinated phenoxyacetic acid, chloroacetamide, and auxin-like. The carbamate pesticides Aldicarb, Aldicarb sulfone, Aldicarb sulfoxide, Methomyl, 3-hydroxycarbofuran, MBC, Benomyl, Isolan, Baygon™, Carbofuran™, Mobam, Carbaryl, Betanal, α-Napthol, Propham, Landrin, Banol, Mesurol, Zectran™, Chloropropham, Bux, Captafol, Barban, Eptam™, Pebulate, Vernolate, Eurex, Butylate, Avadex™, and Avadex™ BW were separated on a μ-Bondapak™ C_{18} column using acetonitrile:water.[161] The carbamate pesticides Carbendazim, Aminocarb, Propoxur, Carbofuran™, Carbaryl, Propham, Captan, Chloropropham, Barban, Benomyl, and Butylate were separated on a C_8 or C_{18} column using acetonitrile:water buffered with phosphate.[162] Linuron, i.e., 3-(3,4-dichlorophenyl)-1-methoxy-1-methylurea, was separated on a μ-Bondapak™ C_{18}/Nova-Pak® column pair with column switching.[163]

4.3.11   Pigments, inks, and dyes

Many pigments, inks, and dyes are very hydrophobic compounds, and the matrices in which they occur may be complex. Direct Blue 6 (2,7-naphthalenedisulfonic acid, 3,3'-[[1,1'-biphenyl]-4,4'-diylbis(azo)]*bis* [5-amino-4-hydroxy-tetrasodium salt) and Direct Blue 15 (2,7-naphthalenedisulfonic acid, 3,3'-[(3,3'-dimethoxy[1,1'-biphenyl]-4,4'-diylbis(azo)]*bis* [5-amino-4-hydroxy-tetrasodium salt) were separated from numerous impurities on C_{18} columns, using methanol:water buffered with triethylamine phosphate.[164] The marker dye Solvent Yellow 124 — N-ethyl-N-[2(-1-isobutoxyethoxy)ethyl]-(4-phenylazophenyl)amine — was analyzed in an acetonitrile:water phase containing dimethyloctylamine.[165] The histological xanthene dye components — fluorescein, 4',5'-dibromofluorescein, Eosin Y, ethyleosin, 2',7'-dichlorofluorescein, 4,4, 6,7-tetrachlorofluorescein, 4',5'-diiodofluorescein, erythrosin B, and phloxine B — were separated on a polymeric Hamilton™ PRP®-1 column in acetonitrile:water containing trimethylammonium hydroxide.[166] Disperse Red II, Solvent Blue II, 1-hydroxy-4-anilino-anthraquinone, Solvent Violet 13, and 1,8-di-*p*-toluidino-anthraquinone were chromatographed on Zipax/HCP in ethanol:water.[167] Contaminants of FD&C Yellow No. 6, including aniline, benzidine, 4-aminobiphenyl, and 4-amino benzidine, were diazotized and coupled to disodium 3-hydroxy-2,7-naphthalenedisulfonate, then chromatographed on a Microsorb™ C_{18} column in acetonitrile:water containing ammonium acetate.[168]

4.3.12   Polymers and polymer additives

Naturally occurring polymers include proteins, polynucleotides, and polysaccharides. Proteins and polynucleotides usually occur as definable chemical species. Polysaccharides and synthetic polymers normally occur as distributions varying widely in the number of constituent units. The number of constituent species is so great as to defy resolution by separation into individual species, so polymers are often analyzed as an unresolved distribution. Gel permeation chromatography, which is described in Chapter 6 of this book, is generally the method of choice. Many synthetic polymers are, in addition, so hydrophobic or strongly hydrogen bonding as to be difficult to chromatograph in RPLC or NPLC. Still, some remarkable separations have been accomplished on RPLC. Synthetic polymers (epoxy ester polymer) from paint was separated from the tertiary-amine modified polymer on Lichrosorb® RP-8 in THF:water, using butane sulfonic acid and acetic acid as modifiers.[169] A mixture containing polyether polyol used in reaction injection molding was chromatographed on Hypersil®-ODS in acetonitrile:water.[170] Degradation products of polyTame, a poly[(peptide ester)phosphazene], including the O-methyl ester of trialanine, dialanine, the O-methyl ester of alanine, diglycine, diglycine amide, and glycine ethyl ester, were separated in acetonitrile:water containing trifluoroacetyl octane sulfonate on a Zorbax® C_8 column.[171] Polymer additives — Irganox-245, -259, -565, -1010, -1035; Anox-3114; and Tinuvin-P, -234, -320, -326, -327, -and -328 — were separated on a Capcell™ Pak C_{18} in methanol:water.[172] Two toxic contaminants of polyurethane — diphenylmethane-4,4'-diisocyanate and toluene diocyanate — were reacted with 9-(methyl aminomethyl)-anthracene and determined as the urea derivative on a Hypersil® RP-18 C_{18} column in acetonitrile:water buffered with triethylamine phosphate.[173]

4.4   Case history: optimizing a chemical synthesis by RPLC

4.4.1   Overview

A discovery group transferred the synthesis of RWJ-26240 (Figure 7)[174] to the Chemical Development Department (R. W. Johnson Pharmaceutical

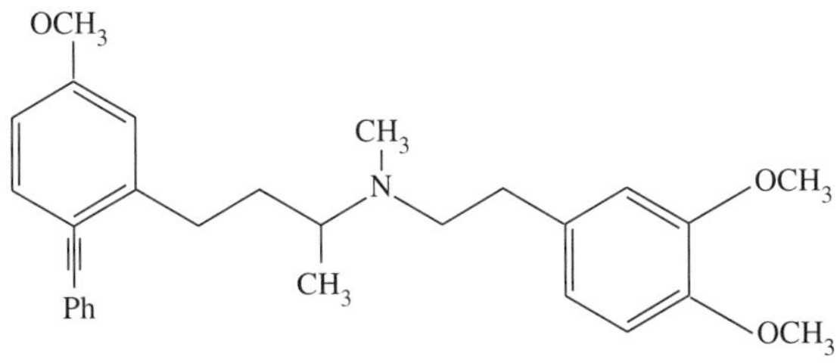

Figure 7   Target compound — RWJ-26240 (McN-5691).

Research Institute; Spring House, PA) for optimization. To develop the small-scale synthetic route used in the discovery phase[175] into a method that could be increased in scale for possible production, Chemical Development scrutinized each reaction step. Analytical techniques, including thin layer chromatography (TLC), gas chromatography (GC), nuclear magnetic resonance (NMR), infrared spectroscopy (IR), and HPLC were developed to analyze each reaction step for purity and yield. During the course of this investigation, we invented a new reaction called ligand-assisted nucleophilic addition (LANA).[176] Each step of the LANA reaction was supported by use of an HPLC technique. Below, we describe the requirements for devising and optimizing a chemical process suitable for production scale. The role of analytical chemistry in process development is described. Both GC and HPLC were key analytical techniques. GC was key in optimizing the small-scale route used in discovery, especially the first three steps of that route. HPLC was the key analytical technique for the LANA process. The need for HPLC methods to monitor reactions in order to minimize byproducts and optimize the yield of intermediates and the pure drug substance is discussed. We summarize by emphasizing the importance of analytical chemistry in process development.

4.4.2   Requirements of a production process

In the discovery phase, a reaction route is developed to allow synthesis of a maximum number of analogues for pharmacological testing. Since the focus is on synthetic flexibility, issues of scale are not central. Once a lead compound exhibits a useful pharmacological activity and is identified as a candidate for further development, larger scale synthesis is required to evaluate stability, bioavailability, toxicity, physicochemical properties, and other compound properties. The Chemical Development Department is usually involved in the preparation of supplies for these activities.

The first step on transfer of the sythesis is to evaluate the discovery route, looking particularly at overall yield and purity, as well as parameters such as cost of production (cost of starting materials, solvents, labor and overhead, and disposal of waste stream), ease of removal of impurities or catalyst from products, and the degree of hazard associated with solvents, reactants, intermediates, and products. The route used in discovery is adopted or modified as appropriate. We require that a production process provide high purity and stable drug substance, using the least hazardous and most cost-effective synthesis possible.

4.4.3   Evaluation of the preliminary route

The route used to synthesize RWJ-26240 (McN-5691) in drug discovery is depicted in scheme 1 (Figure 8). Chemical Development began by repeating the entire reaction route and examining each step for yield; purity; cost; hazardous nature of chemicals, intermediates, and byproducts; choice of solvents and/or catalyst; ease of removal of catalyst from the product; and total time of drug development. In evaluating reaction steps, TLC is useful as a fast, qualitative technique to evaluate the progress of conversion of starting material to product. For very fast reactions, a reaction can be conducted and monitored in the NMR tube. HPLC and/or GC are particularly useful when the reaction is somewhat slower and precise quantitative data is required. Elemental analysis, mass spectral (MS) analysis (including TLC/MS, HPLC/MS, and GC/MS), IR, and NMR are useful in verifying the structures of compounds observed chromatographically.

HPLC is a very powerful technique for qualitative and quantitative analysis. In the support of process development, HPLC plays an important role in monitoring a reaction, since each reaction component can be quantitated. In this role, the HPLC method must be fast, rugged, and specific, capable of separating all reactants, products, and byproducts. Development of appropriate analytical methods is often a rate-limiting step in process development.

After preliminary evaluation of the discovery route (scheme 1), we concluded that the overall yield of RWJ-26240 should be improved and that the use of $NaCNBH_3$ should be eliminated, since it produces a waste stream containing HCN or NaCN. Replacement of the expensive silver reagent, silver acetate, would permit significant cost reduction. The copper-catalyzed palladium coupling step would lead to palladium as a contaminant in the final product.[177] Since a drug substance containing palladium would not be acceptable, this step would also have to be revised.

4.4.4   Modification of the discovery route

With these goals in mind, we developed a modified route as outlined in scheme 2 (Figure 9).

The improvements to the first three steps of scheme 1 were accomplished using GC as a major analytical tool. A capillary GC internal standard method, described above, was used to monitor the first three steps of scheme 1. Figure 10 is a typical chromatogram of the internal standard method for step 1 of scheme 1. To follow a reaction, a known amount of internal standard was added to the reaction vessel. Aliquots were withdrawn at intervals and analyzed on GC. A graph of yield vs. reaction time was prepared to determine the optimum time for completion of the reaction.

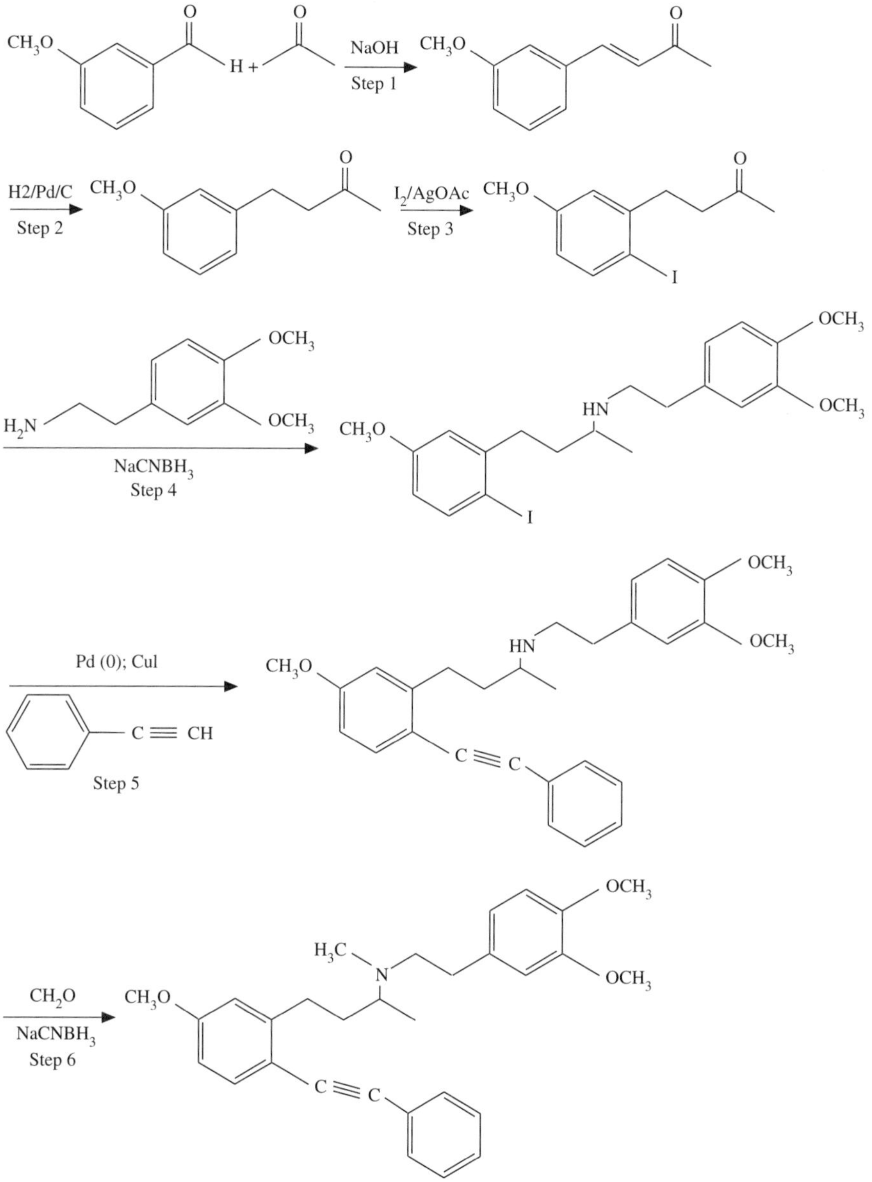

Figure 8 Scheme 1 — RWJ-26240 drug discovery route.

The GC conditions developed to follow steps 1 to 3 of scheme 1 were

- Column: 12 m × 0.2 mm i.d. × 0.3 μm thickness HP-1 capillary (Hewlett-Packard®)
- Detector and injector temperature: 300°C
- Internal standard: decane

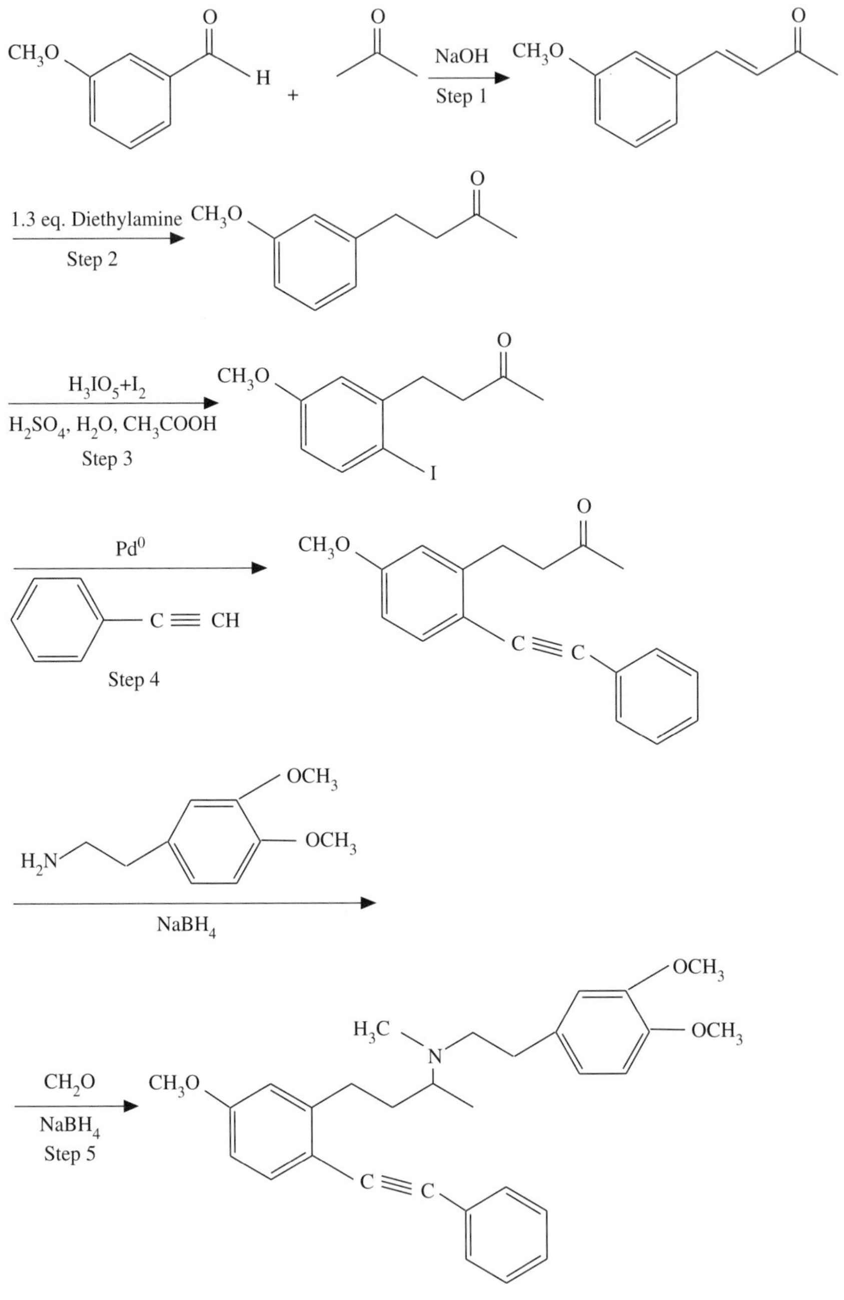

Figure 9 Scheme 2 — RWJ-26240 modified route.

This method allowed us to explore reaction conditions by measuring the conversion of starting material to product as a function of time. Reaction parameters were varied to maximize yield of the desired intermediate products.

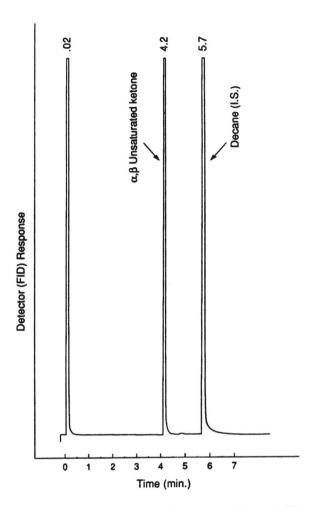

Figure 10 GC Internal Standard Method for step 1, scheme 1. Chromatographic conditions were column: 12 m × 0.2 mm × 0.3 mm HP-1; injector temperature: 300°C; detector temperature: 300°C; column temperature: 75°C for 3 min, then +10°C/min to 280°C.

In step 1, only minor changes in time and temperature were made. In step 2, it was found that the addition of diethylamine led to a decrease in dimeric byproducts. An improved ratio of the intermediate (with the iodide moiety para rather than ortho to the methoxy group) was attained with a reagent far less expensive than the silver acetate used in the preliminary synthetic route.

Next, reductive amination (step 4 in scheme 1) was exchanged with copper catalyzed palladium coupling (step 2 in scheme 1). Atomic absorption analysis for palladium in RWJ-26240 samples prepared by scheme 2 indicated that the level of palladium was reduced to an acceptable level. This improvement may be due to the two reduction steps subsequent to the use of palladium in scheme 2.[177] The final major modification to the reaction scheme was the substitution of $NaBH_4$ for $NaBH_3CN$. The yield of product (60%) was determined by HPLC (Method 2). Reductive alkylation with formalin/$NaBH_4$ afforded a pharmaceutically acceptable drug substance.

Each reaction step was monitored qualitatively by TLC using hexane:ethyl acetate as the developing solvent and quantitatively by GC. Impurity peaks were identified by GC/MS. An HPLC external standard method (Method 2) was developed and used to determine the purity of the final isolated product (RWJ-26240). The following rugged HPLC method was developed to optimize scheme 1, step 6:

Method 2: HPLC Method for RWJ-26240

Instrument:	Waters™ HPLC system
Column:	25 cm × 4.6 mm Zorbax® phenyl, 6-μ particle size (Du Pont)
Mobile phase:	58% A: 39% B: 3% C
	A = 10 mM KH_2PO_4 + 3.0 ml TEA, pH 4.2
	B = acetonitrile
	C = 2-propanol
Flow:	2.0 ml/min
Column temperature:	50°C
Detector:	UV-VIS, 254 nm
Detector sensitivity:	1 AUFS
Sample solvent:	30% B: 70% C
Sample concentration:	40 μg/ml
Injection size:	20 μl
Retention time, RWJ-26240:	14.5 min

The principal goals of process modification had been met. An environmentally safe waste stream was assured by the replacement of $NaCNBH_3$ with $NaBH_4$. The product was essentially free of palladium. The expensive silver acetate had been replaced with a less expensive reagent. The overall yield of RWJ-26240 had been increased by 60%.

4.4.5   Ligand-assisted nucleophilic addition route

To further reduce the cost of production and the number of reaction steps, an entirely new reaction step was developed (scheme 3 in Figure 11).[176]

Using the relatively inexpensive *p*-benzoquinone as a starting material, ligand-assisted nucleophilic addition (LANA) allows the preparation of RWJ-26240 by scheme 4 (Figure 12). Lithium phenylacetylide is added to benzoquinone at −10 to 0°C, followed by addition of a Grignard reagent of the ketal to generate intermediate 2. Aromatization of compound 2 to a

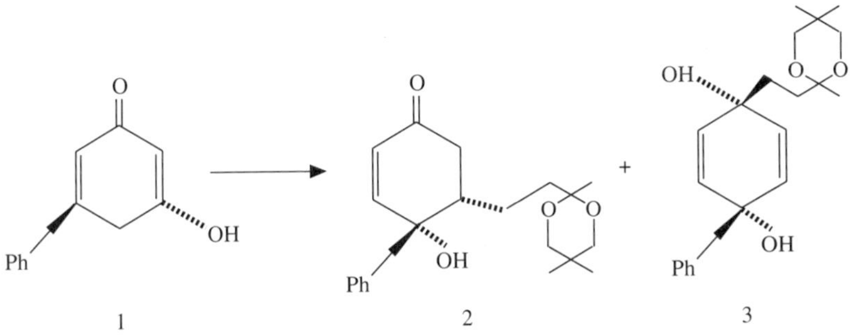

Figure 11 Scheme 3 — ligand-associated nucleophilic reaction.

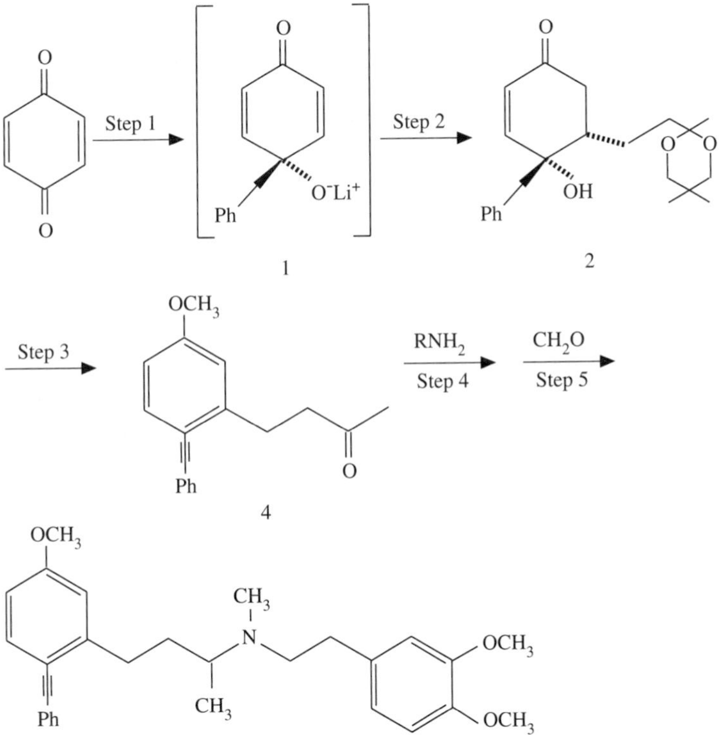

Figure 12 Scheme 4 — LANA route for RWJ-26240.

phenol, followed by alkylation with dimethyl sulfate, or conversion in one step using pyridinium hydrochloride in methanol resulted in high yields of compound 4 of scheme 4. Conversion of compound 4 to the final product was accomplished by the same steps 5 and 6 as in scheme 2. Note that product 2 is formally a 1,2-addition of phenylacetylide to a benzoquinone carbonyl, followed by a 1,4-addition of the Grignard reagent to the enone system, and is thus dubbed the 1, 2/1, 4-addition product, while product 3 is formally a 1,2-addition of phenylacetylide to a benzoquinone carbonyl, followed by a 1, 2-addition of the Grignard reagent to the remaining carbonyl, and is thus dubbed the 1, 2/1, 2-addition product.

A complication of the LANA step (Figure 11, scheme 3) is that in addition to intermediate 2 (by 1, 4-addition of the Grignard reagent to the α, β unsaturated ketone in a Michael fashion), 1,2/1,2-addition to the carbonyl is observed to afford compound 3 (Figure 11).[178] In order to maximize the yield of compound 2 relative to compound 3 (Figure 11), experimental conditions were validated by monitoring the reaction sequence by HPLC. A variety of Grignard reagents, solvents, and temperatures were tried, and the results were monitored by HPLC.

4.4.6 HPLC methods to support the LANA process

For the development of the LANA route, analytical techniques such as GC, TLC, HPLC, NMR, and GC/MS were used. GC methods were developed to monitor formation of the Grignard reagent. Since all of the components of the LANA route are unstable to the elevated temperatures of GC, HPLC and TLC techniques were chosen for qualitative and quantitative analysis of reaction samples, to monitor reaction progress, and to determine the purity of intermediates and final product. Because the process development time was limited and the LANA process was entirely dependent on HPLC analysis, we set criteria for the development of HPLC methods:

- The components of the reaction must be stable in the mobile phase.
- The method must be fast, rugged, and universal for the reaction products. The response factors of all components of interest should be equivalent to permit quantitation of all components without the use of extensive standardization.
- The method must be linear for all starting materials, products, and byproducts over a large concentration range to permit sensitive detection of the appearance of products and the disappearance of starting material.
- The method should be readily adaptable for each step of the process.
- The method development time should be minimal, since the process development time is short.

A reversed phase method (Method 3) was used for the optimization of the LANA reaction scheme (scheme 5; Figure 13). With slight modification of the mobile phase composition, it was also used for steps 1 to 3 of the LANA route (Figure 12).

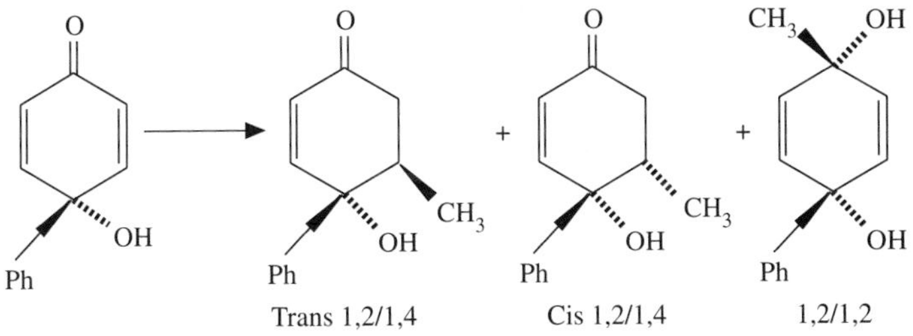

Figure 13 Scheme 5 — key step of a LANA reaction.

Method 3: General HPLC Method for the LANA Process

Instrument:	Waters™ HPLC system
Column:	25 cm × 4.6 mm Zorbax® C$_8$, 5-μ particle size (Du Pont, Wilmington, DE)
Mobile phase:	55% A: 45% B A = 100 mM KH$_2$PO$_4$, pH 6.5 B = acetonitrile
Flow:	1.0 ml/min
Column temperature:	35°C
Detector:	UV-VIS, 220 nm
Detector sensitivity:	1 AUFS
Sample solvent:	50% A: 50% B
Sample concentration:	5.0 mg/ml
Injection size:	10 μl
Retention time:	4.4 min, benzoquinone

The solvent in which a sample is dissolved can play a very important role in HPLC analysis. Immiscibility, precipitation, decomposition, and system peaks are all problems potentially caused by a sample solvent incompatible with the analysis. Ideally, the mobile phase should be identical to the reaction solvent. The addition of an internal standard permits a kinetic analysis to be conducted.

During the LANA process, we observed the decomposition of 1,2/1,4 ketal (compound 2) in 2 hours in HPLC grade acetonitrile. A noticeable color change was observed. Since previous experience with sample decomposition led us to suspect that oxygen played a role in the decomposition reaction, a sample was prepared in acetonitrile with or without degassing. After 1 hour, the apparent pH was found to be 1.3 in acetonitrile that had not been degassed, but 6.2 in the degassed sample. Both samples were analyzed by Method 4, using well degassed mobile phase to exclude oxygen. A single peak was observed in the sample protected from oxygen, representing the 1,2/1,4 ketal, while the sample that had not been protected from oxygen had a new peak at 3.9 minutes, representing the carbinol (compound 1), as

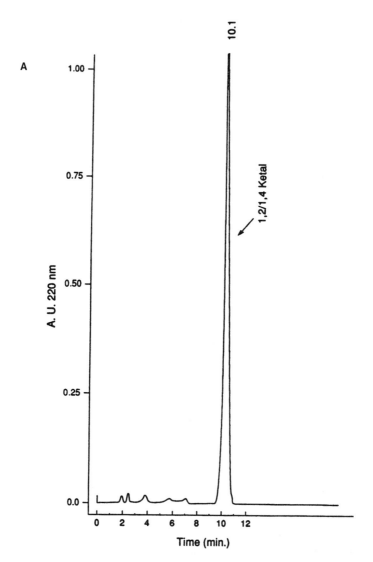

Figure 14   Separation of 1,2/1,4 ketal with and without protection from oxidative degradation. Chromatographic conditions were column: 25 cm × 4.6 mm Zorbax® C₈ (5-µm) column; mobile phase: 100 mM KH₂PO₄ (pH 6.5):acetonitrile (50:50); flow rate: 1.0 ml/min; column temperature: 35°C; detector wavelength: 220 nm. **(A)** Acetonitrile degassed. **(B)** Acetonitrile not degassed.

determined by MS. Figures 14A and B depict degassed and nondegassed samples, respectively.

   To further investigate the stability of the sample to acid and base, a stock solution at 106 µg/ml in degassed acetonitrile was prepared in an acid-washed or a base-washed flask. The samples were analyzed after 2 hours by Method 4. Chromatograms of acid-exposed and base-exposed samples are presented

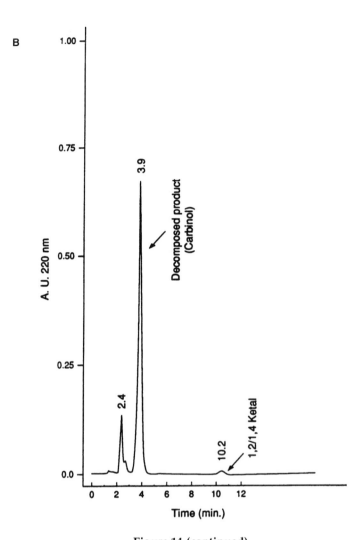

Figure 14 (continued)

in Figures 15A and B, respectively. The decomposition product was again found to be the carbinol (compound 1). The shifts in retention times from those observed in Figure 14 are typical observations in, and a principal limitation of, isocratic chromatography. Since retention times are extremely sensitive to slight differences in mobile phase composition, day-to-day variations in mobile phase preparation lead to retention shifts. Day-to-day variations in flow rate, check valve efficiency, or mixing solenoid performance (in binary, ternary, or quaternary pumping systems) can also contribute to retention shifts. Therefore, compound identification should be performed only by spiking with a known standard or by direct identification with, for example, mass spectral analysis.

Qualitative HPLC methods, using area percent, are used to monitor the disappearance of starting material and the formation of byproduct. Without

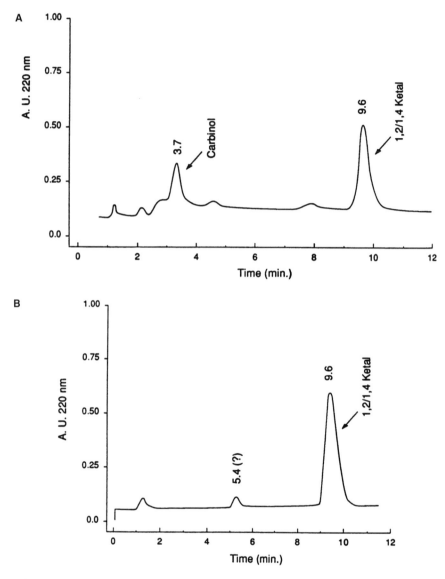

Figure 15 Separation of 1,2/1,4 ketal exposed to acidic or basic conditions. Chromatographic conditions were column: 25 cm × 4.6 mm Zorbax® C_8 (5-µm) column; mobile phase: 100 mM KH_2PO_4 (pH 6.5):acetonitrile (50:50); flow rate: 1.0 ml/min; column temperature: 35°C; detector wavelength: 220 nm. **(A)** Sample exposed to acid-washed flask. **(B)** Sample exposed to alkali-washed flask.

the inclusion of an internal standard and the calculation of response factors, it is not possible to establish with certainty whether all of the starting material can be accounted for. An internal standard must be stable in the reaction mixture, must not co-elute with any of the components, and must be stable in the mobile phase. Ideally, the internal standard has a retention time about half that of the total analysis time. Internal standardization is extremely useful for kinetic studies. Added to the reaction vessel, samples that are withdrawn at various times will contain identical concentrations of internal standard, and chromatograms can be directly compared or adjusted to identical scales to correct for variation in injection volume.

Method 4: HPLC Method for 1,2/1,4 Ketal

Instrument:	Waters™ HPLC system
Column:	25 cm × 4.6 mm Zorbax® C_{18}, 5-μ particle size (Du Pont)
Mobile phase:	50% A: 50% B
	A = 100 mM KH_2PO_4, pH 6.5
	B = acetonitrile
Flow:	1.0 ml/min
Column temperature:	35°C
Detector:	UV-VIS, 220 nm
Detector sensitivity:	1 AUFS
Sample solvent:	50% A: 50% B
Sample concentration:	5.0 mg/ml
Injection size:	10 μl
Retention times:	3.9 min, carbinol
	10.2 min, 1,2/1,4 ketal

Method 3 was modified to an internal standard method into Method 5 by changing the bonded phase and the mobile phase composition. Biphenyl was used as an internal standard added into the reaction. Aliquots were withdrawn, diluted with degassed acetonitrile, and analyzed according to Method 5. This internal standard method, Method 5, was helpful in the optimization of the desired *cis*-1,2/1,4 product of the key step of the LANA reaction (scheme 5).

Method 5: HPLC Internal Standard Method

Instrument:	Waters™ HPLC system
Column:	25 cm × 4.6 mm Zorbax® C_{18} (Du Pont), 5-μ particle size
Mobile phase:	45% A: 55% B
	A = 100 mM KH_2PO_4, pH 6.5
	B = acetonitrile
Flow:	1.0 ml/min
Column temperature:	Ambient
Detector:	UV-VIS, 254 nm
Detector sensitivity:	1 AUFS
Sample solvent:	50% A: 50% B
Sample concentration:	5.0 mg/ml
Injection size:	10 μl
Retention times:	5.8 min, *cis*-1,2/1,4
	17.6 min, biphenyl (IS)

Table 3   Product Yield (*cis*-1,2/1,4 of Scheme 5) as
Weight Percent Using Internal Standard Method

Sample no.	Time (h)	Wt%
1	0.0	0.0
2	0.5	20.0
3	1.0	36.7
4	2.0	38.1
5	19.0	38.7

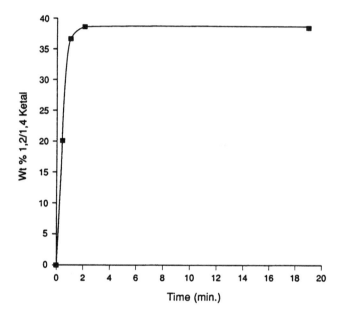

Figure 16   Reaction rate determination of 1,2/1,4 ketal in LANA reaction (scheme
5). Yield, graphed on the y-axis vs. time on the x-axis, was estimated by RPLC on
Zorbax® C_{18}. Column: 25 cm × 4.6 mm (5 μ). The mobile phase was 100 mM KH_2PO_4
(pH 6.5):acetonitrile (45:55) at 1.0 ml/min. The column temperature was 35°C, and
detection was at 254 nm.

The response factor of the product was determined by preparing a concentration series and analyzing as above using Method 5. The weight percent
yield of the *cis*-1,2/1,4 product as shown in scheme 5 calculated as a function
of time and is shown in Table 3 and Figure 16. The product is formed in
almost 40% yield within 3 minutes of the start of the reaction.

Under some conditions, it is difficult to incorporate an internal standard into a method. If the chromatogram is very complex, an internal
standard may interfere with quantitation of a peak of interest. The development of highly precise sample transfer techniques, including modern
autoinjectors, reduces the dependence of the experimentalist on the use of

an internal standard to correct for effects of dilution and transfer losses. In many cases, external standardization can be used effectively. The weight percent purity is determined by comparing the area of each peak in a chromatogram with those generated by separately injected pure standards of known concentration.

To determine the weight percent of compounds 2 and 3 of the LANA reaction presented in scheme 3, Method 6 was developed as an external standard technique. Figure 17 shows a typical chromatogram.

Method 6: HPLC External Standard Method

Instrument:	Waters™ HPLC system
Column:	25 cm × 4.6 mm Supelcosil™ LC-DB-18 (Supelco; Bellefonte, PA), 5-μ particle size
Mobile phase:	45% A: 55% B
	A = 100 mM KH$_2$PO$_4$, pH 6.5
	B = acetonitrile
Flow:	2.0 ml/min
Column temperature:	Ambient
Detector:	UV-VIS, 254 nm
Detector sensitivity:	1 AUFS
Sample solvent:	50% A: 50% B
Sample concentration:	5.0 mg/ml
Injection size:	5 μl
Retention times:	3. 5 min, carbinol 1
	5.5 min, 1,2/1,2 ketal 3
	12.5 min, 1,2/1,4 ketal 2

To determine the weight percent of each compound in the reaction sample, a standard and sample of known weight concentration were prepared and analyzed. The weight percent of each component purity was determined as described in the External Standard Method.

4.5   Conclusions

In the pharmaceutical industry, it is essential to produce pure drug substance, suitable for human consumption, in a cost-effective manner. The purity of a drug substance can be checked by separation techniques such as GC, TLC, and HPLC. Both techniques tend to be more sensitive and specific than spectroscopic methods. HPLC has an advantage over GC as an analytical technique, since analytes need be neither volatile nor extremely stable to elevated temperatures. Highly accurate, almost universal detectors, such as the UV-VIS, make quantitation far easier than with TLC. Standardization, either by external or internal standard techniques, makes it possible to correct for structurally dependent differences in detector response. The development and optimization of the novel LANA reaction used in the synthesis of RWJ-26240 was greatly facilitated by using HPLC for kinetic analysis of

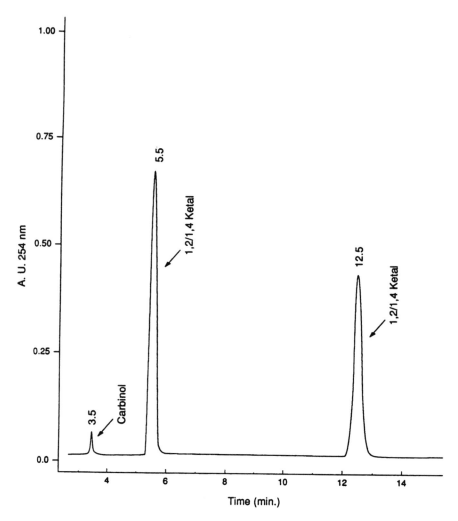

Figure 17 A typical chromatogram of LANA reaction (scheme 3). Chromatographic conditions — 45:55 100 mM KH_2PO_4:acetonitrile at 2.0 ml/min; 25 cm × 0.46 mm Supelcosil™ LC-DB-18 (5-μ) RPLC column; column temperature: ambient; detector wavelength: 254 nm.

reaction progress. Reaction conditions were varied, and the consequences monitored by fast, specific, quantitative HPLC methods.

Acknowledgments

The authors thank Ahmed F. Abdel-Magid, David A. Cherry, Joseph H. Childers, Robert J. Duda, John T. Hortensine, and John E. Mills for their support in providing standards and samples. We thank Dennis C. Liotta and his students, who worked on the LANA process, and David A. Cherry and

John E. Mills for their contributions to the LANA process development. We thank Daksha M. Desai and Robin C. Stanzione for their helpful discussion of the principles of reversed phase liquid chromatography.

References

1. Karger, B. L., Snyder, L. R., and Horváth, Cs, *An Introduction to Separation Science*, John Wiley & Sons, New York, 1973, 121–167.
2. Snyder, L. R. and Kirkland, J. J., *Introduction to Modern Liquid Chromatography*, Second ed., John Wiley & Sons, New York, 1979, chap. 2.
3. Glajch, J. L., Kirkland, J. J., Squire, K. M., and Minor, J. M., Optimization of solvent strength and selectivity for reversed-phase liquid chromatography using an interactive mixture-design statistical technique, *J. Chromatogr.*, 199, 57, 1980.
4. Daignault, L. G., Jackman, D. C., and Rillema, D. P., The case of the elusive t_m values in HPLC, *Chromatographia*, 27, 156, 1989.
5. Bidlingmeyer, B. A. and Warren, Jr., F. V., Column efficiency measurement, *Anal. Chem.*, 56, 1583A, 1984.
6. Antia, F. D. and Horváth, Cs., High-performance liquid chromatography at elevated temperatures: examination of conditions for the rapid separation of large molecules, *J. Chromatogr.*, 435, 1, 1988.
7. Warren, Jr., F. V. and Bidlingmeyer, B. A., Influence of temperature on column efficiency in reversed phase liquid chromatography, *Anal. Chem.*, 60, 2821, 1988.
8. Grushka, E., Snyder, L. R., and Knox, J. H., Advances in band spreading theories, *J. Chromatogr. Sci.*, 13, 25, 1975.
9. van Deemter, J. J., Zuiderweg, F. J., and Klinkenberg, A., Longitudinal diffusion and resistance to mass transfer as causes of nonideality in chromatography, *Chem. Eng. Sci.*, 5, 271, 1956.
10. Giddings, J. C. and Mallik, K. L., Theory of gel filtration (permeation) chromatography, *Anal. Chem.*, 38, 997, 1966.
11. Giddings, J. C., *Dynamics of Chromatography. Principles and Theory*, Marcel Dekker, New York, 1965.
12. Rocca, J. L., Higgins, J. W., and Brownlee, R. G., Peak variance as a function of HPLC column length and diameter, *J. Chromatogr. Sci.*, 23, 106, 1985.
13. Yau, W. W., Characterizing skewed chromatographic band broadening, *Anal. Chem.*, 49, 395, 1977.
14. Kirkland, J. J., Yau, W. W., Stoklosa, H. J., and Dilks Jr., C. H., Sampling and extra-column effects in high-performance liquid chromatography: influence of peak skew on plate count calculations, *J. Chromatogr. Sci.*, 15, 303, 1977.
15. Sánchez, F. C., Toft, J., van den Bogaert, B., and Massart, D. L., Orthogonal projection approach applied to peak purity assessment, *Anal. Chem.*, 68, 79, 1996.
16. Pauls, R. E. and McCoy, R. W., Testing procedures for liquid chromatographic columns, *J. Chromatogr. Sci.*, 24, 66, 1986.
17. Ahuja, S., High resolution liquid chromatography, *J. Liq. Chromatogr.*, 10, 1841, 1987.
18. Simpson, R. C., Brown, P. R., and Schwartz, M. K., Evaluation of HPLC column performance for clinical studies, *J. Chromatogr. Sci.*, 23, 89, 1985.

19. Cooke, N. H. C. and Olsen, K., Some modern concepts in reversed-phase liquid chromatography on chemically bonded alkyl stationary phases, *J. Chromatogr. Sci.*, 18, 512, 1980.
20. Colin, H. and Krstulovic, A. M., The theory of retention in reversed-phase high-performance liquid chromatography, in *Liquid Chromatography in Pharmaceutical Development*, Wainer, I. W., Ed., Aster Publishing, 1985, 171.
21. Dallas, G. and Abbott, S. D., New approaches to the analysis of low-molecular-weight polymers, in *Liquid Chromatographic Analysis of Food and Beverages*, Vol. 2, Charalambous, G., Ed., Academic Press, New York, 1979, 509.
22. Schoenmakers, P. J., Billiet, H. A. H., Tijssen, R., and de Galan, L., Gradient selection in reversed-phase liquid chromatography, *J. Chromatogr.*, 149, 519, 1978.
23. Snyder, L. R., Dolan, J. W., and Gant, J. R., Gradient elution in high-performance liquid chromatography. I. Theoretical basis for reversed-phase systems, *J. Chromatogr.*, 165, 3, 1979.
24. Stadalius, M. A., Gold, H. S., and Snyder, L. R., Optimization model for the gradient elution separation of petide mixtures by reversed-phase high-performance liquid chromatography. Verification of retention relationships, *J. Chromatogr.* 296, 31, 1984.
25. Aguilar, M. -I., Hodder, A. N., and Hearn, M. T. W., High-performance liquid chromatography of amino acids, peptides and proteins. LXV. Studies on the optimisation of the reversed-phase gradient elution of polypeptides: evaluation of retention relationships with β-endorphin-related polypeptides, *J. Chromatogr.*, 327, 115, 1985.
26. Snyder, L. R. and Dolan, J. W., Initial experiments in high-performance liquid chromatographic method development. I. Use of a starting gradient run, *J. Chromatogr. A*, 721, 3, 1996.
27. Lewis, J. A., Snyder, L. R., and Dolan, J. W., Initial experiments in high-performance liquid chromatographic method development. II. Recommended approach and conditions for isocratic separation, *J. Chromatogr. A*, 721, 15, 1996.
28. Quarry, M. A., Grob, R. L., and Snyder, L. R., Measurement and use of retention data from high-performance gradient elution. Contribution from "nonideal" gradient equipment, *J. Chromatogr.*, 285, 1, 1984.
29. Koyama, J., Nomura, J., Shiojima, Y., Ohtsu, Y., and Horii, I., Effect of column length and elution mechanism on the separation of proteins by reversed-phase high-performance liquid chromatography, *J. Chromatogr.*, 625, 217, 1992.
30. Antia, F. D. and Horváth, Cs., Dependence of retention of the organic modifier concentration and multicomponent adsorptive behavior in reversed-phase chromatography, *J. Chromatogr.*, 550, 411, 1991.
31. Poppe, H. and Kraak, J. C., Mass loadability of chromatographic columns, *J. Chromatogr.*, 255, 395, 1983.
32. Guo, D, Mant, C. T., Taneja, A. K, Parker, J. M. R., and Hodges, R. S., Prediction of peptide retention times in reversed-phase high-performance liquid chromatography. I. Determination of retention coefficients of amino acid residues of model synthetic peptides, *J. Chromatogr.*, 359, 499, 1986.
33. Guo, D., Mant, C. T., Taneja, A. K., and Hodges, R. S., Prediction of peptide retention times in reversed-phase high-performance liquid chromatography II. Correlation of observed and predicted petide retention times and factors influencing the retention times of peptides, *J. Chromatogr.*, 359, 519, 1986.

34. Hanai, T., Structure-retention correlation in liquid chromatography, *J. Chromatogr.*, 550, 313, 1991.

35. Sasagawa, T. and Teller, D. C., Prediction of peptide retention times in reversed-pahase HPLC, in *CRC Handbook of HPLC for the Separation of Amino Acids, Peptides, and Proteins*, Vol. II, Hancock, W. S., Ed., CRC Press, Boca Raton, FL, 1984, 53.

36. Kaliszan, R., Quantitative structure-retention relationships applied to reversed-phase high-performance liquid chromatography, *J. Chromatogr. A*, 656, 417, 1993.

37. Jandera, P., Correlation of retention and selectivity of separation in reversed-phase high-performance liquid chromatography with interaction indices and with lipophilic and polar structural indices, *J. Chromatogr. A*, 656, 437, 1993.

38. Kuronen, P., Identification using retention indices on gradient HPLC, in *Retention and Selectivity in Liquid Chromatography*, Smith, R. M., Ed., Elsevier Science, Amsterdam, 1995, chap. 6.

39. Bolck, A. and Smilde, A. K., Multivariate characterization of RP-HPLC stationary phases, in *Retention and Selectivity in Liquid Chromatography*, Smith, R. M., Ed., Elsevier Science, Amsterdam, 1995, chap. 12.

40. Lochmüller, C. H., Hsu, S.-H., and Reese, C., Prediction of the retention behavior of ionizable compounds in reversed-phase LC using factor-analytical modeling, *J. Chromatogr. Sci.*, 34, 77, 1996.

41. Lochmüller, C. H., Moebus, M. A., Liu, Q., and Jiang, C., Temperature effect on retention and separation of poly(ethylene glycol)s in reversed-phase liquid chromatography, *J. Chromatogr. Sci.*, 34, 69, 1996.

42. Crego, A. L., Díez-Masa, J. C., and Dabrio, M. V., Preparation of open tubular columns for reversed-phase liquid chromatography, *Anal. Chem.*, 65, 1615, 1993.

43. Hanson, M., Kurganov, A., Unger, K. K., and Davankov, V. A., Polymer-coated reversed-phase packings in high-performance liquid chromatography, *J. Chromatogr. A*, 656, 369, 1993.

44. Cox, G. B., The influence of silica structure on reversed-phase retention, *J. Chromatogr. A*, 656, 353, 1993.

45. Pesek, J. J. and Williamsen, E. J., Comparison of novel stationary phases, in *Retention and Selectivity in Liquid Chromatography*, Smith, R. M., Ed., Elsevier Science, Amsterdam, 1995, chap. 11.

46. Wilson, T. D. and Simmons, D. M., A particle size distribution analysis of used HPLC column packing material, *Chromatographia*, 35, 295, 1993.

47. Kirschbaum, J., Perlman, S., and Poet, R. B., Anomalies in HPLC, *J. Chromatogr. Sci.*, 20, 336, 1982.

48. Kirschbaum, J., Perlman, S., Adamovics, J., and Joseph, J., Anomalies in HPLC, III, *J. Chromatogr. Sci.*, 23, 493, 1985.

49. Kirschbaum, J., Perlman, S., Joseph, J., and Adamovics, J., Ensuring accuracy of HPLC assays, *J. Chromatogr. Sci.*, 22, 27, 1984.

50. Li, Z., Rutan, S. C., and Dong, S., Wetting of octadecylsilylated silica in methanol-water eluents, *Anal. Chem.*, 68, 124, 1996.

51. Jandera, P., Predictive calcluation methods for optimization of gradient elution using binary and ternary solvent gradients, *J. Chromatogr.*, 485, 113, 1989.

52. Engelhardt, H. and Jungheim, M., Comparison and characterization of reversed phases, *Chromatographia*, 29, 59, 1990.

53. Aiken, R. L., Fritz, G. T, Marmion, D. M., Michel, K. H., and Wolf, T., An evaluation of quantitative precision in high performance liquid chromatography, *J. Chromatogr. Sci.*, 19, 338, 1981.

54. McCoy, R. W., Aiken, R. L., Pauls, R. E., Ziegel, E. R, Wolf, T., Fritz, G. T., and Marmion, D. M, Results of a cooperative study comparing the precision of peak height and area measurements in liquid chromatography, *J. Chromatogr. Sci.*, 22, 425, 1984.

55. Kipiniak, W., A basic problem — the measurement of height and area, *J. Chromatogr. Sci.*, 19, 332, 1981.

56. Knoll, J. E. and Midgett, M. R, The area overlap method for determining adequate chromatographic resolution, *J. Chromatogr. Sci.*, 20, 221, 1982.

57. Karnes, H. T., Shiu, G., and Shah, V. P., Validation of bioanalytical methods, *Pharm. Res.*, 8, 421, 1991.

58. Inman, E. L. and Tenbarge, H. J., High-low chromatography: estimating impurities in HPLC using a pair of sample injections, *J. Chromatogr. Sci.*, 26, 89, 1988.

59. Gilpin, R. K., Jaroniec, M., and Lin, S., Dependence of the methylene selectivity on the composition of hydro-organic eluents for reversed-phase liquid chromatographic systems with alkyl bonded phases, *Chromatographia*, 30, 393, 1990.

60. Colin, H., Guiochon, G., Yun, Z., Diez-Masa, J. C., and Jandera, J., Selectivity for homologous series in reversed-phase LC. III. Investigation of non-specific selectivity, *J. Chromatogr. Sci.*, 21, 179,1983.

61. Smith, R. M. and Finn, N., Comparison of retention index scales based on alkyl aryl ketones, alkan-2-ones and 1-nitroalkanes for polar drugs on reversed-phase high-performance liquid chromatography, *J. Chromatogr.*, 537, 51, 1991.

62. Issaq, H. J. and Jaroniec, M., Enthalpy and entropy effects for homologous solutes in HPLC with alkyl chain bonded phases, *J. Liq. Chromatogr.*, 12, 2067, 1989.

63. Thruston Jr., A. D., High-pressure liquid chromatography techniques for the isolation and identification of organics in drinking water extracts, *J. Chromatogr. Sci.*, 16, 254, 1978.

64. Swadesh, J. K., Stewart, Jr., C. W., and Uden, P. C., Comparison of liquid chromatographic methods for analysis of homologous *n*-alkyl esters of biphenyl-4,4'-dicarboxylic acid, *Analyst*, 118, 1123, 1993.

65. Farinotti, R., Siard, Ph., Bourson, J., Kirkiacharian, S., Valeur, B., and Mahuzier, G., 4-bromomethyl-6, 7-dimethoxycoumarin as a fluorescent label for carboxylic acids in chromatographic detection, *J. Chromatogr.*, 269, 81, 1983.

66. Possanzini, M. and Di Palo, V., Improved HPLC determination of aliphatic amines in air by diffusion and derivatization techniques, *Chromatographia*, 29,151, 1990.

67. Kunugi, A. and Tabei, K., HPLC of ketone semicarbazones: relation of the capacity factors of ketone semicarbazones to carbon numbers, *J. HRC&CC*, 11, 600, 1988.

68. Wan, Q.- H., Davies, M. C., Shaw, P. N., and Barrett, D. A., Retention behavior of ionizable isomers in reversed-phase liquid chromatography: a comparative study of porous graphitic carbon and octadecyl bonded silica, *Anal. Chem.*, 68, 437, 1996.

69. Yang, W.-H., Chen, I-L., and Wu, D.-H., Chemically bonded phenylsilicone stationary phases for the liquid chromatographic separation of polycyclic aromatic hydrocarbons and cyclosiloxanes, *J. Chromatogr. A*, 722, 97, 1996.

70. Gupta, P. K. and Nikelly, J. G., Determination of UV-transparent compounds by liquid chromatography using the UV detector, *J. HRC&CC*, 9, 572, 1986.

71. Katayama, M., Masuda, Y., and Taniguchi, H., Determination of alcohols by high-performance liquid chromatography after pre-column derivatization with 2-(4-carboxyphenyl)-5,6-dimethylbenzimidazole, *J. Chromatogr.*, 585, 219, 1991.

72. Geyer, R. and Saunders, G. A., Determination of phthalic anhydride in workplace air using reverse phase high performance liquid chromatography, *J. Liq. Chromatogr.*, 9, 2281, 1986.

73. Dong, M. W. and Greenberg, A., Liquid chromatographic analysis of polynuclear aromatic hydrocarbons with diode array detection, *J. Liq. Chromatogr.*, 11, 1887, 1988.

74. Liberti, A. and Ciccioli, P., High resolution chromatographic techniques for the evaluation of atmospheric pollutants, *J. HRC&CC*, 9, 492, 1986.

75. Dong, M. W. and DiCesare, J. L, Analysis of priority pollutants by very high-speed LC, *J. Chromatogr. Sci.*, 20, 517, 1982.

76. Verzele, M. and Dewaele, C., Liquid chromatography in packed fused silica capillaries or micro-LC: a repeat of the capillary gas chromatography story?, *J. HRC&CC*, 10, 280, 1987.

77. Trisciani, A. and Andreolini, F, Evaluation of a micro-HPLC system dedicated to packed capillary column liquid chromatography, *J. HRC & CC*, 13, 270, 1990.

78. Šlais, K., Small bore LC/EC, *J. Chromatogr. Sci.*, 24, 321, 1986.

79. Lee, D. P., Reversed-phase HPLC from pH 1 to 13, *J. Chromatogr. Sci.*, 20, 203, 1982.

80. Kafil, J. B. and Last, T. A., Liquid chromatography with voltammetric detection for quantitation of phenolic acids, *J. Chromatogr.*, 348, 397, 1985.

81. Nowack, B., Karl, F. G., Hilger, S. U., and Sigg, L., Determination of dissolved and adsorbed EDTA species in water and sediments by HPLC, *Anal. Chem.*, 68, 561, 1996.

82. Franco, C. M., Fente, C. A., Vazquez, B., Cepeda, A., Lallaoui, L., Prognon, P., and Mahuzier, G., Simple and sensitive high-performance liquid chromatography-fluorescence method for the determination of citrinin. Application to the analysis of fungal cultures and cheese extracts, *J. Chromatogr. A*, 723, 69, 1996.

83. Gennaro, M. C., Bertolo, P. L., and Marengo, E., Determination of aromatic amines at trace levels by ion interaction reagent reversed-phase high-performance liquid chromatography. Analysis of hair dyes and other water-soluble dyes, *J. Chromatogr.*, 518, 149, 1990.

84. Gennaro, M. C., Giacosa, D., and Abrigo, C., The role of pH of the mobile-phase in ion-interaction RP-HPLC, *J. Liq. Chromatogr.*, 17, 4365, 1994.

85. Gao, C. -X., Chou, T. -Y., Colgan, S. T., Krull, I. S., Dorschel, C., and Bidling-meyer, B., New polymeric benzotriazole reagents for off-line derivatizations of amines and polyamines in HPLC, *J. Chromatogr. Sci.*, 26, 449, 1988.

86. Bartók, T., Börsök, G., and Sági, F., RP-HPLC separation of polyamines after automatic FMOC-Cl derivatization and precolumn sample clean-up using column switching, *J. Liq. Chromatogr.*, 15, 777, 1992.

87. Romanová, D. and Novotný, L. Chromatographic properties of cytosine, cytidine and their synthetic analogues, *J. Chromatogr. B*, 675, 9, 1996.
88. Roston, D. A. and Rhinebarger, R. R., Evaluation of HPLC with pulsed-amperometric detection of an aminosugar drug substance, *J. Liq. Chromatogr.*, 14, 539, 1991.
89. Dominguez, L. M. and Dunn, R. S., Analysis of OPA-derived amino sugars in tobacco by high-performance liquid chromatography with fluorimetric detection, *J. Chromatogr. Sci.*, 25, 468, 1987.
90. Díaz, J., Liberia, J. L., Comellas, L., and Broto-Puig, F., Amino acid and amino sugar determination by derivatization with 6-aminoquinolyl-N-hydroxysuccinimidyl carbamate followed by high-performance liquid chromatography and fluorescence detection, *J. Chromatogr. A*, 719, 171, 1996.
91. Antonopoulou, S., Andrikopoulos, N. K., and Demopoulos, C. A., Separation of the main neutral lipids into classes and sepcies by PR-HPLC and UV detection, *J. Liq. Chromatogr.*, 17, 633, 1994.
92. Balestrieri, C., Camussi, G., Giovane, A., Iorio, E. L., Quagliuolo, L., and Servillo, L., Measurement of platelet-activating factor acetylhydrolase activity by quantitative high-performance liquid chromatography determination of coumarin derivatized 1-O-alkyl-2-sn-lysoglyceryl-3-phosphorylcholine, *Anal. Biochem.*, 233, 145, 1996.
93. Ruiz-Gutiérrez, V. and Barron, L. J. R., Methods for the analysis of triacylglycerols, *J. Chromatogr. B*, 671, 133, 1995.
94. Di Corcia, A., Marchetti, M., Samperi, R., and Marcomini, A., Liquid chromatographic determination of linear alkylbenzenesulfonates in aqueous environmental samples, *Anal. Chem.*, 63, 1179, 1991.
95. Schreuder, R. H., Martijn, A., Poppe, H., and Kraak, J. C., Determination of the composition of ethoxylated alkylamines in pesticide formulations by high-performance liquid chromatography using ion pair extraction detection, *J. Chromatogr.*, 368, 339, 1986.
96. Kawase, J., Ueno, H., and Tsuji, K., Analysis of amphoteric surfactants by liquid chromatography with post-column detection. I. Mono- and dialanine surfactants, *J. Chromatogr.*, 264, 415, 1983.
97. Bogusz, M., Franke, J. P., de Zeeuw, R. A., and Erkens, M., An overview of the standardization of chromatographic methods for screening analysis in toxicology by means of retention indices and secondary standards. Part II. High performance liquid chromatography, *Fresenius J. Anal. Chem.*, 347, 73, 1993.
98. Caliguri, E. J, Capella, P., Bottari, L., and Mefford, I. N., High-speed microbore liquid chromatography with electrochemical detection using 3 μ C_{18} packing, *Anal. Chem.*, 57, 2423, 1985.
99. Mitsui, A., Nohta, H., and Ohkura, Y., High-performance liquid chromatography of plasma catecholamines using 1,2-diphenylethylenediamine as precolumn fluorescence derivatization reagent, *J. Chromatogr.*, 344, 61, 1985.
100. Mishin, V. M., Koivisto, T., and Lieber, C. S., The determination of cytochrome P450 2E1-dependent *p*-nitrophenol hydroxylation by high-performance liquid chromatography with electrochemical detection, *Anal. Biochem.*, 233, 212, 1996.
101. Gfeller, J. C., Frey, G., and Frei, R.W., Post-column derivatization in high-performance liquid chromatography using the air segmentation principle: application to digitalis glycosides, *J. Chromatogr.* 142, 271, 1977.

102. Lobell, M. and Schneider, M. P., 2,3,4,6-tetra-O-benzoyl-β-D-glucopyranosyl isothiocyanate: an efficient reagent for the determination of amino acids, β-adrenergic blockers and alkyloxiranes by high-performance liquid chromatography using standard reversed-phase columns, *J. Chromatogr.*, 633, 287, 1993.

103. Dong, M. W. and DiCesare, J. L., Rapid analysis of some commonly abused drugs by LC, *J. Chromatogr. Sci.*, 20, 330, 1982.

104. Adams, R. F., Schmidt, G. J., and Vandemark, F. L., A micro liquid column chromatography procedure for twelve anticonvulsants and some of their metabolites, *J. Chromatogr.*, 145, 275, 1978.

105. Liang, G. S., Zhang, Z., Baker, W. L., and Cross, R. F., Formation and verification of the structure of the 1-fluorenylmethyl chloroformate derivative of sulfamethazine, *Anal. Chem.*, 68, 86, 1996.

106. Salomies, H. and Salo, J.-P., An HPLC stdy of tinidazole hydrolysis, *Chromatographia*, 36, 79, 1993.

107. Beran, M. and Zima, J., Determinations of Monensins A and B in the fermentation broth of *Streptomyces cinnamonensis* by high performance liquid chromatography, *Chromatographia*, 35, 206, 1993.

108. Shihabi, Z. K., Review of drug analysis with direct serum injection on the HPLC column, *J. Liq. Chromatogr.*, 11, 1579, 1988.

109. Simonella, A., Torreti, L., Filipponi, C., Falgiani, A., and Ambrosil, L., Simultaneous determination of aflatoxins G_1, B_1, G_2, B_2 in animal feedstuffs by HP-TLC and RP-HPLC, *J. HRC&CC*, 10, 626, 1987.

110. Sharman, M. and Gilbert, J., Automated aflatoxin analysis of foods and animal feeds using immunoaffinity column clean-up and high-performance liquid chromatographic determination, *J. Chromatogr.*, 543, 220, 1991.

111. Scholten, A. H. M. T. and Frei, R. W., Identification of ergot alkaloids with a photochemical reaction detector in liquid chromatography, *J. Chromatogr.*, 176, 349, 1979.

112. Tscherne, R. J. and Umagat, H., Determination of isophenindamine in phenindamine tartrate using an argentated high-performance liquid chromatographic mobile phase, *J. Pharm. Sci.*, 69, 342, 1980.

113. Irth, H., De Jong, G. J., Brinkman, U. A. Th., and Frei, R. W., Determination of disulfiram and two of its metabolites in urine by reversed-phase liquid chromatography and spectrophotometric detection after post-column complexation, *J. Chromatogr.*, 424, 95, 1988.

114. Amin, M., Simultaneous determination of prostaglandins (PG) E_2, A_2 and B_2 and stability studies of PGE_2 in pharmaceutical preparations by ion-pair reversed phase HPLC, *Pharm. Acta. Helv.*, 64, 45, 1989.

115. McGuffin, V. L. and Zare, R. N., Femtomole analysis of prostaglandin pharmaceuticals, *Proc. Natl. Acad. Sci. U.S.A.*, 82, 8315, 1985.

116. Segelman, A. B., Adusumalli, V. E., and Segelman, F. H., Automated liquid chromatographic determination of ranitidine in microliter samples of rat plasma, *J. Chromatogr.*, 535, 287, 1990.

117. Kimura, M., Kanehira, K., and Yokoi, K., Highly sensitive and simple liquid chromatographic determination in plasma of B_6 vitamers, especially pyridoxal 5'-phosphate, *J. Chromatogr. A*, 722, 295, 1996.

118. Danielson, N. D., Targove, M. A., and Miller, B. E., Pre- and postcolumn derivatization chemistry in conjunction with HPLC for pharmaceutical analysis, *J. Chromatogr. Sci.*, 26, 362, 1988.

119. Ahuja, S., Chemical derivatization for the liquid chromatography of compounds of pharmaceutical interest, *J. Chromatogr. Sci.*, 17, 168, 1979.
120. Larsen, B. R. and West, F. G., A method for quantitative amino acid analysis using precolumn *o*-phthaladehyde derivatization and high performance liquid chromatography, *J. Chromatogr. Sci.*, 19, 259, 1981.
121. Hervé, C., Beyne, P., Jamault, H., and Delacoux, E., Determination of tryptophan and its kynurenine pathway metabolites in human serum by high-performance liquid chromatography with simultaneous ultraviolet and fluorimetric detection, *J. Chromatogr. B*, 675, 157, 1996.
122. Heinrikson, R. L. and Meredith, S. C., Amino acid analysis by reverse-phase high-performance liquid chromatography: precolumn derivatization with phenylisothiocyanate, *Anal. Biochem.*, 136, 65, 1984.
123. Alegría, A., Barberá, R., Farré, R., Ferrerés, M., Lagarda, M. J., and López, J. C., Isocratic high-performance liquid chromatographic determination of tryptophan in infant formulas, *J. Chromatogr. A*, 721, 83, 1996.
124. Chang, J. -Y., Knecht, R., and Braun, D. G., Amino acid analysis at the picomole level, *Biochem. J.*, 199, 547, 1981.
125. Cooper, J. D. H., Ogden, G., McIntosh, J., and Turnell, D. C., The stability of the *o*-phthaldehyde/2-mercaptoethanol derivatives of amino acids: an investigation using high-pressure liquid chromatography with a precolumn derivatization technique, *Anal. Biochem.*, 142, 98, 1984.
126. Bartók, T., Szalai, G., Lőrincz, Zs., Börcsök, G., and Sági, F., High-speed RP-HPLC/FL analysis of amino acids after automated two-step derivatization with *o*-phthaldialdehyde/3-mercaptoproprionic acid and 9-fluorenylmethyl chloroformate, *J. Liq. Chromatogr.*, 17, 4391, 1994.
127. Stocchi, V., Palma, F., Piccoli, G., Biagiarelli, B., Magnani, M., Masat, L., and Cucchiarini, L., Analysis of amino acids as DABS-derivatives with a sensitivity to the femtomole using RP-HPLC narrow-bore columns, *Amino Acids*, 3, 303, 1992.
128. Swadesh, J. K., High-speed tryptic fingerprinting, *BioTechniques*, 9, 626, 1990.
129. Swadesh, J. K., Tryptic fingerprinting on a poly(styrene-divinyl benzene) reversed-phase column, *J. Chromatogr.*, 512, 315, 1990.
130. Lau, S. Y. M., Taneja, A. K., and Hodges, R. S., Effects of high-performance liquid chromatographic solvents and hydrophobic matrices on the secondary and quaternary structure of a model protein. Reversed-phase and size exclusion high-performance liquid chromatography, *J. Chromatogr.*, 317, 129, 1984.
131. Swadesh, J. K., unpublished, 1991.
132. Henderson, D. E. and Mello, J. A., Physicochemical studies of biologically active peptides by low-temperature reversed-phase high-performance liquid chromatography, *J. Chromatogr.*, 499, 79, 1990.
133. Gough, T. A. and Baker, P. B., Identification of major drugs of abuse using chromatography, *J. Chromatogr. Sci.*, 20, 289, 1982.
134. Funasaki, N., Hada, S., and Neya, S., Conformational effects in reversed-phase liquid chromatographic separation of diastereomers of cyclic dipeptides, *Anal. Chem.*, 65, 1861, 1993.
135. Zhou, N. E., Mant, C. T., and Hodges, R. S., Effect of preferred binding domains on peptide retention in reversed-phase chromatography; amphipathic α-helices, *Peptide Res.*, 3, 8, 1990.

136. Lockhart, K. L., Kenley, R. A., and Lee, M. O., Comparing electrochemical, fluorescence, and electrochemical detectors for HPLC analysis of the decapeptide, nafarelin, *J. Liq. Chromatogr.*, 10, 2999, 1987.

137. Pearson, J. D., Lin, N. T., and Regnier, F. E., The importance of silica type for reverse-phase protein separations, *Anal. Biochem.*, 124, 217, 1982.

138. Pearson, J. D., Mahoney, W. C., Hermodson, M. A., and Regnier, F. E., Reversed-phase supports for the resolution of large denatured protein fragments, *J. Chromatogr.*, 207, 325, 1981.

139. Wilson, K. J., Van Wieringen, E., Klauser, S., Berchtold, M. W., and Hughes, G. J., Comparison of the high-performance liquid chromatography of peptides and proteins on the 100- and 300-Å reversed-phase supports, *J. Chromatogr.*, 237, 407, 1982.

140. Griehl, C. and Merkel, S., Synthesis and separation of pprotected tripeptide epimers by RP-HPLC, *Intl. J. Peptide Protein Res.*, 45, 217, 1995.

141. James, S. and Bennett, H. P. J., Use of reversed-phase and ion-exchange batch extraction in the purification of bovine pituitary peptides, *J. Chromatogr.*, 326, 329, 1985.

142. Zhao, Q., Sannier, F., Garreau, I., Lecoeur, C., and Piot, J. M., Reversed-phase high-performance liquid chromatography coupled with second-order derivative spectroscopy for the quantitation of aromatic amino acids in peptides: application to hemorphins, *J. Chromatogr. A*, 723, 35, 1996.

143. Wimley, W. C. and White, S. H., Quantitation of electrostatic and hydrophobic membrane interactions by equilibrium dialysis and reverse-phase HPLC, *Anal. Biochem.*, 213, 213, 1993.

144. Rohrer, J. S., Cooper, G. A., and Townsend, R. R., Identification, quantification, and characterization of glycopeptides in reversed-phase HPLC separations of glycoprotein proteolytic digests, *Anal. Biochem.*, 212, 7, 1993.

145. Williams, R. C., Vasta-Russell, J. F., Glajch, J. L., and Golebiowski, K., Separation of proteins on a polymeric fluorocarbon high-performance liquid chromatography packing, *J. Chromatogr.*, 371, 63, 1986.

146. McKern, N. M., Edskes, H. K., and Shukla, D. D., Purificatiion of hydrophilic and hydrophobic peptide fragments on a single reversed phase high performance liquid chromatographic system, *Biomed. Chromatogr.*, 7, 15, 1993.

147. Randall, C. S., Malefyt, T. R., and Sternson, L. A., Approaches in the Analysis of Peptide in *Peptide and Protein Drug Delivery*, Lee, V. H., Ed., Marcel Dekker, New York, 1991, chap. 5.

148. Titani, K., Sasagawa, T., Resing, K., and Walsh, K. A., A simple and rapid purification of commercial trypsin and chymostrypsin by reverse-phase high-performance liquid chromatography, *Anal. Biochem.*, 123, 408, 1982.

149. Berchtold, M. W., Heizmann, C. W., and Wilson, K. J., Ca^{2+}-binding proteins: a comparative study of their behavior during high-performance liquid chromatography using gradient elution in reverse-phase supports, *Anal. Biochem.*, 129, 120, 1983.

150. Cohen, K. A., Schellenberg, K., Benedek, K., Karger, B. L., Grego, B., and Hearn, M. T. W., Mobile-phase and temperature effects in the reversed phase chromatographic separation of proteins, *Anal. Biochem.*, 140, 223, 1984.

151. Sadler, A. J., Micanovic, R., Katzenstein, G. E., Lewis, R. V., and Middaugh, C. R., Protein conformation and reversed-phase high-performance liquid chromatography, *J. Chromatogr.*, 317, 93, 1984.

152. Benedek, K., Dong, S., and Karger, B. L., Kinetics of unfolding of proteins on hydrophobic surfaces in reversed-phase liquid chromatography, *J. Chromatogr.*, 317, 227, 1984.

153. Zale, S. E. and Klibanov, A. M., Why does ribonuclease irreversibly inactivate at high temperature?, *Biochemistry*, 25, 5432, 1986.

154. Terabe, S., Nishi, H., and Ando, T., Separation of cytochromes c by reversed-phase high-performance liquid chromatography, *J. Chromatogr.*, 212, 295, 1981.

155. Szczerba, T. J., Baehr, D. N., Glunz, L. J., Perry, J. A., and Holdoway, M. J., New packing and column for fast protein high performance liquid chromatography, *J. Chromatogr.*, 458, 281, 1988.

156. Swadesh, J. K., Huang, I.-Y., and Budzynski, A. Z., Purification and characterization of hementin, a fibrinogenolytic protease from the leech *Haementeria ghilianii*, *J. Chromatogr.* 502, 359, 1990.

157. Hanson, M., Unger, K. K., Mant, C. T., and Hodges, R. S., Polymer-coated reversed-phase packings with controlled hydrophobic properties. I. Effect on the selectivity of protein separations, *J. Chromatogr.*, 599, 65, 1992.

158. Čoupek, J. and Vinš, I., Hydroxyethyl methacrylate-based sorbents for high-performance liquid chromatography of proteins, *J. Chromatogr. A*, 658, 391, 1994.

159. Regnier, F. E. and Gooding, K. M., High-performance liquid chromatography of proteins, *Anal. Biochem.*, 103, 1, 1980.

160. Benedek, K. and Swadesh, J. K., HPLC of proteins and peptides in the pharmaceutical industry, in *HPLC in the Pharmaceutical Industry*, Fong, G. W. and Lam, S. K., Eds., Marcel Dekker, New York, 1991, chap. 11.

161. Sparacino, C. M. and Hines, J. W., High-performance liquid chromatography of carbamate pesticides, *J. Chromatogr. Sci.*, 14, 549, 1976.

162. Marvin, C. H., Brindle, I. D., Singh, R. P., Hall, C. D., and Chiba, M., Simultaneous determination of trace concentrations of benomyl, carbendazim (MBC) and nine other pesticides in water using an automated on-line pre-concentration high-performance liquid chromatographic method, *J. Chromatogr.*, 518, 242, 1990.

163. Cessna, A. J., The determination of the herbicide Linuron in Saskatoon berries using HPLC with column switching, *J. Liq. Chromatogr.*, 11, 725, 1988.

164. Shan, A., Harbin, D., and Jameson, C. W., Analyses of two azo dyes by high-performance liquid chromatography, *J. Chromatogr. Sci.*, 26, 439, 1988.

165. Henricsson, S. and Westerholm, R., Liquid chromatographic method for analysing the colour marker Solvent Yellow 124, N-ethyl-N-[2-(1-isobutoxyethoxy)ethyl](4-phenylazophenyl)amine, in diesel fuels, *J. Chromatogr. A*, 723, 395, 1996.

166. van Liedekerke, B. M. and de Leenheer, A. P., Analysis of xanthene dyes by reversed-phase high-performance liquid chromatography on a polymeric column followed by characterization with a diode array detector, *J. Chromatogr.*, 528, 155, 1990.

167. Passarelli, R. J. and Jacobs, E. S., High-pressure liquid chromatography analysis of dyes and intermediates, *J. Chromatogr. Sci.*, 13, 153, 1975.

168. Richfield-Fratz, N., Bailey, Jr., J. E., and Bailey, C. J., Determination of unsulphonated aromatic amines in FD&C Yellow No. 6 by the diazotization and coupling procedure followed by reversed-phase high-performance liquid chromatography, *J. Chromatogr.*, 331, 109, 1985.

169. Folonari, C. V. and Garlasco, R., Ion-paired HPLC characterization of cathodic electrodeposition paint polymers, *J. Chromatogr. Sci.*, 19, 639, 1981.

170. Noël, D. and VanGheluwe, P., High-performance liquid chromatography of industrial polyurethane polyols, *J. Chromatogr. Sci.*, 25, 231, 1987.

171. Eickhoff, W. M., Liversidge, G. G., and Mutharasan, R., Liquid chromatographic analysis of a potential polymeric-pendant drug delivery system for peptides. Application of high-performance size-exclusion chromatography, reversed-phase high-performance liquid chromatography and ion chromatography to the evaluation of biodegradable poly[(chloromethoxytrialanine methyl ester) phosphazenes], *J. Chromatogr.*, 536, 255, 1991.

172. Jinno, K. and Yokoyama, Y., Retention prediction for polymer additives in reversed-phase liquid chromatography, *J. Chromatogr.*, 550, 325, 1991.

173. Rastogi, S. C., Analysis of diisocyanate monomers in chemical products containing polyurethanes by high pressure liquid chromatography, *Chromatographia*, 28, 15, 1989.

174. Carson, J. R., U.S. Patent appl. 665, 684, 1984, U.S. Patent 4,661,635, 1987.

175. Carson, J. R., Almond, H. R., Brannan, M. D., Carmosin, R. J., Flaim, S. F., Gill, A., Gleason, M. M., Keely, S. L., Ludovici, D. W., Pitis, P. M., Rebarchak, M. C., and Villani, F. J., 2-ethynylbenzenealkanamines: a new class of calcium entry blockers, *J. Med. Chem.*, 31, 630, 1988.

176. Liotta, D. C., Maryanoff, C. A., and Paragamian, V., U.S. Patent appl. 27, 762, 1987, U.S. Patent 4,772,755, 1988.

177. Maryanoff, C. A., Mills, J. E., Stanzione, R. C., and Hortestine, Jr., J. T., Catalysis from the perspective of an organic chemist: common problems and possible solutions, in *Catalysis of Organic Reactions*, Rylander, P. N., Greenfield, H., and Augustine, R. L., Eds., Marcel Dekker, New York, 1988, 359–379.

178. Solomon, M., Jamison, W. C. L., McCormick, M., Liotta, D. C., Cherry, D. A., Mills, J. E., Shah, R. D., Rodgers, J. D., and Maryanoff, C. A., Ligand assisted nucleophilic additions. Control and face attack of nucleophiles on 4-oxidoenones, *J. Am. Chem. Soc.*, 110, 3702, 1988.

chapter five

Ion exchange chromatography

Joel Swadesh

0-8493-2682-6/97/$0.00+$.50
© 1997 by CRC Press, Inc.

5.1   Introduction

Ion exchange chromatography (IEC) has been in use for over half a century.[1,2] The first high performance applications are substantially more recent;[3-7] however, IEC tends to be less dependent on particle size and column packing technique than reversed phase liquid chromatography. There is, therefore, value in reviewing some of the older separations, many of which are remarkable in light of the simplicity of the technology. Advances in separation speed, minimum sample size, and detection have been astonishing.

The principle of IEC is that a charged analyte is bound to the stationary phase by means of electrostatic attraction. Ions from the mobile phase selectively displace components of the analyte from adsorption to the stationary phase. Some examples of so-called *ion-pairing* chromatography on reversed phase may actually be ion exchange chromatography. In ion pairing, a counter-ion binds to a charged analyte to form a neutral complex. However, if the ion pairing agent binds more strongly to the reversed phase than to the analyte, a noncovalent IEC phase can be said to have formed. Since hydrogen bonds may also have some degree of ionic character, some separations that are termed *normal phase* chromatography also may be considered ion exchange separations. Likewise, some *ligand exchange* separations could be considered to be ion exchange separations. Ethylene glycol dimethacrylate-methacrylic acid copolymeric phases imprinted with L-phenylalanine anilide (which is then extracted) were used to form a chiral ion exchange material.[8] Finally, *ion exclusion*,[9] which relies on the effect of ion exchange on the Donnan potential, might be termed an ion exchange method. Ions with the same charge as the ion exchanger are excluded from binding to a degree proportional to the analyte charge. Strongly ionized species elute before weakly ionized species.

There is often mixed-mode adsorption of the ligand to the stationary phase in ion exchange chromatography. Typically, the analyte and the stationary phase have some hydrophobic character, so selectivity may be modified by adding an organic component to the mobile phas. There has been recent progress in interpreting selectivity in ion exchange chromatography.[10,11] As long as there are not multiple equilibria involved in either the analyte or the eluent, selectivity is readily defined.[12,13] Changes in pH, however, alter the equilibrium between the species of a polyprotic ion as well as the properties of some ion exchange materials. The role of pH complicates the interpretations of many ion exchange separations.

Detection in IEC is often problematic. Of the substances commonly analyzed by ion chromatography, many simple inorganic ions, sugars, and amino acids lack strong absorbance in the UV, while many mobile phases absorb strongly. In some applications, however, the difference in absorbance between analyte and mobile phase is exploited as an indirect or vacancy detection. Refractive index detection has generally been regarded to be unsuitable for gradient applications in analysis, although recent instrumentation has begun to make such applications possible.[14] Electrochemical detection, such as conductivity, amperometry, coulometry, and voltammetry are therefore widely used in ion exchange applications.[15-19] The use of conductivity may require the selective removal of mobile phase ions prior to detection.

The chromatographic pumps and flow path used in IEC must be resistant to corrosion. For this reason, polymers such as poly(etheretherketone) (PEEK®, ICI Americas; Wilmington, DE) have entered into widespread usage in ion chromatography. Electrochemical detectors may also be subject to corrosion by certain ions. This chapter will review the chromatographic materials, detectors, and applications of ion exchange chromatography. For some classes of compounds, where reversed phase or normal phase alternatives may have been developed, alternative separation techniques will be presented.

5.2   Ion exchange stationary phases

5.2.1   General

Some of the earliest materials used as stationary phases in ion exchange were based on zeolites and clays.[20] Ion exchange materials include amines and alkyl amines, which bind anions, and carboxylates and sulfonates, which bind cations. Quaternized amines and sulfonates are called strong ion exchangers, while carboxylates and incompletely alkylated amines such as diethylaminoethylated (DEAE) phases are termed weak ion exchangers. Quaternized amines are incapable of titration. Sulfonates titrate only at extremely low values of pH, and the weak ion exchange materials titrate at intermediate values of pH. Crown ethers have also been used as ion exchange phases for inorganic cations. Functional groups that chelate metals, such as iminodiacetate, have also found considerable use in what is termed metal interaction chromatography. Silica itself acts as an ion exchange material.[21] Polymers, particularly polymers of styrene and divinyl benzene, have been the preferred solid support for ion exchange. Polymers of methacrylate are also widely used. Carbohydrate polymers, such as carboxymethylcellulose and cross-linked dextrans, have excellent pH stability but are less suited for high performance use because of instability to elevated pressure. As methods to stabilize silica, alumina, zirconia, and other inorganic materials to elevate pH are perfected, these materials are finding increased utility as supports for ion exchange.[22] Silica, first treated with triethoxyvinylsilane, was coated with styrene or glycidyl methacrylate and divinyl benzene and polymerization initiated to form the support for a cation exchange material stable to elevated pH.[23] Silica has been treated with tetramethylcyclotetrasiloxane and allyl glycidyl ether prior to derivatization with an anion exchange material to generate a column stable even at pH 11.[24] Aminopropylated silica was formed from the epoxy silica, then cross-linked with formaldehyde. While the chromatographic applications shown for this aminopropyl phase were in predominantly organic phases and might be regarded as normal phase separations, a strong dependence on salt concentration in the separation of nucleotides suggested the operation of an ion exchange mechanism. Small[2,25] and Arshady[26] have contributed informative descriptions of the process of fabrication of polymeric supports, and further information is available in Chapter 6.

To illustrate the difference between strong and weak ion exchange, consider the example of a stationary phase formed of a fully alkylated amine $-NR_3^+$ vs. one formed of a free amine $-NH_2$, which can be protonated to form $-NH_3^+$. As noted above, the former stationary phase is termed a strong ion exchange material, because it is charged at any value of pH. The latter stationary phase is called a weak ion exchange material, because at sufficiently basic pH, it will be fully or partially deprotonated. Anions, whether of the mobile phase or the analyte, interact with a charged cationic stationary phase to form a complex. Some anions bind more strongly than others. If the anion of the analyte binds more strongly than the anion of the mobile phase, the analyte will be retained on the column and can be eluted by increasing the ionic strength of the mobile phase.

5.2.2   Theory of ion binding

Ion exchange is actually a far more complicated process than the simple exchange of one cation for another or one anion for another, particularly for complex polyions such as proteins and oligonucleotides. Each ion of the stationary or mobile phase is surrounded by a cluster of counterions and water. When an ion in the mobile phase is transferred to the stationary phase, the cluster of counterions and water surrounding it is partially or completely displaced. A similar event occurs at the binding site of the stationary phase. Ions displaced from the stationary phase undergo the converse process. The overall equilibrium for exchange of a polyanion Z^{-n} (bound to an ion exchange material at m sites) with a univalent anion X^- on an anion exchange material such as a quaternized amine can be represented as

$$\{[\text{Stationary phase-NR}_3^+]_m \, [Z^{-n}\cdots(n-m)Y^+\cdots xH_2O]\} + [mX^-\cdots mY^+\cdots yH_2O] \rightleftarrows$$

$$m \, \{[\text{Stationary phase-NR}_3^+] \, [X^-\ldots(x+y-r)/m \, H_2O]\} + [Z^{-n}\ldots nY^+\ldots rH_2O]$$

where n, m, x, y, and r are integers, $n \geq m$. Even this representation is oversimplified. Some polyanions include both positive and negative charges, and there are often interactions between the analyte and stationary phase other than the simple electrostatic interactions described above. Further, many macromolecular materials can undergo conformational changes which affect elution behavior. One experimental study of the diffusion and binding of unstirred Rhodamine B in 16 to 20 μ MCI-GEL ion exchange particles showed a biphasic process, with diffusion in the 2-μ thick surface layer requiring about 1 hour and diffusion into the center of the particle requiring on the order of days.[27]

Polyions can bind at several sites on an ion exchange material. The process of exchange involves a re-arrangement of the cluster of counterions surrounding both exchanger and analyte, with the concomitant rehydration of both exchanger and analyte. Proteins and peptides are typically ampholytes, i.e., possessed of both positive and negative charges. In addition, proteins and peptides tend to have hydrophobic and hydrogen-bonding character and may undergo conformational changes as a function of pH, temperature, or solvent conditions.

Several theoretical approaches have been developed to explain the behavior of proteins and other polyions, such as oligonucleotides on ion exchange. The stoichiometric displacement model[28-31] calculates the number of effective binding sites of the protein on the ion exchange material, Z. The capacity factor k' is linearly related to D_o^{-Z}, where D_o is the concentration of the eluent. An alternative model, based on a linearized Poisson-Boltzmann equation, proposes that the protein interacts with the stationary phase in the way that one charged slab interacts with another in classical electrostatic theory.[32] A model based on the Manning ion condensation model[33] has also been presented.[34] This model incorporates the reduction in electrolyte activity caused by ion release when the protein binds; ion release is entropy driven. In simple terms, when a polyelectrolyte binds to an ion exchange surface, counterions around both the polyelectrolyte and the ion exchange surface are displaced into free solution. Water molecules are also displaced from around the polyelectrolyte and the ion exchange surface. Therefore, the activities of the electrolytes in free solution are altered. Neglecting co-ion release and a change in hydration, the capacity factor k' is given as

$$k' = K_o^T + \ln\varphi + \zeta\xi^{-1}\ln\delta\gamma_\pm - \zeta\psi \, \ln m$$

where K_o^T is the thermodynamic equilibrium constant, ζ is the number of binding sites on the analyte, ξ is a dimensionless charge density parameter equal to $e^2/bDkT$, φ is the phase ratio, b is the effect of the mobile phase salt concentration on the mean salt activity coefficent $\gamma_\pm$, δ is 0.33b in aqueous solution near 25°C, ψ is the thermodynamic binding constant equal to $1 - (2\zeta)^{-1}$, k is the Boltzmann factor, D is the bulk dielectric constant, T is the absolute temperature, and m is the mobile phase salt concentration.[34]

The number of binding sites can be determined in this model by a plot of d lnk′/dlnm at constant temperature, pH, and ion valency. To do that, it may be assumed that dln$\gamma_\pm$/dlnm is approximately zero. The actual value is -0.04 for 0.1 to 0.5 *M* sodium chloride and less at lower concentrations. To a first approximation, the stoichiometry of water molecules released by binding protein could be determined from the slope of the plot of dlnk′/dlnm vs. m. However, especially at low salt concentration and near the isoelectric point, the slope of such plots is nonlinear. The nonlinearity may be due to hydrophobic interaction between stationary phase and protein or a large change of ionic hydration on binding.[34]

5.3   *Suppliers of ion exchange materials*

Suppliers (and trademarks) of ion exchange materials include Alltech (Universal), Astec, Dionex (Sunnyvale, CA; IonPac® and CarboPac™, Dionex Corp.), Bio-Rad Laboratories (Hercules, CA; Aminex®), EM Separations (Polyspher IC), Hamilton (Reno, NV; PRP®), Interaction® Chemicals (San Jose, CA; ACT™, AA, Hydrophase™, CHO, ION), Metachem, Mitsubishi Kasei (MCI Gel), PerSeptive Biosystems (Cambridge, MA; Poros®), Phenomenex™ (Torrance, CA; Rezex™, Spherex™, Selectosil™, BioSep™, W-Porex™), Polymer Laboratories, Supelco (SigmaChrom), Tosoh (Tokyo, Japan; TSK-Gel™ and ToyoPearl®), and Waters (Milford, MA; Protein-Pak™, IC-Pak™).[35,36] Polymeric supports include poly(styrene/divinyl benzene), polymethacrylate, poly(hydroxymethacrylate), polyvinyl copolymer, and poly(ethylvinyl) polymers. Very high speed separations have been achieved on macroporous poly(styrene-divinylbenzene) phases.[37] Stationary phase ligands for anion exchange include ammonium, methyl quaternary ammonium, ethyl quaternary ammonium, alkyldimethylammonium, and diethylamino ethyl ammonium groups. Cation exchange materials include sulfonates, carboxylates, and alkyl sulfonates. Mixed-mode chromatography, combining controlled hydrophobic interactions with ion exchange, has also been used to control selectivity.[38]

5.3   *Electrochemical detectors*

5.3.1   *Overview of electrochemical detectors*

Electrochemical detection is extremely selective and is consequently useful for determination of known components in complex mixtures. The nomenclature to describe electrochemical techniques is somewhat more complicated than would seem to be necessary for a science that consists of measuring current, potential, and resistance. To review the basic terminology, *current* is the flow of charge per unit time. The *potential* (or, more accurately, the potential difference) is the work that must be done to move a charge from one point to another. *Resistance* is defined as the potential divided by the current. *Resistivity* is the resistance multiplied by the area of the path of conduction and divided by the length of the path. *Conductivity* is the inverse of the resistivity. The *specific conductance* is the path length of the current flow divided by the resistivity. The *equivalent conductance* is 1000 × specific conductance/C, where C is the concentration of an electrolyte in equivalents/L.[2]

In measurements of conductivity, no electrochemical reactions occur. Differences in conductivity are due to differences in the ionic strengths of solutions. An alternating potential is applied to the solution at a known potential. The current is measured and the conductivity in Siemens/cm calculated.[16] In potentiometry, the analyte is presumed to undergo no electrochemical reaction. The potential at the electrode changes due to changes in potential across the surface of the membrane in a membrane electrode or at the electrode surface of a solid electrode. The most familiar example of a potentiometric electrode is the pH electrode. In amperometry, current does flow, due to reduction or oxidation of the substance being analyzed.

5.3.2   Conductivity

The resistance of pure water is extremely high, but drops quickly with ionic content. Therefore, conductivity detectors are most sensitive when the concentration of ionized species in the mobile phase (from the eluting salt and buffer) is extremely low. Analyte concentrations are often in the ppm-ppt range, while eluent ion concentrations are often in the high ppt range. Several strategies have been developed to address this problem. One approach is to place a second ion exchange column, called the "suppressor", in line after the separator.[2] The suppressor serves as a reservoir of hydroxide or hydronium ions to neutralize the counterion of the basic or acidic eluent. This approach traditionally has suffered from a number of drawbacks, especially the need to regenerate the resin. A new device that claims to have solved the problems of packed-bed suppressors is available from Alltech.[39] Post-column membrane suppressors are micro-dialysis devices used to remove eluent ions from the column effluent prior to conductivity detection.[16] Consider a separation of anions performed with sodium hydroxide as the eluent. On one side of the membrane circulates an acidic regenerant, while the analyte and eluent are on the other side of the membrane. The membrane carries a negative charge, so anions do not readily cross the membrane. Hydrogen ion is exchanged for the sodium ion of the eluent and the cation of the analyte. The hydrogen ion neutralizes the hydroxide, reducing the conductivity of the solution to the point that the sensitivity of the conductivity detector to the analyte is increased. The drawback is that large volumes of acidic regenerant are required. This problem has been overcome. An electrolytically self-regenerating device has been described.[40] This device, however, requires large volumes of deionized water for operation. An evaluation and comparison of the Alltech solid phase SPCS™ (Alltech Assoc.; Deerfield, IL) suppressor and the Dionex AMMS™ (Dionex; Sunnyvale, CA) anion micromembrane suppressor have been published.[41] Other factors, such as temperature control and pump pulsation were also examined.

Conductivity detection is also sensitive to temperature, with the conductance changing by about 2% per degree Celsius.[42] With proper controls, conductivity detection is linear over 5 orders of magnitude.[43] To obtain universal detection, a microelectrodialytic NaOH generator is installed post-suppressor.[44] Given the acid dissociation value of a weak acid, dual detection under neutral and alkaline conditions can be used for concentration calibration. Peak integrity, i.e., the coelution of a second analyte, may also be established using dual detection.

5.3.3   Voltammetry

Voltammetry is the term applied to measurement of current as a function of potential. If the potential is continuously varied in a cyclic manner, the technique is called *cyclic voltammetry*. At certain values of the potential, no electrochemical reaction of analyte occurs, so no current flows. As the magnitude of the potential is raised, the analyte may be oxidized at more positive potentials or reduced at more negative ones. For a perfectly reversible chemical reaction in a static solution, the current would return to its original value with the return of the potential to its original value. Since electrochemical reactions often are highly irreversible, voltammetric experiments to optimize current flow are done with a stirred solution or a rotating electrode or by injection of fresh solution, a technique called *hydrodynamic voltammetry*.[45] At sufficient high or low potentials, the solvent and the electrode surface may also oxidize or be reduced. Also, at a sufficiently large potential, the analyte is consumed as rapidly as it can approach the electrode. The current that is generated under conditions in which diffusion of analyte limits the current flow is called the *diffusion-limited current*. Above this limit, detector response is flat. *Polarography* refers to voltammetry using a dropping mercury electrode.[46]

5.3.4   Potentiometry

Potentiometry is the measurement of the potential at an electrode or membrane electrode, so the detector response is in units of volts. The potentiometric response tends to be slow, so potentiometry is used infrequently in analysis.[47] One example is the use of a polymeric membrane impregnated with ionophores for the selective detection of potassium, sodium, ammonium, and calcium.[48] In process chromatography, potentiometry may be used to monitor selected ions or pH as these values change over the course of the gradient.

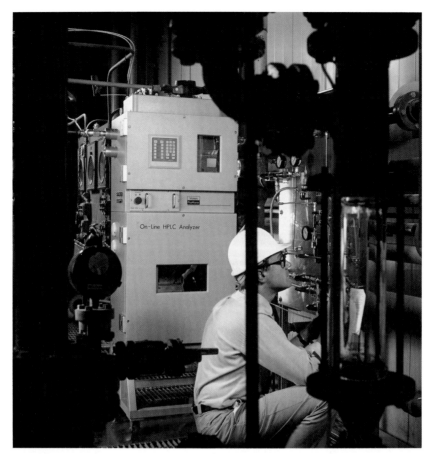

CHAPTER 2, PLATE 1A. Waters™ ProMonix On-Line HPLC Analyzer. The upper compartment door contains a keypad for programming and operation of the analyzer. The upper window allows viewing of indicator lights and a liquid crystal display which provides the operator-analyzer interface, programmed parameters, and instrument status results. The lower chamber contains the pumps, valves, injector, and detector(s) requited for the chromatographic separation. The sample conditioning plate for online process monitoring is to the right of the analyzer. This is a typical process HPLC. (From Cotter, R.L. and Li, J.B., *Lab. Rob. Autom.,* 1, 251, 1989. With permission of VCH Publishers.)

CHAPTER 2, PLATE 1B. Purged enclosure for operation of laboratory HPLC equipment in process area.

5.3.5   *Amperometry*

Amperometry measures the current flowing from an electrode at a given potential, with the detector response in units of amperes; therefore, amperometry might be regarded as a form of hydrodynamic voltammetry in which the potential has been selected and the species being measured are changing. *Pulsed amperometry* implies that the potential is changed, with measurement typically being made at a single potential value. The upper and lower limits for the applied potential typically are limited by the tendency of the solvent to be oxidized or reduced. Therefore, solvents used in electrochemical detection must be relatively electrochemically inert.

The analytes that can be determined electrochemically are those that are readily oxidized or reduced, such as sugars and other alcohols, amines, and sulfur compounds. Aryl nitrates are also detected by this method.[49,50] DNA, as well as nucleotides, were detected by amperometry at the 0.1-ng level using amperometry on glassy carbon.[51] An example of amperometric detection in a reversed phase liquid chromatographic system is shown in Figure 1.[50] The figure illustrates the selectivity of detection possible with differential pulse amperometry. Another example involving the detection of amines and amphetamines, phenols, phenothiazines, resorcinols, and other biologically relevant materials with amperometry has been reported.[52] Carbamate pesticides such as Benomyl, Baygon® (Bayer-Aktiengsellschaft; Lever Kusen, Germany), Sevin® (Union Carbide; New York, NY), Propham, Landrin, Nesurol, Chloropropham, and Barban have also been detected amperometrically.[53] Isoproteronol, a bronchodilator, was detected in a reversed phase system using a glassy carbon electrode.[54]

The use of non-inert and chemically modified electrodes and other strategies for the detection of species that are difficult to analyze with the normal electrode materials have been reviewed.[55] Photosensitization prior to amperometric detection is another tactic that has proved useful for the analysis of substances that are normally considered to be electrochemically inert.[56] The use of pulsed amperometry has recently been reviewed.[57]

Oxidation or reduction of an analyte at an electrode surface tends to foul the surface, leading to a change in detector response. It is therefore generally necessary to clean the surface frequently. This may be accomplished *in situ* by reversing the potential on the electrode. In the case of gold and platinum electrodes, commonly used in electrochemical detection, a strong positive potential is used to convert the surface to metal oxides, from which the products of analyte decomposition are desorbed. Then, a strong negative potential regenerates the metal electrode. In chromatographic systems, a small positive or negative potential is used for detection. Periodically, a strong positive pulse may be applied to clean the electrode and a strong negative potential pulse to regenerate the surface. If detection is performed at regular intervals between cleaning and regeneration, this detection scheme is called *pulsed amperometry.* Since electrode cleaning is not infinitely rapid, there is a limitation on how frequently the pulse may be repeated. To reduce

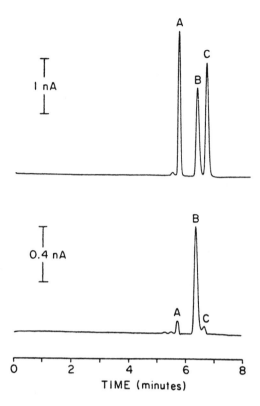

Figure 1 Electrochemical detection of catechol, acetaminophen, and 4-methyl catechol, demonstrating the selectivity of differential pulse detection vs. constant potential detection. **(A)** Catechol, **(B)** acetaminophen, and **(C)** 4-methylcatechol were
separated by reversed phase liquid chromatography and detected by amperometry
on a carbon fiber electrode. In the upper trace, a constant potential of +0.6 V was
used. In the lower trace, a base potential of +425 mV and a pulse amplitude of +50
mV were used. An Ag/AgCl reference electrode was employed. Note that acetaminophen responds much more strongly than catechol or 4-methylcatechol under the
differential pulse conditions, allowing highly selective detection. (Reproduced with
permission from St. Claire, III, R. L. and Jorgenson, J. W., *J. Chromatogr. Sci.* 23, 186,
1985. Preston Publications, A Division of Preston Industries, Inc.)

the noise, the sampling period may be extended, a technique which is called
integrated amperometry.

The amount of current that flows is dependent not only on the condition
of the electrode, but also on temperature, pH, and ionic strength of the
solvent. Therefore, careful control of the conditions of detection is essential.
A reduction of the slope of the baseline in gradient elution is often performed
by post-column addition of a solution of strong alkali. Flow is also an important variable,[58] and pump fluctuations may be an important source of noise.[59]
At very high flow rates, amperometric response can decrease depending on

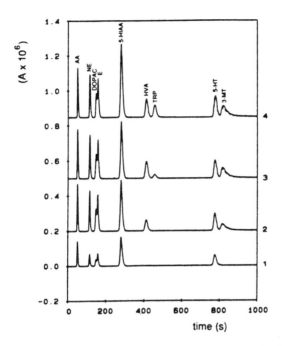

Figure 2 Selective electrochemical detection of a mixture on multielectrode amperometry. AA = Ascorbic acid, NE = norepinephrine, DOPAC = 3-4-dihydroxy-phenylacetic acid, E = epinephrine bitartrate, 5-HIAA = 5-hydroxyindole-3-acetic acid, HVA = homovanillic acid, TRP = tryptophan, 5-HT = 5-hydroxytryptamine, and 3-MT = 3-methoxytyramine (separated by RPLC). Detection was with a 4-electrode glassy carbon array, with electrode 1 at 500 mV, electrode 2 at 700 mV, electrode 3 at 900 mV, and electrode 4 at 1100 mV. Note that at electrode 1, HVA, TRP, and 3-MT are not seen. At electrode 2, only TRP is not seen. A standard calomel electrode was used as reference. (Reprinted with permission from Hoogvliet, J. C., Reijn, J. M., and van Bennekom, W. P., *Anal. Chem.,* 63, 2418, 1991. ©1991 Analytical Chemistry.)

cell geometry.[59] One of the interferences in amperometry is oxygen which must be removed by sparging and the re-entry of which must be limited by avoiding gas-permeable plastic tubing.[19] For very high sensitivity, it may be advisable to de-gas samples, as well, as dissolved oxygen can enter and elute as a plug, producing a peak. Nonelectroactive substances have also been found to be interferences in amperometry.[60] The effect seems to be caused by a drop in pH.

Multiple electrodes have been used to obtain selectivity in electrochemical detection. An early example involved the separation of catecholamines from human plasma using a Vydac® (The Separation Group; Hesperia, CA) SCX cation exchange column eluted with phosphate-EDTA.[61] A sensor array using metal oxide-modified surfaces was used with flow injection to analyze multicomponent mixtures of amino acids and sugars.[62] An example of the selectivity provided by a multi-electrode system is shown in Figure 2.[63]

5.3.6   Coulometry

Coulometric detection describes amperometry in which the reaction of the analyte proceeds to completion.[59] The extent of reaction is a function of cell geometry, flow rate, and analyte concentration, so all cells can be made coulometric by stopping the flow.[59] Several designs for coulometric cells that are independent of flow rate have been described.[46]

5.3.7   Vacancy detection

If the analyte is electrochemically inactive, it is possible to add a background of an electrochemically active material to the eluent post-column. If dispersion is kept a minimum, the analyte elutes as a well defined band with a lower electrochemical response than the background, forming a vacancy peak.

5.3.8   Post-column reactors

Post-column reactors are also widely used in ion exchange detection. One advantage of post-column detection is the insensitivity to most buffer components. Another significant advantage is that, to a first approximation, response is proportional to the concentration of a particular functional group. This property of the response is particularly useful in purity determinations, since the accuracy of the analysis does not rely heavily on calibration. Finally, post-column detection allows for extremely selective detection. As an example, a post-column system for detection of either sulfhydryls or sulfhydryls plus disulfides greatly simplified interpretations in protein peptide mapping.[64] Phosphorylated sugars were separated on a silica-based strong anion exchange detected using a europium additive for ultraviolet (UV) detection.[65] The europium was believed to complex with the phosphate group. A post-column fluorescence detector using periodate oxidation and guanidinium derivatization was devised for the detection for the detection of nonreducing saccharides.[66] A similar approach has been used for reducing saccharides.[67] Additional post-column detection schemes, particularly the ninhydrin and *o*-phthaldehyde systems popular in amine and amino acid analysis, are described below.

5.3.9   Other detectors

Light scattering (nephelometry) was used as a detection system for glycosaminoglycans from urine, eluted from a DEAE Sephadex™ (Pharmacia Biotechnology; Uppsala, Sweden) A-25 column.[68] This technique has been more recently applied to protein characterization.[69] Interferometry was used for analysis of dextran eluted from a size exclusion column.[70] One of the problems of electrochemical detection is that it is relatively insensitive to polymers. Because many of the materials discussed below (DNA, proteins, and polysaccharides) are polymeric, a brief mention of some alternative

techniques is appropriate. A more extensive discussion of polymer characterization is to be found in Chapter 6, while absorbance and refractive index
detection are discussed in Chapter 1.

5.4 Applications

5.4.1 Inorganic ions

For many applications, spectroscopic techniques such as atomic absorption
spectroscopy are preferred for the determination of simple inorganic ions.
For complex mixtures, for samples derived from complex matrices, or in
cases when differentiation of oxidation states is required, ion exchange is
usually superior. Prevention is critical for a number of industrial applications
of ion exchange chromatography for inorganic ions, particularly in nuclear
power. Applications are also to be found in electronics, agriculture, the food
industry, mining,[71] and biomedical science. Conductivity detection is sufficient for most simple ions. Preconcentration is useful for very dilute samples
of low ionic strength. By this means, it has been possible to detect ppb levels
of chloride, nitrate, and sulfate in rainwater.[72] A typical application of anion
exchange is the monitoring of oilfield waters, in which brine content may
be high. Measurement of bromide, sulfate, and nitrate in the presence of
large excesses of sodium chloride was performed on a Dionex IonPac®
AS9-SC column with bicarbonate and carbonate as eluents.[73] Similarly, in
natural gas, the content of hydrogen sulfide may be a corrosion concern, so
an amine solution may be used as a scrubber. To examine the regeneration
of the scrubber solution, the Dionex IonPac® AS9-SC was used to resolve
nitrate, hydrogen sulfide, sulfate, oxalate, and thosulfate.[74] A zwitterionic
detergent was found useful as an ion pairing agent when pure thiosulfate
is used as the water mobile phase, which eliminates the need for post-column
suppression.[75] An application demonstrating simultaneous separation of
inorganic and organic ions at high resolution and excellent sensitivity is
shown in Figure 3.[76] Representative ion exchange separations for small ions
are shown in Table 1.[71,77-90]

A problem with using amperometry in the determination of electroactive
ions in the presence of large quantities of nonelectroactive ions is that the
detector may respond due to effects on the pH.[60] Another approach to detection is to form complexes of the metal with a chelating agent, such as oxalate,
and to perform the separation on a reversed phase column in a mobile phase
containing an alkyl ammonium salt.[91] In samples of high organic content,
pre-column UV photolysis has been found useful in destroying interfering
carboxylic acids to permit the detection of transition metals and anions
separated by ion chromatography.[92] UV photolysis was also found to be
useful in destroying hydrogen peroxide or ammonia from solutions used in
cleaning or stripping photoresist from silicon wafers used in electronics.[93]
Direct analysis of anions and transition metals has been performed on the

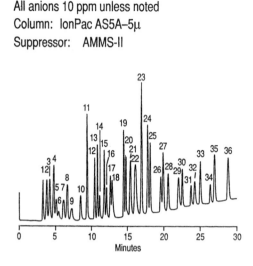

All anions 10 ppm unless noted
Column: IonPac AS5A–5μ
Suppressor: AMMS-II

Peaks:

1. F⁻ (1.5 ppm)	19. HPO_3^{2-}
2. α–Hydroxybutyrate	20. SeO_3^{2-}
3. Acetate	21. Br⁻
4. Glycolate	22. NO_3^-
5. Butyrate	23. SO_4^{2-}
6. Gluconate	24. Oxalate
7. α–Hydroxyvalerate	25. SeO_4^{2-}
8. Formate (5 ppm)	26. α–Ketoglutarate
9. Valerate	27. Fumarate
10. Pyruvate	28. Phthalate
11. Monochloroacetate	29. Oxalacetate
12. BrO_3^-	30. PO_4^{3-}
13. Cl⁻ (3 ppm)	31. AsO_4^{3-}
14. Galacturonate	32. CrO_4^{2-}
15. NO_2^- (5 ppm)	33. Citrate
16. Glucouronate	34. Isocitrate
17. Dichloroacetate	35. cis–Aconitate
18. Trifluoroacetate	36. trans–Aconitate

Figure 3   Gradient separation of anions using suppressed conductivity detection. Column: 0.4 × 15 cm AS5A, 5 μ latex-coated resin (Dionex). Eluent: 750 μM NaOH, 0–5 min., then to 85 mM NaOH in 30 min. Flow: 1 ml/min. 1: fluoride, 2: α-hydroxybutyrate, 3: acetate, 4: glycolate, 5: butyrate, 6: gluconate, 7: α-hydroxyvalerate, 8: formate, 9: valerate, 10: pyruvate, 11: monochloroacetate, 12: bromate, 13: chloride, 14: galacturonate, 15: nitrite, 16: glucuronate, 17: dichloroacetate, 18: trifluoroacetate, 19: phosphite, 20: selenite, 21: bromide, 22: nitrate, 23: sulfate, 24: oxalate, 25: selenate, 26: α-ketoglutarate, 27: fumarate, 28: phthalate, 29: oxalacetate, 30: phosphate, 31: arsenate, 32: chromate, 33: citrate, 34: isocitrate, 35: *cis*-aconitate, 36: *trans*-aconitate. (Reproduced with permission of Elsevier Science from Rocklin, R. D., Pohl, C. A., and Schibler, J. A., *J. Chromatogr.*, 411, 107, 1987.)

complex solutions from the manufacture of printed circuit boards.[94] Destruction of hydrogen peroxide and ammonia was necessary for analysis of trace impurities by ion chromatography. Ion exchange is one of the few chromatographic techniques capable of isotopic separation. A review of determination of inorganic anions in food is available.[95] Simultaneous determination of anions and cations was performed on a PRP®-X100 (Hamilton; Reno, NV) anion exchange column eluted with phthalic acid, using conductivity to detect anions and flame emission spectrometry to detect chromatographically unresolved cations.[96]

A PRP®-1 (Hamilton; Reno, NV) reversed phase column was coated with cetylpyridinium and eluted with tetramethylammonium salicylate:acetonitrile:water.[89] The separation was comparable to that observed on conventional ion exchange. Coated phases were also used to separate oxalate complexes of manganese, cobalt, copper, and zinc.[90] Reversed phase silica supports were coated with poly(N-ethyl-4-vinylpyridinium bromide), poly(dimethydiallylammonium chloride), poly(hexamethyleneguanidinium

Table 1 Representative Separations of Simple Ions

Manufacturer	Column	Mobile phase	Ions
Dionex	CS10[77]	HCl/2,3-diaminoproprionic acid	Li+, Na+, NH4+, K+, Rb+, Mg2+, Ca+2, Sr2+, Ba2+
	CS2[78]	HCl/2,3-diaminoproprionic acid	Na+, K+, NH4+
	AS4[78]	NaHCO3	Cl-
	AS4A[71]	NaHCO3 or NaHCO3-tetrabutyl ammonium hydroxide-acetonitrile	SO_3^{2-}, SO_4^{2-}, $S_2O_6^{2-}$
	AS10[79]	Sodium-4-cyano phenolate	Cl-, NO3-, Br-, NO2-, HPO_4^{2-}, SO_4^{2-}, citrate, nucleotides
	Transition metal[80]	4-(2-pyradylazo)-resorcinol, NaCl, LiOH, Na2SO3, Na2SO4, ascorbic acid, methanol	Cu, Ni, Zn, Co, Mn, Fe2+
	CS12[81]	Methane sulfonic acid	Na+, NH4+, K+, Mg2+
Alltech	Universal cation[82]	HNO3	Li+, Na+, NH4+, K+, Mg2+, Ca+2
Hamilton	PRP®-X100[83,84]	p-Hydroxy benzoic acid	F-, Cl-, NO2-, Br-, NO3-, HPO_4^{2-}, SO_4^{2-}
Interaction Chemicals	ION-100[85]	Potassium hydrogen phthalate	F-, Cl-, NO2-, Br-, NO3-, SO_4^{2-}
Waters™	IC-Pak™[86]		F-, Cl-, HPO_4^{2-}, SO_4^{2-}
Other	Tris(2,6)-dimethoxy phenyl phosphine[87]	NH3OH-NH4Cl	Au3+, Cu2+, Pt4+
	Cetyl dimethyl ammonium-coated C18 [88,89,90]		F-, Cl-, and organic ions

hydrochloride), and 2,5-ionene, a polymeric quaternary amine used for sep-arations of many common anions.[97] Alkali and alkaline metal earth ions such as sodium, lithium, potassium, calcium, and magnesium were separated on the Metrohm Cation 1-2, a poly(butadiene-maleic acid) cation exchange col-umn.[98] Alkyllead and alkylmercury compounds were separated by precon-centrating the mercaptoethanol complexes onto RPLC and eluting with an organic solvent.[99] Stability constants for the complexation of Zn^{+2}, Ni^{+2}, Co^{+2}, Cd^{+2}, Mn^{+2}, and Pb^{+2} with oxalate, tartrate, malonate, and pyridine-2,6-dicarboxylate anions were measured by ion exchange chromatography on a reversed phase column coated with sodium dodecyl sulfate.[100] A review of some applications is available.[101]

Since many ion exchange columns exhibit mixed-mode interactions with analytes, factor analysis has been found to be useful in optimization.[84] A 3-year, comprehensive review of inter-laboratory errors in determinations of the anions chloride, nitrate, and sulfate and the cations sodium, potassium, magnesium, and calcium suggested that multipoint calibration is essential and nonlinear calibration desirable.[102] The need for nonlinear calibration was confirmed by an extended quality assurance study of chloride, sulfate, and nitrate in rainwater.[103]

5.4.2   Amines

Perhaps the majority of chromatographic applications include compounds containing a basic amino functionality; proteins, peptides, amino acids, and amino sugars are discussed in separate sections below. However, amines are also present in many pharmaceutical compounds, polymer initiators, pH buffers, surfactants, and other materials. Many amines have some hydro-phobic character and can be analyzed by reversed phase liquid chromatog-raphy. Primary and secondary amines can be derivatized with a hydrophobic nucleus to increase the hydrophobicity, making RPLC feasible. Pre-column derivatization, discussed in more detail in the sections on amines and amino acids of Chapter 4, can also be used to enhance detectability. Polyamines are more difficult to analyze, since a derivatization reaction must be driven to completion at all of the sites to generate a homogeneous material for analysis. Multiple derivatization with a hydrophobic nucleus may make the polyamine too hydrophobic for convenient RPLC analysis. Therefore, in some amine separations, ion exchange chromatography remains the method of choice. It should also be noted that derivatization inherently makes an analytical method more complex and more susceptible to variability. When extremely high precision is required, ion exchange should be considered as the separation method of choice.

In the analysis of isocyanuric acid, a stabilizer used in swimming pools, ion exchange separation on an Omnipac PCX-500 was used to separate isocyanuric acid from ammelide, ammeline, and melamine (Figure 4).[104] Since ammelide has one primary amine, ammeline two, melamine three, and isocyanuric acid none, derivatization and RPLC would have been problematic.

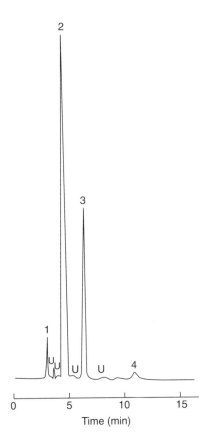

Figure 4 Ion chromatogram of crude isocyanuric acid and its impurities. Column: Omnipac PCX-500. Eluent: 100 mM KCL-200 mM HCl-5% acetonitrile. Flow: 1.0 ml/min. Detection: UV at 215 nm. The following peaks were resoved from crude isocyanuric acid: (1) Isocyanuric acid, (2) ammelide, (3) ammeline, (4) melanine, and (U) unknowns. (Reproduced with permission of Elsevier Science from Debowski, J. K. and Wilde, N. D., *J. Chromatogr.*, 639, 338, 1993.)

In Figure 5 is shown a separation of aspirin, paracetamol, phenacetin, caffeine, phenylephrine, and salbutamol on a Merckosorb SI-60-SCX strong cation exchange column.[105] 2-Amino-2-methyl propan – 1-ol and 5-aminopentan – 1-ol, used to inhibit corrosion in steam generators, were analyzed by a CS2 column (Dionex) eluted with 8 mM HCl.[106] Conductivity detection permitted measurement with a precision greater than 1% at the high ppb level, even though a nonlinear calibration curve was observed. A poly(butadiene-maleic acid)-coated phase, eluted with dilute nitric acid, EDTA, and an organic modifier, was used to simultaneously separate alkyl amines or alkanolamines and alkali metal salts.[107] Detection was by conductivity. Over 50 biogenic amines, including serotonin and norepinephrine, were separated on Amberlite® (Rohm & Haas Co.; Spring House, PA) CG-50 using pyridinium acetate.[108] Aliphatic amines were separated on Hitachi 2632

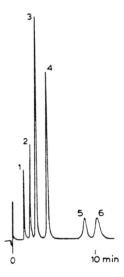

Figure 5 Separation of pharmaceuticals, including amines, on strong cation exchange. Column: 0.46 × 15 cm Merckosorb SI-60-SCX, 5 μ. Eluent: 50 mM aqueous ammonium formate-10% ethanol, pH 4.8. Flow: 1 ml/min. Temperature: 50°C. The peaks are (1) aspirin, (2) paracetamol, (3) phenacetin, (4) caffeine, (5) phenylephrine, (6) salbutamol. (Reproduced with permission of Elsevier Science from Cox, G. B., Loscombe, C. R., Slucutt, M. J., Sugden, K., and Upfield, J. A., *J. Chromatogr.*, 117, 269, 1976).

anion exchange resin.[109] A Vydac® strong cation exchanger was used to separate L-dopa, norepinephrine, epinephrine, and dopamine from human plasma.[61] Nonaqueous solvents containing methane sulfonic acid were used in the ion exchange separation of homologous *n*-alkyl amines on a home-made sulfonated poly(styrene-divinyl benzene) resin to eliminate the effects of hydrophobic interactions.[110] Detection was by conductivity. Amperometric detection is very useful in determination of amines.[52]

   The polyamines putrescine, cadaverine, spermidine, and spermine, which are seen at elevated levels in some victims of cancer, were separated on a Technicon® (The Technicon Company; Chauncey, NY) TSM Amino Acid Analyzer packed with an 8% divinylbenzene-co-polystyrene sulfonated resin with post-column ninhydrin detection.[111] Amines such as ethanolamine, noradrenaline, hexamethylene diamine, methoxytryptamine, spermine, and spermidine were separated from amino acids on a DC-4A cation exchange resin.[112] A similar approach, using a Beckman Model 121M amino acid analyzer equipped with an AA-20 column, was also successful.[113] A Polyaminpak strong cation exchange column (JASCO) was eluted with a citrate buffer for the detection of putrescene, spermine, cadaverine, and 1,5-diaminohexane from rat thymus.[114] A post-column *o*-phthaldehyde detection system was used.

5.4.3 Amino acids

Amino acids are the building blocks of proteins and are therefore encountered by the analyst in the food and pharmaceutical industries. Amino acid analysis was first automated by Hirs, Stein, and Moore,[115,116] and has subsequently undergone substantial technical improvement. Amino acid analysis is commonly used in the food industry and in biopharmaceutical analysis of proteins. In the food industry, sample size is rarely limiting. In biopharmaceutics, however, a typical sample might contain 0.1 mg of protein for hydrolysis. In the hydrolyzate, certain amino acids typically will be abundant, while others may constitute only a single residue per hundred residues. The entire sample may contain only a few nmol of the least abundant amino acid; therefore, sample preparation and handling usually are more critical to analysis than the chromatographic methodology.[117,118] In most industrial applications, method ruggedness and throughput are primary considerations. Polystyrene-based supports are useful in amino acid analysis because of their chemical stability and ease of cleaning. Either strong anion or cation exchange materials have been used, but the cation exchangers are more prevalent. Microbore systems have been in use for several decades.[119]

Reversed phase chromatography following pre-column derivatization with phenyl isothiocyanate, fluorenyl methoxycarbonyl, or *o*-phthaldehyde has become a popular technique. Ion exchange with post-column derivatization with ninhydrin,[115,116] fluorescamine,[120] or *o*-phthaldehyde[121] remain popular methods because retention times are extremely stable. Electrochemical detection has also been used.[62] None of these methods are specific to amino acids, but they do detect all amines, including amino sugars and peptides. Retention time stability becomes extremely important in making correct identifications with complex chromatograms, particularly on automated data systems.

Higher speed separations were obtained by the use of packings with smaller particle diameter,[4,5] and the major improvement in ion exchange separations of amino acids has been to reduce the time of analysis appreciably. In Figure 6 is shown an analysis with a medium performance packing, requiring 6 hours. In Figure 7 is shown an analysis with high performance materials, requiring 40 minutes and permitting the simultaneous determination of amino sugars. Analyses as short as approximately 20 minutes have been reported.[122]

Simultaneous determination of amino sugars and other amines may be particularly important for samples derived from biological sources. The length of the glycan chain in bacterial cell wall peptidoglycans, for example, could be estimated by simultaneous determination of muramicitol, muramic acid, glutamic acid, glycine, alanine, 2,6-diaminopimelic acid, isoglutamine, 5-hydroxy-4-aminopentanoic acid, glucosamine, glucosaminitol, and alaninol from the acid hydrolyzate of borohydride-reduced peptidoglycan on an LKB Aminex® A-6 column.[123] Pre-column dabsylation and separation on

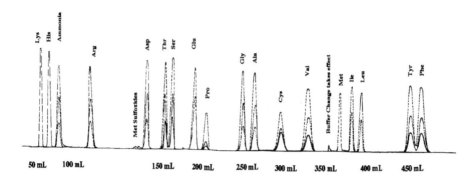

Figure 6 Separation of amino acids on conventional ion exchange: Beckman 120B Amino Acid Analyzer. Column: 15 cm. Eluent: 0.35 M sodium citrate buffer, pH 5.28. Flow rate: 30 ml/hr. Temperature: 50°C. Note that the separation requires approximately 6 hours. Compare to a modern separation shown in Figure 7. (Reproduced with permission from Beckman Instruments; Fullerton, CA.)

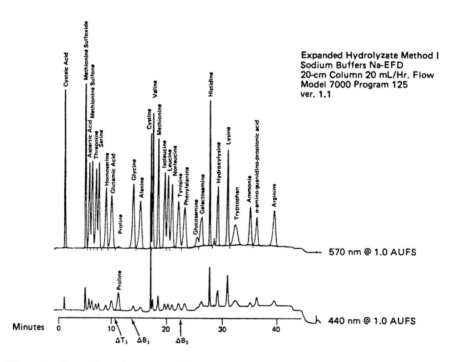

Figure 7 Separation of amino acids by high performance ion exchange: Beckman Model 6300/7300 Amino Acid Analyzer. Column: 20 cm sodium high performance column. Mobile phase: Na-E, Na-F, Na-D, and Na-R regenerant, with buffer changes occurring at ΔB_1 and ΔB_2. Flow rate: 20 ml/hr. Detection: Ninhydrin post-column (absorbance at 440 nm and 570 nm). Temperature: 49°C rising to 79°C at ΔT_1. Note that the separation requires about 40 min. Compare to Figure 6. (Application Note #A6300-AN007, reproduced with permission from Beckman Instruments; Fullerton, CA.)

RPLC was used for the simultaneous determination of amino acids and amino sugars, including galactosamine and glucosamine, in glycopeptide analysis.[124]

A strong cation exchanger, IRICA SCX-1005, was used to separate S-pyridylethylated amino acid hydrolyzates.[125] Furosine [(ε-N-(2-furoyl-methyl)-L-lysine], which is formed on the Maillard reaction of lysine with sugars, was separated on MCI Gel CK10F.[126] Maillard adducts are important in studying protein aging. A 0.46×12-cm AA503 column was used to separate amino acids by ion exchange, followed by post-column *o*-phthalde-hyde.[85] The separation required 90 minutes, five discrete buffers, and a change of temperature during the run. Physiological fluids were separated on Beckman M82 and M72 resins[127] and on a Hitachi #2618 Custom Resin. [128] The isomers of hydroxyphenlyalanine, including *o*-tyrosine, *m*-tyrosine, 2,4-DOPA, and β-phenylalanine, were separated on a Hitachi #2612 resin with post-column ninhydrin detection.[129] Amino acids from leukocytes were analyzed on a Chromo-Beads C_2 cation exchange resin.[130] A Type W-1 cation exchange resin was used for the separation of 32 amino acids.[131] The use of an organic modifier, 1-propanol, has been examined.[132] An early application of separation on a pH gradient generated with a citrate-borate buffer allowed rapid separatio.[133] A split-stream system was devised to allow sample collection.[134] A *m*-divinylbenzene column exhibited unusually high resolution, an observation that was ascribed to perfusive flow through the bead.[135] Sulfur-containing amino acid derivatives such as cysteine, cystine, methionine, and the reduced and oxidized forms of glutathione were separated on a Dionex OmniPac PCX-500 column and detected using a special amperometric waveform.[136] Since most amino acids react with bromine, a post-column reactor was combined with a platinum working electrode to create an electrochemical detector for amino acids and proteins separated on a SynchroPak® (Micra Scientific; Northbrook, IL) SAX-300 strong anion exchanger.[137]

5.4.4   Organic acids and phenols

Organic carboxylic acids are commonly found in foods, in the adipate process stream, and as pollutants. Fatty acids are the lipophilic portion of glycerides and a major component of the cell membrane. Phenols are widely used in polymers, as wood preservatives, and as disinfectants. Chlorophenols such as 4-chlorophenol, two isomeric dichlorophenols, 2,4,6-trichlorophenol, three isomeric tetrachlorophenols, and pentachlorophenol were separated on a Dowex® (The Dow Chemical Co.; Midland, MI) 2-X8 anion exchange resin using an acetic acid-methanol gradient.[138]

An early example of a separation of carboxylates is shown in Figure 8.[6] The figure demonstrates that very high speed separations originated in the late 1960s. Figure 9 shows a separation of mono-, di-, and trichloracetate on a sulfonated poly(styrene-divinyl benzene) column, using suppressed conductivity detection.[139] The figure shows that high-sensitivity, on-column conductivity detection was practical in the mid-1970s. A 0.32×2.5-cm Vydac®

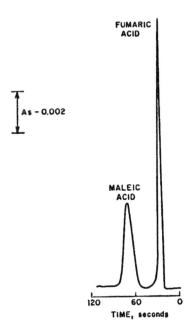

Figure 8   Separation of isomeric acids (maleic and fumaric acid) by controlled surface porosity anion exchange chromatography. Column: Sulfonated fluoropolymer coated onto a 50-μ glass bead. Average pore size about 1000 Å. Flow rate: 2.73 ml/min. Eluant: 10 mM HNO$_3$. Temperature: 60°C. Detection: absorbance. (Reproduced from Kirkland, J. J., *J. Chromatogr. Sci.*, 7, 361, 1969. By permission of Preston Publications, A Division of Preston Industries, Inc.)

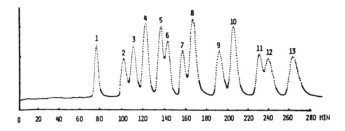

Figure 9   A synthetic mixture of water-soluble carboxylic acids separated by anion-exchange chromatography. Column: 0.3 cm × 300 cm Diaoion CA 08, 16–20 μ (Mitsubishi Kasei Kogyo). Eluant: 200 mM HCl. Detection: reaction with Fe^3-benzohydroxamic acid-dicyclohexy carbodiimide-hydroxylamine perchlorate-triethyl amine with absorbance at 536 nm. Analytes: (1) aspartate, (2) gluconate, (3) glucuronate, (4) pyroglutamate, (5) lactate, (6) acetate, (7) tartrate, (8) malate, (9) citrate, (10) succinate, (11) isocitrate, (12) *n*-butyrate, (13) α-ketoglutarate. (Reprinted with permission from Kasai, Y., Tanimura, T., and Tamura, Z., *Anal. Chem.*, 49, 655, 1977. ©1977 Analytical Chemistry).

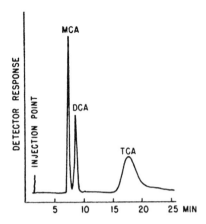

Figure 10 Separation of monochloroacetate, dichloroacetate, and trichloroacetate on a sulfonated poly(styrene-divinyl benzene) column with suppressed conductivity detection. Column: 2% cross-linked sulfonated poly(styrene-divinyl benzene); capacity 0.02 meq/g. Flow rate: 64 ml/hr. Eluant: 15 mM sodium phenate. Suppressor: 0.28 × 25 cm Dowex® 50W X8 column (200–400 mesh). Detector: Chromatronix conductivity cell connected to a Dow conductivity meter. (Reprinted with permission from Small, H., Stevens, T. S., and Bauman, W. C., *Anal. Chem.*, 47, 1801, 1975. ©1975 Analytical Chemistry.)

SCX column was used to separate oxalate and urate.[140] In this separation, differential pulse and DC amperometric detection were compared. Differential pulse detection was found to allow better selectivity in detection. Anion exchange on Diaion CA08 was used to separate 20 carboxylic acids in the analysis of white wine, as shown in Figure 10.[141] Because many carboxylic acids have a relatively weak absorbance, detection is difficult. The colorimetric detection scheme shown in the figure may be useful in some applications.

Malate, tartrate, and citrate were separated from fruit juices on Dowex® 1-X8 eluted with an acetic acid-acetate buffer.[142] Naphthoate ions were separated on the TSK-Gel™ IC-Anion-PW or -SW columns.[143] Ascorbic acid and sulfite from beer were separated on either a Fast Acid or an Aminex® HPX 87H column.[144] An Aminex® HPX-87H[145,146] was eluted with dilute sulfuric acid to separate 11 carboxylic acids. Formic and acetic acids, atmospheric pollutants that are relatively difficult to quantitate by gas chromatography, were separated on a Hamilton PRP®-X300 anion exclusion column with dilute sulfuric acid.[147] Organic acids in honey, including butyric, acetic, formic, lactic, succinic, pyroglutamic, malic, citric, and gluconic acids, were captured on a Dowex® 50 column.[148] Benzoic acid, phenylacetic acid, cyclohexylacetic acid, 3-phenylpropanoic acid, and cyclohexylpropanoic acid were separated from the corresponding alcohols using an Aminex® A-5 column and eluted with dilute acid.[149] In this separation, the adsorption mechanism involved hydrophobic as well as hydrogen bonding. A cation exchange column, PA-28 (Beckman; Fullerton, CA) was used to separate carboxylic acids, phenols, and aldehydes.[150] A 0.65 × 10-cm ARH-601 ion exchange resin eluted with dilute sulfuric acid was used to separate oxalic, citric, shikimic, fumaric, butyric, homoprotocatechuic, gallic, protocatechuic, gentisic, *p*-hydroxybenzoic, benzoic, and salicylic acids.[85] Ion exclusion and partitioning mechanisms were believed to be operative.

In the more recent literature, the Dionex AS-10 column was used to separate acetate, propionate, and formate in an amine scrubber solution for sour gas.[74] Short chain organic acids such as β-hydroxybutyrate acetate, glycolate, formate, α-ketobutyrate, and pyruvate were determined in drinking water using an AS-10 column from Dionex.[151] Temperature was found to decrease retention of phenyl-substituted alcohols on a PRP®-X100 trimethylammonium anion exchange column, while retention of phenyl-substituted carboxylates showed little correlation with temperature.[152] This was interpreted to mean that there is a change in retention mechanism for the carboxylates, possibly due to mixed-mode adsorption (solvophobic assistance) in retention. Inclusion of a hydrophilic linker near the surface of the support was used to form a hydrophilically shielded cation- and anion-exchange phases for the separation of chloramphenicol, trimethoprim, and propanolol from serum.[153] Proteins are too large to penetrate the hydrophilic layer surrounding the surface of the hydrophilically shielded phase and are eluted at the void volume, while charged molecules bind to the shielded ion exchange sites. For systems involving mass spectrometric detection, the non-volatile mobile phases used in ion exchange chromatography pose special problems. In the separation of carboxylates using an ion-pairing system of cetyltrimethylammonium bromide on cyano silica, a trapping system to remove the ion pairing agent prior to mass spectrometry was required.[154] Applications of separations of carboxylates using ion exchange are available in the literature.[7]

5.4.5 Nucleotides

DNA and RNA are formed of nucleotides. Each *nucleotide* or *nucleoside* is composed of a purine or pyrimidine base linked to the 1-position of a ribose sugar in the case of RNA and a 2′-deoxyribose sugar in the case of DNA.[155] The 5′ position is phosphorylated in the case of a *nucleotide*, while the *nucleoside* is not phosphorylated; therefore, nucleotides are nucleoside phosphates. Phosphorylation can include one, two, or three phosphate groups. The most familiar example of a phosphorylated nucleotide is phosphorylated adenosine, which occurs as the mono-, di-, and triphosphate (AMP, ADP, and ATP, respectively) and is a principal means of energy storage in biological systems.

Nucleotides can be linked together into oligonucleotides through a phosphate bridge at the 5′ position of one ribose unit and the 3′ position of another. The purine bases, adenine and guanine, have two heterocyclic rings, while the pyrimidines cytosine, thymine, and uracil have one. The structure of adenosine monophosphate is shown in Figure 11.

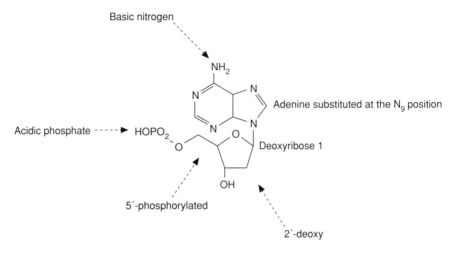

Figure 11 Structure of adenosine monophosphate.

Nucleotides are particularly well suited to separation by IEC, since the strong negative charge of the phosphate promotes adsorption to a positively-charged resin. Nucleotides also have a strong hydrophobic character due to the purine or pyrimidine rings. Adenine, guanine, and cytosine include basic nitrogen moieties that can interact with packing materials. All of the bases except adenine additionally have carbonyl moieties that can serve as proton acceptors. As an example of a packing that exploits mixed mode interactions, one of the most widely used phases in nucleotide analysis was RPC-5, a phase incorporating both hydrophobic and anionic character. Mixed ligand phases, incorporating both anion exchange and hydrophobic interactions have been reported.[156,157] Uracil, aminouracil, aminopropyltriethoxysilane, and 2-thiobarbituric acid were separately bonded to silica to form several phases for separation of nucleic acids.[158] Retention occurred by various mechanisms, including hydrophobic interaction, ligand binding, and ion exchange.

Nucleotide separations on classical ion exchange materials were reported at least as long ago as 1950.[159] An examination of pH gradient separation of ribonucleotides on Dowex®-1, including a clear description of the interaction of ionic strength and pH, was presented in 1960.[160] Complexation of nucleic acids with borate, followed by cation exchange separation on Aminex® A-6 (Bio-Rad Laboratories; Hercules, CA) was used to separate ribonucleosides, 2'-deoxyribonucleosides, and arabinonucleosides.[161] However, classical ion exchange materials continued to have insufficient resolving power for many of the complex separations required.[162] The first high performance separation was achieved in 1967.[3] A poly(styrene-divinyl benzene) copolymer was formed on the surface of a 50-μ glass bead, then this was chloromethylated and reacted with dimethylbenzylamine to form a pellicular anion exchange material. The common ribonucleotides were resolved in only 30 minutes.

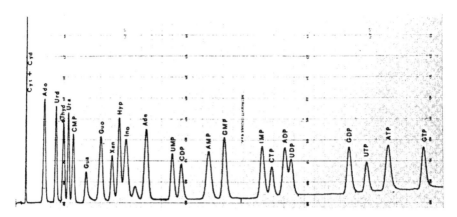

Figure 12 Gradient separation of bases, nucleosides and nucleoside mono- and polyphosphates. Column: 0.6 × 45 cm. Aminex® A-14 (20 ± 3 μ) in the chloride form. Eluent: 0.1 M 2-methyl-2-amino-1-propanol delivered in a gradient from pH 9.9–100 mM NaCl to pH 10.0–400 mM NaCl. Flow rate: 100 ml/hr. Temperature: 55°C. Detection: UV at 254 nm. Abbreviations: (Cyt) cytosine, (Cyd) cytidine, (Ado) adenosine, (Urd) uridine, (Thyd) thymidine, (Ura) uracil, (CMP) cytidine monophosphate, (Gua) guanine, (Guo) guanosine, (Xan) xanthine, (Hyp) hypoxanthine, (Ino) inosine, (Ade) adenosine, (UMP) uridine monophosphate, (CDP) cytidine diphosphate, (AMP) adenosine monophosphate, (GMP) guanosine monophosphate, (IMP) inosine monophosphate, (CTP) cytidine triphosphate, (ADP) adenosine diphosphate, (UDP) uridine monophosphate, (GDP) guanosine diphosphate, (UTP) uridine triphosphate, (ATP) adenosine triphosphate, (GTP), guanosine triphosphate. (Reproduced with permission of Elsevier Science from Floridi, A., Palmerini, C. A., and Fini, C., *J. Chromatogr.*, 138, 203, 1977.)

The peak capacity of the early high performance separations, however, was inadequate for the complexity of most biological samples. By 1977, separation of a number of unusual ribonucleosides and deoxyribonucleosides, such as pseudouridine and 5-methyl uridine, was achieved on 10 to 12 μ Beckman M-71 strong cation exchange resin in 3 hours.[163] An Aminex® A-14 column was used to achieve a separation of the common mono-, di-, and triphosphates of nucleotides in about 4 hours, as is shown in Figure 12.[164] By 1985, mixtures of nucleotides, nucleosides, 3'- and 5'-nucleotides, and 2':3' cyclic nucleotides were separated isocratically at acidic or neutral pH in ammonium acetate on a 30-cm Nucleosil™ (Macherey-Nagel; Düren, Germany) 10SB column in 20 to 40 minutes.[165] A decade after the first high-performance separation of nucleotides, a weak ion exchange material formed from polyethyleneimine resolved the 5'-mononucleotides in 5 minutes.[166] There is continuing interest in nucleic acid analysis, because alkylated, arylated, and oxidized nucleic acids are formed due to environmental damage.[167] Reversed phase techniques, however, presently dominate the field of nucleotide separations.

Table 2 Representative Separations of Macromolecules

Manufacturer	Column	Mobile phase	Analyte
Pharmacia	Mono Q® [175]	NaCl, pH 8.2 (20–60°C)	DNA restriction fragments
	Mono Q® [208]	NaCl, pH 7.6	Malaria antigen
	Mono S® [210]	Borate-mannitol-NaCl ascending pH gradient	IgG isoforms
Toyo Soda	DEAE 5PW™ [176]	Ammonium acetate or ammonium formate, 35–80°C	Single-stranded oligodeoxyribonucleotides
	DEAE 5PW™ [202]	NaCl or sodium acetate, pH 6–8.5	Superoxide dismutase
Macherey-Nagel	Nucleogen® DMA [178]	KCl-potassium phosphate/urea	Viroid DNA and RNA
Dionex	Nucleopak PA-100 [185]	Formamide-LiBr	Phosphorothionate oligonucleotides
	CarboPak™ PA-1 [249]	Sodium acetate-NaOH	Dextrin digestion products
Synchrom, Inc.	SynchroPak® Q-300 [216]	NaCl, pH 5–9	Carbonic anhydrase, soybean trypsin inhibitor, and ovalbumin

For higher-molecular-weight oligomers of nucleic acids, classical ion exchange phases used include DEAE[168] and Amberlite® IR-120 in the aluminum salt form.[169-172] The medium-performance phases available in the mid-1980s were reviewed.[173] Zorbax®-NH2 (Du Pont; Wilmington, DE), Ultrasil-NH2 (Altex/Beckman), and Nucleosil™-NH2 were capable of separating oligoriboadenylates up to a degree of polymerization of 22. Aminex® A-28 was used for separation of tRNA.[174]

High-molecular-weight oligonucleotides, particularly DNA and RNA, require special conditions for separation. Self-aggregation, acid depolymerization,[169] protein adsorption,[170] conformational effects,[171] and pore exclusion are common problems in the analysis of oligonucleotides. Electrophoretic separation of DNA restriction fragments was compared with HPLC on RPC-5, with a DEAE 5PW™ (LKB) weak ion exchanger, with Nucleogen® 500 and 4000 weak ion exchangers (Macherey-Nagel), and with a Mono Q® (Pharmacia) strong ion exchanger.[175] This and other macromolecular separations are presented in Table 2.

Since separation by size, rather than by base composition, is a desired goal in restriction fragment analysis, the RPC-5 resin was found to be less useful than the less hydrophobic DEAE and Mono Q® phases. The chromatogram from the DEAE 5PW™ column showed unexplainable complexity, possibly due to pore exclusion. Salt and temperature effects were also examined. Single-stranded oligodeoxyribonucleotides were separated on a DEAE 5PW™ anion exchange column (ToyoSoda) using a gradient of ammonium

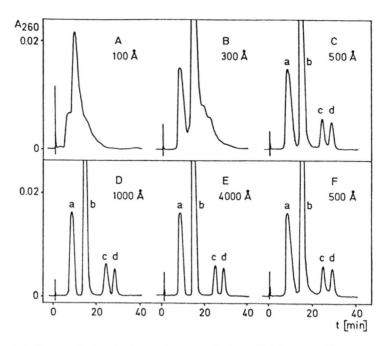

Figure 13   Pore exclusion in the gradient resolution of high-molecular-weight RNA from anion exchange. Column: 0.62 × 5 cm Nucleogen®-DMA column (Panels A–E) or a DEAE column (Panel F). Eluent: 20 mM potassium phosphate, 5 M urea, pH 6.7, operated as a gradient from 400 mM to 1 M KCl in 50 min. Flow rate: 1 ml/min. Detection: absorbance at 260 nm. A mixture of (a) tRNA, (b) 5S RNA, (c) 7S RNA, and (d) viroid RNA was chromatographed. tRNA has a molecular weight of 25 kD, 5S RNA has a molecular weight of 35 kD, 7S RNA has a molecular weight of 100 kD, and viroid RNA has a molecular weight of 120 kD. Below a porosity of 500 Å, resolution is poor, presumably due to exclusion of the high-molecular-weight materials from the pores. At very high pore size, some resolution is lost in oligonucleotide separation. (Reproduced with permission of Elsevier Science from Colpan, M. and Riesner, D., *J. Chromatogr.*, 296, 339, 1984.)

acetate or ammonium formate at temperatures from 30 to 80°C, and pH values of 5 to 7.[176] A review of separations of RNA pointed out that the popular RPC-5 phase, coated onto a hydrophobic, nonporous poly(chlorotrifluoroethylene) support called Plascon 2300, suffered from bleeding of the trimethylammonium coating.[177]

One of the most important considerations with porous chromatographic materials in ion exchange of high-molecular-weight oligonucleotides is that the pore size should be substantially larger than the analyte. As is illustrated in Figure 13, viroid DNA and RNA of 7S size were evidently excluded from the pores of a 300-Å Nucleogen®-DMA (Macherey-Nagel) column, while there was no exclusion from a 500-Å material.[178] In this potassium phosphate (pH 6.7) mobile phase, urea was used to disrupt aggregation between looped

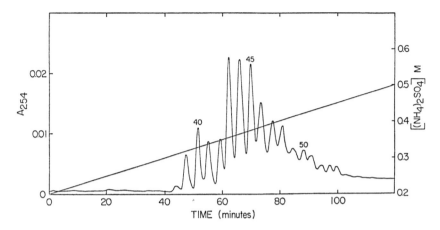

Figure 14 Fractionation of 40–60-base oligodeoxyadenylates. Column: 0.41 × 5 cm column packed with cross-linked and methylated PEI on Hypersil®, 3 μ. Eluent: 50 mM potassium phosphate, 15% acetonitrile, pH 5.9 with a gradient from 200–500 mM ammonium sulfate. Flow rate: 0.5 ml/min. Oligomers of deoxyadenylic acid were fractionated up to a degree of polymerization of 60.[180] (Reproduced with permission of Academic Press from Drager, R. R. and Regnier, F. E., *Anal. Biochem.*, 145, 47, 1985.)

regions, and EDTA was used to prevent metal bridging of backbone phosphates. Elevated temperature was not found to be useful to improve resolution. A 4000-Å pore size Nucleogen® DEAE was used to separate yeast RNA, yeast tRNA, *Escherichia coli* rRNA, *E. coli* and salmon DNA, and supercoiled cccDNA markers.[179] A variety of mobile phases were evaluated at neutral pH, and the use of formamide or urea to denature the analytes was examined. In Figure 14 is shown the separation of oligodeoxyadenylates on a cross-linked polyethyleneimine phase, packed into a 5 × 0.42-cm column.[180] The oligoadenylates had the terminal phosphates removed to increase the ionic difference between homologs. A mobile phase containing acetonitrile or methanol and using an ascending gradient of ammonium sulfate was used for elution. The methyl-quaternized strong ion exchange material and the unquaternized weak ion exchange material were compared. Resolution on the quaternized ion exchange material was far better than on the unquaternized material. Alkali denaturation, followed by anion exchange chromatography on a modified RPC-5 phase called NACS 37 (Bethesda Research Laboratories), was useful in differentiating supercoiled DNA from cellular DNA, RNA, and nicked circular molecules.[181] Neosorb® LC (Roquette Freres; Lestrem, France), a material similar to RPC-5, was applied to separations of oligodeoxyribonucleotides of 20 to 160 units.[182]

Understanding the detailed mechanism of ion exchange adsorption may be critical to devising phases that provide resolving power and speed comparable to competing techniques. A review of computer-assisted prediction

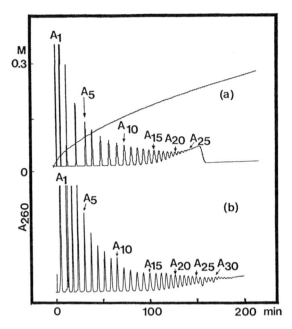

Figure 15 Comparison of theory and experiment for the fractionation of oligoade-nylates on ion exchange materials. **(a)** Simulated chromatogram. **(b)** Observed chro-matogram. An example of how theory is being used to attempt to optimize performance of ion exchange materials. The curve in (a) shows the nonlinear gradient development with a convex curvature. (Reproduced with permission of Elsevier Science from Baba, Y., Fukuda, M., and Yoza, N., *J. Chromatogr.*, 458, 385, 1988.)

of retention and peak width of oligonucleotides is available, with deviations between predicted and observed retention averaging under 8%.[183] A com-parison of theory and experiment is shown in Figure 15.[184]

One of the most important pharmaceutical applications of ion exchange separation of nucleotides is in anti-sense therapy, in which an oligodeoxy-nucleotide sequence complementary to a particular sequence of RNA is used to block protein synthesis and thereby shut down a particular biochemical pathway.[185] To reduce the susceptibility to degradation by ribonuclease, a sulfur is used to replace a nonbridging oxygen in the phosphate group. An example of a high-speed separation of phosphorothionate oligonucleotides and oxidation products is shown in Figure 16.[185] A 0.9×25-cm Dionex Nucleopak PA-100, operated at 1 ml/min, 30% formamide-25 mM Tris, pH 8.5, 70°C, and a gradient from 200 mM to 1 M LiBr, was used for the separation.

5.4.6   Peptides and proteins

Polymers of amino acids, like polymers of nucleic acids, have both ionic and hydrophobic character. Unlike the polymers of nucleic acids, amino acid polymers may carry either a negative or a positive (or zero) net charge, are far more subject to irreversible denaturation and oxidation, and are labile to

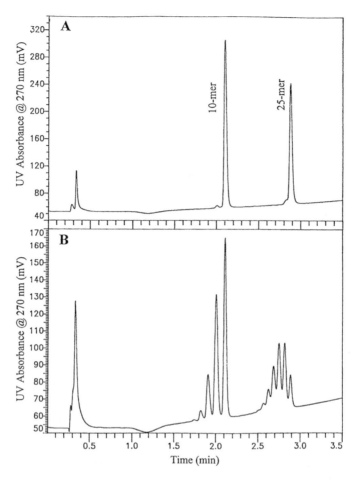

Figure 16 High-speed gradient anion exchange HPLC of phosphorothioates oligo-mers (SODNs) and SODNs with an oxygenated defect. Column: 0.4 × 5 cm Nucleopak PA-100. Eluent: 30% formamide-25 mM Tris, pH 8.5, operated on a gradient from 200 mM to 1 M LiBr. Flow rate: 1 ml/min. Temperature: 70°C. **(A)** Standards of degree of polymerization 10 and 25. **(B)** Standards after heating in glycidol at 100°C to replace a sulfur with an oxygen. Detection: UV absorbance at 270 nm. (Reproduced with permission of Elsevier Science from Bourque, A. J. and Cohen, S., *J. Chromatogr.*, 617, 43, 1993.)

depolymerization under basic as well as acidic conditions. Under physio-logical conditions, aspartic acid and glutamic acid carry a negative charge, while lysine and arginine carry a positive charge. Histidine is protonated slightly below neutral pH, while cysteine is deprotonated slightly above neutral pH. Tyrosine, serine, and threonine may be phosphorylated, intro-ducing a negative charge. Asparagine, serine, and threonine may be glyco-sylated, introducing a positive charge from deacetylated amino sugars, or a negative charge from glycuronates or sulfonated sugars. Asparagine and glutamine may also deamidate to negatively charged aspartate and glutamate. Asparagine may also rearrange to α- or β-aspartate.[186] Ion exchange is particularly useful for characterization of proteins, since charge microheterogeneity due to phosphorylation, sulfation, methylation, N-terminal formylation or acetylation, and deamidation is readily determined.

Ion exchange chromatography is a technology critical to the production and analysis of biopharmaceutical proteins. Because biopharmaceutical compounds are extremely biologically potent, there may be no clear distinction between analysis and production in the early stages of development. In production schemes, ion exchange is commonly used as the first step of chromatographic purification.[187] Both ion exchange chromatography and the related technique, chromatofocusing, are extremely useful in differentiating protein charge variants. For peptides, the use of RPLC is more common than ion exchange. The ease of mobile phase preparation, mobile phase transparency to UV detection, reduced instrumental maintenance, and the high peak capacity of modern RPLC materials are considerations in selecting RPLC over IEC. For protein isolation, however, IEC is attractive, since it can be done under nondenaturing conditions. The greater loading capacity of IEC is also a consideration in deciding between the techniques. An example illustrating the gains in speed of separation of proteins is presented in Figure 17.[188] Cation exchange separation of calf lens nuclear γ-crystallins, achieved in about 27 hours on a classical ion exchanger (SP-Sephadex™), was completed with far greater resolution in about 20 minutes on a high-performance resin (SynchroPak® CM300). A very high-speed protein separation is presented in Figure 18.[189] Ribonuclease A, carbonic anhydrase, conalbumin, and bovine serum albumin were separated in about 60 seconds, using a 3.5 × 0.46-cm pellicular polyethylenimine anion exchange column, with very high recovery. Irreversible adsorption has been a problem with the support materials used in ion exchange of proteins. Sulfonated 2-hydroxyethylmethacrylate, known as Spheron, was an early attempt to use a hydrophilic support.[190]

In the area of peptide separations, an example of a separation on the classical anion exchange material, Aminex® A-5, used the tryptic peptides of disulfide-reduced, S-carboxymethylated lysozyme and sperm whale apomyoglobin.[191] Both JEOLCO AR-15 resin and carboxymethylcellulose have been used for separations of the peptides generated from the bacterial ribonuclease, barnase.[192] In the area of high performance separations, angiotensins were separated on a weak anion exchanger.[193] Cation exchange of dynorphin A, β-casomorphin, bradykinin, Met-enkephalin, Leu-enkephalin, methionine sulfoxide and sulfone variants of MEHFKFG, neurotensin, angiotensin I, and numerous other peptides was performed on a sulfoethylaspartamide cation exchange column.[194] Peptides from bovine pituitary extract were fractionated into acidic and basic groups using Sep-Pak® cartridges packed with an ion exchange material.[195] Dipeptides were separated on Nucleosil™ 5 SB and selectively detected electrochemically by post-column addition of Cu^{2+}.[196] RPLC and anion exchange chromatography on TSK

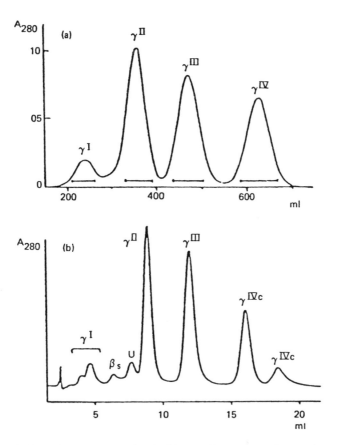

Figure 17 High performance vs. classical ion exchange in cation exchange of crystallins. **(a)** SP-Sephadex™ column, 0.5 ml/min. The separation time was 7 hr. **(b)** SynchroPak® CM300, 1 ml/min. The separation time was 20 min. (Reproduced with permission of Academic Press from Siezen, R. J., Kaplan, E. D., and Anello, R. D., *Biochem. Biophys. Res. Comm.*, 127, 153, 1985.)

DEAE 25W™ were operated in tandem to map the peptides generated on tryptic hydrolysis of very large proteins, such as albumin genetic variants.[197] A useful review that covers many aspects of peptide analysis is available.[198]

As examples of protein separations, macroporous cross-linked polyethyleneimine phases were used in the separation of lipoxygenase I, serum albumin oligomers, and ovalbumin.[199] A sulfopropyl-based cation exchange material, SP-5PW™ (TSK), was used to separate hemoglobin, lipoxidase, β-amylase, fibrinogen, and numerous test proteins.[200] A MemSep 1010 sulfopropyl cation exchange membrane cartridge was used to separate α-lactalbumin, BSA, and BSA dimer at the semi-preparative scale.[201] Superoxide dismutase was separated on DEAE 5PW™ (TSK) at different values of pH and flow rate.[202] Filtration through CM52 was used as a pre-purification step to characterize four myelin basic proteins.[203] One area of intensive

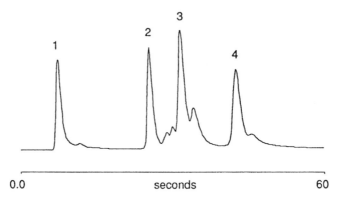

Figure 18 Very-high-speed gradient anion exchange chromatography of proteins. Column: 0.46 × 3.5 cm ZipSep™ AX, 3 μ. Eluent: Tris-HCl, pH 8.0, operated on a gradient from 0–0.5 M NaCl. Flow rate: 2 ml/min. Detection: UV absorbance at 280 nm. (1) Ribonuclease A, (2) carbonic anhydrase, (3) conalbumin, (4) bovine serum albumin. (Reproduced from Hatch, R. G., *J. Chromatogr. Sci.*, 31, 469, 1993. By permission of Preston Publications, A Division of Preston Industries, Inc.)

interest has been in the separation of hemoglobin variants. Separations that might be performed in 10 to 20 hours on classical ion exchange materials[204] can now be done in minutes or even seconds. Hemoglobin variants were separated on a 25 × 0.4-cm SynchroPak® AX300 column.[205] Several poly(aspartic acid) stationary phases, formed from the reaction of poly(succinimide) reacted with aminated silicas, were used to separate hemoglobin variants and other proteins.[206] A succinylated polyethyleneimine anion exchange material, MA7C, was used to separated hemoglobin variants.[207] A Plasmodium falciparum malaria antigen expressed using the baculovirus system was purified on a 0.5 × 5-cm or a 1 × 10-cm Mono Q® column (Pharmacia) using a Tris-HCl buffer (pH 7.6) and a NaCl gradient.[208] A hydrophilically modified poly(styrene-divinyl benzene) was used as the support for a sulfonated cation exchange resin in the separation of lysozyme, cytochrome c, α-chymotrypsinogen A, conalbumin, and myoglobin.[209]

Another area of considerable interest is the separation of antibodies by ion exchange. Charge variants due to glycosylation and the hydrophobicity of some classes of antibodies make method development more difficult. A Mono S® anion exchange column (Pharmacia), eluted with a borate-mannitol-salt buffer was used in the separation of IgG isoforms.[210] Borate is capable of forming a complex with the cis diols of sugars. Because the Mono S® column is a cation exchange material, separation cannot be ascribed to variable neuraminic acid decoration, as is often observed in anion exchange chromatography of glycoproteins. Instead, the separation was apparently based on the number of sugars capable of complexation with borate. The Bakerbond ABx phase was used in separating a fusion protein, CTLA4Ig, which is formed by fusing the extracellular sequence from human CTLA4 receptor and the hinge CH2 and CH3 domains of human IgG 1.[211] The ABx column, a weak cation exchanger with some anion exchange and some hydrophobic character, was operated in a water-methanol gradient and eluted with sodium acetate. Isoform of antibody Fab fragments were separated on a Poros® S/M strong cation exchange column using a pH gradient.[212]

Presumed glycoprotein variants from a hepatoma were separated from other plasma proteins by anion exchange in the presence of detergent on a 7.5 × 0.75-cm TSK DEAE 5PW™ anion exchange column.[213] Seven forms of rat microsomal cytochrome P-450 were separated by sequential chromatography on DEAE 5PW™ anion exchange and SP-5PW™ cation columns (ToyoSoda).[214] The separations were sensitive to buffer, eluent salt, detergent concentration, and detergent type. It was reported that polyanionic exchange resins were useful in the separation of a deamidation variant of recombinant human DNase.[215] Polyethyleneimine or other amines, coated onto silica and cross-linked with epoxy resins, alkyl bromides, or nitro alcohols, formed a a weak ion exchange packing used to separate serum and isoenzymes of rat kidney hexokinase. Peptides were weakly retained.[166] Separations of ovalbumin, carbonic anhydrase, β-lactoglobulin, conalbumin, chymotrypsinogen, α-amylase, ribonuclease, lysozyme, soybean trypsin inhibitor, ferritin, and cytochrome c were performed on a SynchroPak® AX300 weak anion exchange column and on a quaternized Q-300 strong anion exchange column.[216] Retention on the weak ion exchanger was susceptible to changes in pH. The strong anion exchanger had better resolution and exhibited higher recovery. Elevated temperature was found to improve resolution. Recombinant rat cytochrome b_5 was separated at various temperatures from *E. coli* lysate by ion exchange chromatography on Pharmacia Q-Sepharose™.[217] It was concluded that adsorption is an entropically-driven process. Superoxide dismutase was separated.[218] Isoforms of kidney angiotensinase A, differing in many properties but in particular in neuraminic acid content, were separated by anion exchange on Mono Q®.[219]

It is possible to optimize the conditions for ion exchange separations of proteins by mapping the retention as a function of pH. High-speed separations and automated pH mapping have greatly reduced the tedium involved in this approach. Another approach is to perform an electrophoretic titration curve.[220] The sample is run on a pH gradient along one dimension of the gel, then electrophoresis performed along the other dimension of the gel. This generates a two-dimensional representation of the predicted elution pattern as a function of ionic strength and pH. The effects of pH on capacity and retention have been examined using the Mono Q® anion exchange column (Pharmacia) and the SynchroPak® S300 cation exchange column.[221] The capacity is strongly dependent on the pH, with the cytochrome c capacity changing by a factor of about 5 on cation exchange from pH 4 to 8. The capacity for ovalbumin and conalbumin also changed by a factor of about 5 from pH 6 to 10 on anion exchange. Retention was also affected.

Protein separations are also sensitive to the nature of the eluting salt.[222] While specific interactions with proteins are possible, as was observed in the

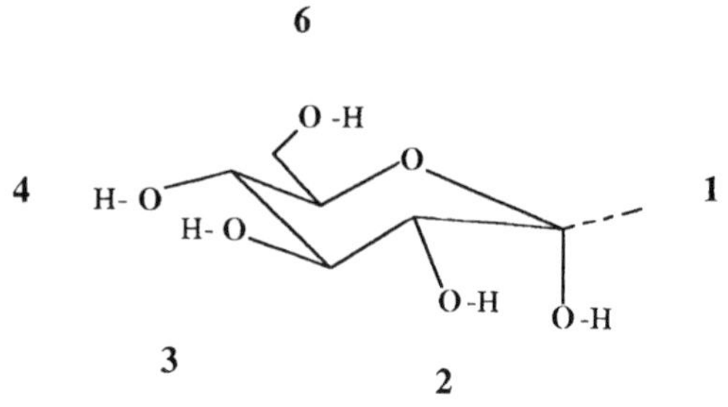

6

Figure 19 α-D-glucose, a simple sugar. The numbers indicate the positions around the ring. In the β-D-anomer, the position of the H- and -OH at position 1 are exchanged.

case of borate complexation with sugars described above, differences in elution were ascribed to nonspecific interactions. Nonspecific interactions were differentiated from specific interactions by determining whether the changes in retention were proportional for all components of a mixture or whether certain peak retention times were affected differently than others.

5.4.7   Simple sugars and lower-molecular-weight oligomers

Sugar determinations are important in the food industry, in analysis of carbohydrates and their polymers, and in pharmaceutical analysis of glycoproteins. A number of column types have been used in the separation of sugars, particularly amino columns and strong anion exchange columns. Amperometric detection is widely used, but post-column detection schemes, UV detection, and refractive index detection are also useful. Sugars are capable of being linked together at a number of locations on the ring, making the determination of linkage and branching one of the most challenging issues in analysis. In Figure 19 is shown α-D-glucose and the linkage points from which oligomers and polymers may be formed.

In Figure 20 is shown an application in clinical chemistry involving the separation of mono- and disaccharides on a Dionex CarboPac™ PA-1 column with post-column 4-aminobenzoylhydrazide detection at 400 nm.[223] In Figure 21 is shown a separation of α-(1,4)-linked oligomers of galacturonic acid, obtained from enzymatic cleavage of polygalacturonate from citrus, on a 25 × 0.46 Rainin Dynamax™-60A weak anion exchange column.[224] The UV absorbance of the carboxylic acid of galacturonate was adequate for detection in phosphate buffer. With the more strongly UV-absorbing acetate buffer, refractive index may be preferable. On a 25 × 2.1-cm column, 300 mg/hr of sample could be processed. Protamine, a basic arginine-containing protein, was found to be useful in the fabrication of an anion-exchange phase for sugar separations.[225] A 15 × 0.46-cm pre-packed silica column was coated with protamine, then eluted with acetonitrile-water to separate monosaccharides, malto-oligosaccharides, disaccharides, and sugar alcohols. A post-column detection system using guanidine-periodate was used.[66,67]

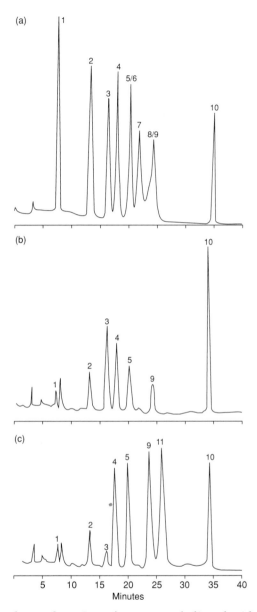

Figure 20 Post-column detection of mono- and disaccharides with 4-amino-benzoylbenzamide. Column: CarboPac™ PA-1. Gradient: 1–10 mm NaOH (0–20 min.), 10–20 mM NaOH (20–35 min). Flow rate: 1 ml/min. Detection: absorbance at 400 nm after reaction with 4-aminobenzoylhydrazide. **(a)** Standard mixture of fucose (1), arabinose (2), galactose (3), glucose (4), xylose and N-acetylglucosamine (5 and 6), allose (7), 3-fucosyllactose (8), fructose (9), lactose (10), Man-β-(1,4)-GlcNac. **(b)** Normal urine. **(c)** Urine from a child with β-mannosidosis. (Reproduced with permission of Academic Press from Peelen, G. O. H., de Jong, J. G. N., and Wever, R. A., *Anal. Biochem.*, 198, 334, 1991.)

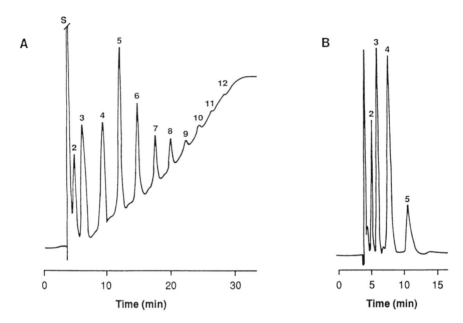

Figure 21 Analytical scale separation of oligomers of α-(1,4)-linked galacturonic acid. Column: 25 × 0.46 cm Dynamax™ NH2. **(A)** Gradient separation, pH 5.9, 0.1–0.4 M phosphate in 25 min. UV detection at 220 nm. **(B)** 0.8 M acetate, pH 5. Refractive index detection. (Reproduced with permission from Elsevier Science from Hotchkiss, Jr., A. T., Hicks, K. B., Doner, L. W., and Irwin, P. L., *Carb. Res.*, 215, 81, 1991.)

In Figure 22 is shown a separation of the sugars generated from hydrolysis of carboxymethylcellulose, a polymer important in foods, coatings, and oil drilling.[226] A CarboPac™ PA-1 anion exchange column, eluted with an acetate buffer, was capable of separating positional isomers of carboxymethyl, dicarboxymethyl, and tricarboxymethylglucose. The monosaccharides obtained from methanolysis and acid hydrolysis of uronic acid-containing polysaccharides were analyzed by ion exchange chromatography on a CarboPac™ PA-1 column with pulsed amperometric detection.[227] A similar approach was used for soil glycuronates.[228,229] A similar separation scheme was used for monosaccharides obtained from the acid hydrolysis of the capsular polysaccharide of *Vibrio vulnificus* M06-24.[230] Almost 100 standard monosaccharides and oligosaccharides were chromatographed on either the CarboPac™ PA-1 or the HPIC AS-6 columns, both strong anion exchangers.[231] The CarboPac™ PA-1 was also used for analysis of mannitol, 3-O-methyl glucose, and lactulose in urine;[232] for sugars derived from the hydrolysis of glycosides in grape musts;[233] and for simple sugars and sugar oligomers in dairy products.[234] The activity of a glucantransferase (debranching) enzyme from yeast was monitored by observing the production of maltotriose and maltose from maltapentaose on a CarboPac™ PA-1 column.[235]

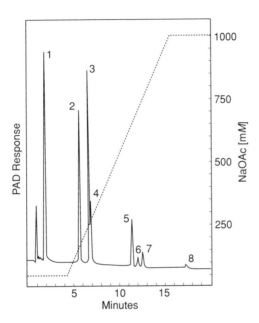

Figure 22 Typical high-pH AEC trace of a carboxymethylcellulose hydrolysate. Column: 25 × 0.9 cm Dionex CarboPac™ PA-1 column. Gradient: 0.1 *M* NaOH-50 m*M* sodium acetate (0–0.3 min) to 0.1 *M* NaOH-1 *M* sodium acetate (15.3 min). Flow rate: 4.0 ml/min. Detection was amperometric. Peaks are D-glucose (1), 6-O-CM-D-glucose (2), 2-O-CM-D-glucose (3), 3-O-CM-D-glucose (4), 2,6-di-O-CM-D-glucose (5), 3,6-di-OCM-D-glucose (6), 2,3-di-O-CM-D-glucose (7), and 2,3,6-tri-O-CM-D-glucose (8). (Reproduced with permission of Elsevier Science from Kragten, E. A., Kamerling, J. P., and Vliegenhart, J. F. G., *J. Chromatogr.*, 623, 49, 1992.)

Amperometric parameters have been studied and optimized for sugar separations.[57,236] The orcinol reaction has also been used as a post-column detection system for sugars.[237]

Another approach is the use of metals such as calcium, lead, and silver, complexed to cation exchange columns, to adsorb sugars or the sugar alcohols generated by borohydride reduction of sugars.[238-240] Amino columns have been used in normal phase mode for sugar separations.[241] Monosaccharides from immunoglobulin M were separated by separation on a 30 × 0.78-cm Aminex® HPX-87H column using a sulfuric acid eluent and refractive index detection.[242] An aminopropylsilica column was operated in normal phase mode to separate simple sugars, sugar alcohols, and oligosaccharides.[24] A DextroPak™ (Waters) column eluted with water and employing pulsed amperometric detection was used to separate methyl glycosides obtained from methanolysis and acetylation of glycoproteins and glycolipids.[243] Monosaccharides were separated on a protamine-coated silica gel support.[225] Ion pair chromatography was used to separate phenyl pyranosides by reversed phase.[244] Ion exclusion chromatography on a Showa Denko

IonPac® KC-811 column separated the labile compounds N-acetylneuraminic acid and N-glycolylneuraminic acid released by mild acid hydrolysis of bovine vitronectin.[245] Sialic acid is extremely labile to conditions of handling and must be released by mild acid hydrolysis.[246] Derivatization with phenylisothiocyanate and separation by reversed phase chromatography was found useful in analysis of hexosamines from gastric mucosa.[247] A review on separation of sugars and other carbohydrates which covers many important aspects is available.[248]

5.4.8   Carbohydrate polymers

Higher oligosaccharides occur as natural homopolymers, such as inulin, dextran, starch, and pullulan, and as heteropolymers, such as pectin and heparin. Complex oligosaccharides are found as N-linked or O-linked prosthetic groups of glycoproteins. As has been noted, analysis of the branching of polysaccharides is one of the most challenging problems in contemporary analysis. One means of approaching that problem is though enzymatic digestion with linkage-specific enzymes. In Figure 23 is shown an isoamylase digest of waxy-β-limit dextrin separated on a CarboPac™ PA-1 column (Dionex) using an acetate gradient and amperometric detection.[249] Isoamylase removes maltosyl and maltotriosyl A-chain stubs, to generate linear polymers, and polymers with enzyme-resistant branches. One point to be aware of in inspecting amperometric analyses of oligosaccharides is that the response factor decreases with increasing degree of polymerization. The peaks labeled 5 and 10 in the figure are about the same height, but peak 10 would contain about twice as much material as peak 5.

Ion exchange with pulsed amperometric detection has been used to characterize the degree of polymerization of malto-oligosaccharides and pullulans. A Dionex CarboPac™ PA-1 column was used for ion exchange separation chromatography of homopolymers of arabinose and xylose and complex oligosaccharides from pectin,[250] and for yeast high-mannose isomers with cores of Manβ(1→4)GlcNacβ(1→4)GlcNAcα,β; Manα(1→6),Manα (1→3)Manβ(1→4)GlcNacα,β; and Manα(1→6), Manα(1→3)Manα(1→4) GlcNacβ(1→4)GlcNacα,β.[251] The same system was applied to the separation of positional isomers of neutral oligosaccharides and desialylated fetuin oligosaccharides released from the tryptic peptide by N-glycanase.[252] The CarboPac™ PA-1 with pulsed amperometric detection also was found to be useful in the characterization of exopolysaccharides from lactic acid bacteria, arabinoxylans, and substituted celluloses.[253] Mass spectroscopic detection was made possible by use of an anion suppressor. The Dionex CarboPac™ PA-1 and PA-100 columns have been used in combination with a micromembrane suppressor and a mass spectrometric detector for analysis of maltodextrins and arabinogalactans.[254] The micromembrane suppressor was essential to remove acetate, which accelerates thermal degradation in the mass spectroscopic (MS) interface. Hyaluronan, a linear (1→3)-O-

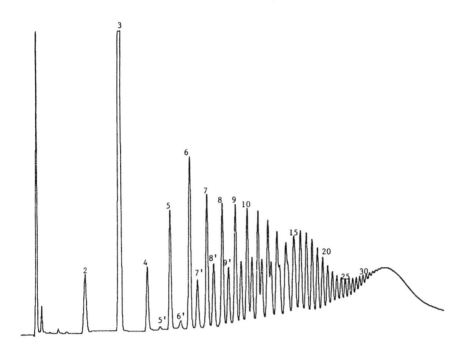

Figure 23 High performance anion exchange chromatogram of isoamylase digest of waxy β-limit dextrin using amperometric detection. Column: 25 × 0.4 cm Carbo-Pac™ PA-1. Eluent: 100 mM NaOH. Gradient: 50–500 mM sodium acetate. Flow rate: 1 ml/min. Detection: pulsed amperometric detection. The numbers over each peak represent the number of repeating α-(1,4)-linked glucose residues. The peaks with the apostrophe also contain α-(1,6)-linked glucose branches. (Reproduced with permission of Elsevier Science from Ammeraal, R. N., Delgado, G. A., Tenbarge, F. L., and Friedman, R. B., *Carb. Res.*, 215, 179, 1991.)

2-acetamido-2-deoxy-β-D-glucopyranosyl-(1→4)-O-β-D-glucopyranosyl polymer of medical significance, was found to be unstable under conditions of alkaline pH;[255] therefore, it was separated on a Dionex CarboPac™ PA-100 column with UV detection under neutral conditions, with oligomers of degree of polymerization 20 being resolved. The oligosaccharides obtained from N-glycanase treatment of glycoproteins such as platelet-derived growth factor and human chorionic gonadotropin were analyzed at both high and low pH on a CarboPac™ PA-1 column with pulsed amperometric detection.[256]

Alkaline conditions are used so frequently in carbohydrate separations on CarboPac™ columns that it should be pointed out that acidic conditions are suitable for separation of acidic sugars. Figure 24 shows the separation of sialic acid containing oligosaccharides.[256] Alkaline conditions were used for neutral milk oligosaccharides and mucin oligosaccharide alditols were characterized similarly.[257] The carbohydrates released from yeast mannoprotein with N-acetyl-β-D-glucosaminidase were also fractionated on CarboPac™ PA-1, as were oligosaccharides derived from treatment with jack bean

α-mannosidase.[258] Oligosaccharides from depolymerized glycosamino gly-
cans such as heparin, chondroitin sulfates A and C, and keratan sulfate were
separated using strong anion exchange on an IonPac® NS1 column (Dionex)
with UV and membrane-suppressed conductivity detection.[259] Endotoxin
was fractionated into immunologically distinct fractions using the Pharmacia
Mono Q® column.[260]

There are alternatives to ion exchange in carbohydrate separations. Carbohydrate polymers may be fractionated according to size, a topic discussed
in far greater detail in Chapter 6. Neutral oligosaccharides from milk have
been separated on a Fractogel® TSK HW 40 (S) column according to size.[261]
Reversed phase may be preferable to ion exchange when mass spectra of
oligosaccharides is desired. Reversed phase has been used in combination
with normal phase chromatography for the separation and UV detection of
p-aminobenzoic acid ethyl ester derivatives of periodate oxidized oligosac-
charaminitols from sea squirt.[262] Oligouronates from alginates were separated by ion-pair chromatography on a reversed phase column with refractive index detection.[263] Reversed phase with amperometric detection has
been used to separate disaccharides such as gentiobiose, trehalose, laminar-
ibiose, nigerose, kojibiose, and sophorose; oligosaccharides such as
underivatized and glucosylated cyclomaltohexaose, cyclomaltoheptaose,
and cyclomaltooctaose; and monosaccharides in their anomeric forms.[264]
Reversed phase with fluorescent and mass spectroscopic detection was used
in the analysis of pyridylated and N-acetylated oligosaccharides released
from immunoglobulin G by hydrazinolysis.[265]

Hydrophilic interaction chromatography on Asahipak® NH2P or Excelpak CHA-P44 with pulsed amperometric detection has been used to frac-
tionate malto-oligosaccharides.[266] The Asahipak® NH2P is a polyvinyl alcohol support with a polyamine bonded phase, and the Excelpak is a
sulfonated polystyrene in the Zn^{+2} form. Amine adsorption of sialic acid-
containing oligosaccharides was performed on a Micropak AX-5 column
(Varian) using acetonitrile-water-acetic acid-triethylamine.[267]

Oligosaccharides from immunoglobulin M were separated on an
Aminex® HPX-42A column (Bio-Rad) at 85°C using a water eluent.[242]
Malto-oligosaccharides were separated in gram quantity on a Waters™
Dynamax™-60A NH2 column in normal phase mode.[268] Hydrophilic interaction chromatography on a Waters™ Protein-Pak™ 60 column eluted with
a mixed organic phase was also used for hydrolysis products of dextrans,
inulins, and starches.[269] Pyridylated oligosaccharides released from urinary
kallidinogenase by digestion with glycopeptidase A and subsequent
hydrolysis with β-galactosidase or α-fucosidase were separated on hydrophilic interaction chromatography on TSK-Gel™ Amide 80 coupled to a
reversed phase system to generate a two-dimensional map.[270] Carbohydrate oligomers of high-fructose corn syrup were separated by reversed
phase and detected with post-column derivatization with 4-aminobenzoic
acid hydrazide.[271]

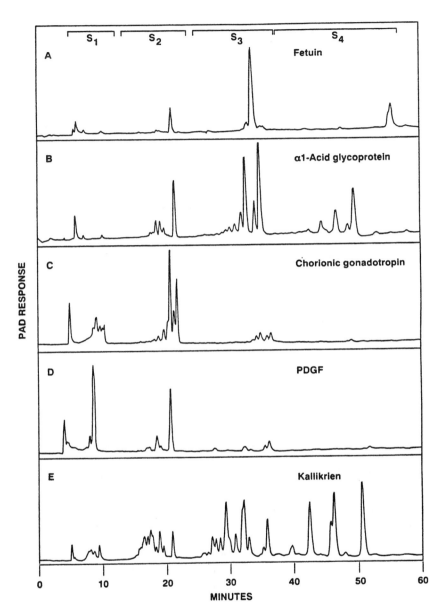

Figure 24 Anion exchange chromatography of reduced oligosaccharides under low pH conditions. Column: 0.46 × 25 cm Dionex CarboPac™ Pa-1. Eluent: 100 mM NaOH. Gradient: 2–300 mM sodium acetate. Flow rate: 1 ml/min. Detection was by amperometry. Enzymatically released sialic acid-containing oligosaccharides, reduced with sodium borohydride, were separated by high performance anion exchange chromatography. **(A)** Fetuin. **(B)** α1-Acid glycoprotein. **(C)** Chorionic gonadotropin. **(D)** Platelet-derived growth factor. **(E)** Kallikrein. (Reproduced with permission of Academic Press from Watson, E., Bhide, A., Kenney, W. C., and Lin, F. K., *Anal. Biochem.*, 205, 90, 1992.)

5.4.9   Vitamins and coenzymes

Reversed phase separations have replaced many ion exchange applications for the analysis of vitamins and their derivatives. Therefore, the bulk of the literature concerns low-performance methodologies. Nicotinamide and thionicotinamide derivatives were separated on Dowex® 1-X8, a weak ion exchanger.[272] Glycerophosphorylcholine, phosporylcholine, betaine, and choline chloride, essential components of cell membranes, were separated on a Dowex® 50W cation exchanger.[273] Thiamine and phosphorylated thiamines were separated on a Dowex® 1-X8 column.[274] One interesting medium performance separation demonstrated the simultaneous separation of Coenzyme A, Coenzyme M, panthetheine, 4'-phosphopantetheine, and numerous other biological thiols on a MBF/SS amino acid analyzer (Dionex) with a AA-10 column (Beckman)using monobromobimane or monobromotrimethylammoniobimane to form a fluorescent derivative.[275] A pellicular cation exchange material, HS Pellionex SCX (Reeve Angel) was used to separate vitamins B_1, B_2, B_6, and nicotinamide from veterinary vitamin formulations.[276] Amperometry of sulfur species has been optimized.[57]

5.4.10   Other compounds

A bisphosphonate, clodronate, was separated from its esters on an HPIC AS7 column (Dionex) using dilute nitric acid as the eluent.[277] Thorium-EDTA-xylenol was used as a postcolumn complexation agent for optical detection at 550 nm. Polyphosphonates and polycarboxylates, used as color preservatives and spoilage retardants in food, were separated on a PRP®-X100 anion-exchange column (Hamilton) using vacancy detection and conductivity detection.[278] Alkylbenzenesulfonates, used as surfactants, were detected as environmental contaminants using a TSK-Gel™ IC-Anion-PW column (Tosoh) eluted with aqueous acetonitrile-sodium perchlorate.[279] The PRP®-X100 and the IonPac® AS11 were compared in the separation of alkane sulfonates, alkyl sulfates, and alkylbenzene sulfonates, with the AS11 being judged more versatile due to better column efficiency and lesser hydrophobic interactions.[280]

Ethanol and glycerol have been separated from beer on an Aminex® HPX 87H or Inores S259H column using dilute sulfuric acid as the eluent.[281] The separation was presumably accomplished by a hydrogen-bonding interaction rather than an ionic interaction. Separation of benzyl and *p*-xylylene adducts of tetrahydrothiophene, used in the preparation of poly(-phenylene vinylene) was unsuccessfully attempted by cation exchange on the Nucleosil™ SA sulfonated silica column, using KCl-HCl-LiNO$_3$ as the eluent.[282] Separation was successfully performed on reversed phase. Phosphate esters of myoinositol, encountered in legumes, corn, and rice, were separated on a PL-SAX strong ion exchanger at pH 4.0 with sodium nitrate.[283] Post-column detection with Fe (III)-sulfosalicylic acid was used with detection at 500 nm. Sugar phosphates from fish muscle were separated on Dowex® 1-X8.[284]

Eleven sugar phosphates, including glucose-6-phosphate, fructose-6-phosphate, dihydroxyacetone phosphate, and 3-phosphoglyceric acid, were separated on a 0.46 × 10-cm Spherisorb® SAX strong anion exchanger (Phase Separations) using mobile phases containing formate and Eu^{+3} as a complexation agent to enhance UV detection.[65] Finally, it may be noted that ion exchange resins have been successfully used in the fractionation of micro-organisms.[285] While this application is not, strictly speaking, analytical IEC, it is an indicator of the versatility of ion exchange.

5. 5 Case history: isolation and partial characterization of hementin, a compound of biopharmaceutical interest

5.5.1 Background

This section describes the isolation of the protease hementin. Hementin has antithrombotic properties that could find application in biopharmaceutics. Found in extracts of the salivary glands of the giant South American leech *Haementeria ghilianii*,[286] hementin digests fibrinogen to fragments resembling plasmic fragments Y, D, and E, rendering the proteolyzed fibrinogen non-clottable.[287,288] As such, this protease was of potential interest in pharmaceutical and biomedical applications, particularly in dissolving platelet-rich clots.[289] However, the organism that produces hementin, *Haementeria ghilianii*, is a rare and endangered species. A leech colony at the University of California at Berkeley, until recently the sole source in the U.S. for extracts containing hementin, collapsed due to the progressive effects of inbreeding. Since the publication of the work reported below, a small colony has been re-established in South Carolina, and hementin has become commercially available.

The scarcity of hementin and the complexity of the sample matrix from which it was extracted made the isolation and full characterization of hementin a considerable challenge. This section illustrates the central role of high performance ion exchange chromatography in the discovery phase of biopharmaceutical development. The precise structure, composition, and molecular weight of hementin are still controversial. This work described in this case history, which was performed at SmithKline&French Research Laboratories, helped to advance the understanding of the physical composition of hementin and was used for a study of the specificity of hementin proteolysis at the Temple University Medical School.[288]

For the work described below, only 600 µl (6 mg) of gland extract, obtained from the *posterior* gland, were available. The active protein was subsequently determined to be present at a concentration of about 0.5%, representing less than 100 µg (1 nmol) of hementin. The remainder of the extract consisted of several hundred inactive proteins and peptides. While the protein composition and cell types of the anterior and posterior glands are very different, the electrophoretic mobility of the active enzyme from anterior or posterior gland sources was identical, implying that the enzyme of both glands is identical.[286]

The goal of the work presented here was to confirm the molecular weight of hementin, determine the N-terminal amino acid sequence, and provide sufficient purified protein for biochemical studies of the fibrinolytic activity. Earlier work on the more abundant anterior gland extract had suggested that the protein stability was poor.[290] Metal-chelating agents such as EDTA abolished the activity. EGTA, a calcium-chelating agent, was partially inhibitory. The optimum of activity was observed at about pH 7 to 9, with almost total loss of activity below pH 5 and above pH 10. Phosphate and Tris buffers were reported to inactivate the enzyme, as were reducing agents. Dialysis and exposure to elevated temperature were also reported to inactivate the enzyme, while reduced temperature was reported to lead to a reversible diminution of activity.

In the original isolation attempt,[290] about 60 mg of total protein were purified by anion exchange on DE-52 diethylaminoethyl cellulose, precipitated by addition of ammonium sulfate, and chromatographed by cation exchange on CM-52 carboxymethylcellulose. The active component was unretained on CM-52. This approach yielded 750 µg of total protein (1.2% of the original), representing a yield of only 18% of the activity, with a net purification factor of only 16. Gel filtration on Ultrogel AcA 34 (LKB) and fibrinogen affinity chromatography resulted in loss of activity. Ultimately, nondenaturing electrophoresis was chosen for the final step of purification. Approximately 25% of the total protein was found in the active band, which would correspond to a purification factor of 64. Both reducing and nonreducing SDS-PAGE showed a predominant band in this isolate corresponding to a molecular weight of 120 kD.

The 120 -kD SDS-PAGE band could not, of course, be assayed for activity because SDS is denaturing. The activity of the precursor was such that 248 pmol of fibrinogen could be cleaved (to nonprecipitable peptides) per second per mg of enzyme. This would correspond to 15 nmol per min per mg hementin. Commercially available hementin is claimed to "decoagulate" 1 µmol of fibrinogen per unit per minute, with 15 units per mg hementin.[291] The claim is presumably intended to indicate that the fibrinogen is rendered incoagulable. This would correspond to 67 nmol fibrinogen per min per mg hementin. Another report claimed a preparation of hementin at an activity of 1860 U/mg,[292] which would correspond to fibrinogenolysis at the rate of 8 µmol fibrinogen per min per mg hementin.

The data described below confirmed that hementin is hydrophobic and acidic, but the molecular weight was found to be about 80 kD rather than the 120 kD reported previously. One key experimental detail to which attention was not drawn in the previously published description of the purification of hementin[293] (described in more detail below) was that activity was not recovered when a C-8 guard column was used in the RPLC. Activity was recovered only when the guard column was omitted. The stability of hementin was found to be far better than anticipated. Also unpublished are the contributions of Dr. Jan Kochansky of the U.S. Department of Agriculture, who synthesized the peptide corresponding to the putative N-terminus of hementin, and Amrut Bhogle of the University of Massachusetts at Amherst, who generated antibodies from Dr. Kochansky's peptide. Confirmation of the N-terminal sequence by affinity chromatography with those antibodies or by other antibody methods remains to be done. The excellent sequencing work of J. G. Gorniak and Dr. J. E. Strickler also bears mention again.

5.5.2   Considerations in designing the isolation

At the outset of the isolation effort reported below,[293] it was presumed that the protein of interest had a molecular weight of 120,000, required neutral pH, moderate temperature, and the presence of calcium to maintain activity. The protein was believed to be acidic, causing it to bind to cation exchange but not to anion exchange materials; the electrophoretic migration on cellulose acetate was consistent with this assumption.[286] The stability was presumed to be poor, and the protein was presumed to be very hydrophobic, leading to losses on hydrophobic materials. The complexity of the sample, which was subsequently estimated to contain several hundred constituents, was daunting, particularly given the small sample size and the need to reserve the majority of protein for biochemical studies. A diagram of the isolation effort is shown in Figure 25.

5.5.3   Notes on protein analytical techniques

Area analysis of chromatograms is familiar to the chromatographer. The assumption is made that the area of a peak is proportional to the mass of the component present. The spectrophotometric detector is set at about 280 nm to detect the aromatic residues or at 210 to 230 nm to detect the peptide bond. The absorbance at 280 may be influenced by peptide composition, while the absorbance near 220 nm is more closely correlated to mass.

In purity analysis by slab gel SDS-PAGE, the analyte mixture is electrophoresed to separate by molecular weight. Under favorable conditions, proteins differing by 1 to 2% in molecular weight can be separated. Then, the separated proteins are stained with either Coomassie Brilliant Blue (CBB) dye or with silver stain. Coomassie staining tends to give a linear response, while the response of silver is highly nonlinear, tending to emphasize trace impurities. Gels can be scanned by an absorption spectrophotometer known as a densitometer, generating a densitometrogram, which resembles a chromatogram. An important strategy in using SDS-PAGE for purity determination is to reduce the disulfide bonds with a mercaptan, such as β-mercaptoethanol, and to compare the result with that obtained from the nonreduced protein. Proteins that have suffered proteolysis may be held together by the disulfide bonds, but on reduction the fragments are released and separated electrophoretically. Even fragments too small to resolve from the marker dye

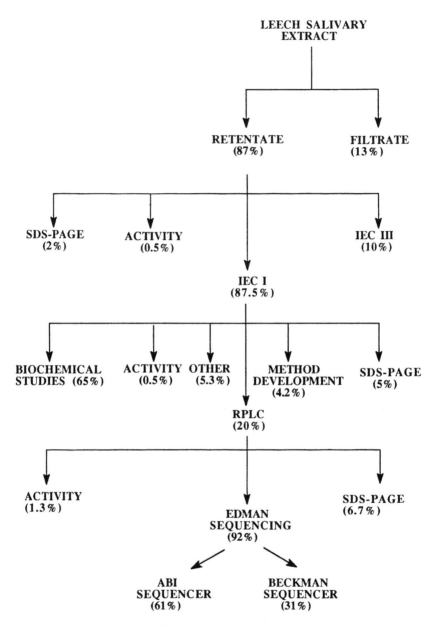

Figure 25   Outline of hementin isolation process.

can be seen by careful inspection of the dye front at the low-molecular-weight end of the gel. Small peptides, however, tend to be washed out of the gel during the staining process.

In Edman sequencing, the N-terminal amino acids are sequentially cleaved and labeled with a hydrophobic chromophore. Then, the labeled amino acids are separated by RPLC, identified by retention, and quantitated by absorbance. The response factors in Edman sequencing vary, depending strongly on the efficiency of the cleavage and losses during sample processing. Each sequencer will tend to recover a few amino acids much less efficiently than the others. There is usually a background of contaminating amino acids, particularly from amino acids derived from incomplete cleavage and from degradation at sites other than the N-terminus. Minor contaminants are often not distinguishable from this background of amino acids. The response is, to a first approximation, proportional to the molar amount of each N-terminus present. For example, a mixture of 10 parts by weight of a protein of molecular weight 100 kD, and 1 part by weight of a protein of 10 kD will show equal quantities of each N-terminus.

5.5.4   The first step in isolation: stability testing

Before beginning the isolation attempt, a simple stability study was conducted. Hementin was ultrafiltered against three volumes of ammonium bicarbonate, pH 7.7, on an Amicon YM 100 membrane filter. The YM 100 has a nominal molecular weight cutoff of 100 kD. The *retentate* was combined with solutions containing calcium chloride, Gly-Trp-Gly with and without calcium chloride, bovine lung trypsin inhibitor with and without calcium chloride, glycerol with and without calcium chloride, and Gly-Trp-Gly with calcium chloride and glycerol. This was a simple factorial design to determine whether calcium, a protease inhibitor, or a hydrophobic peptide could serve to improve stability of the crude extract. Each volume of the *filtrate* was maintained separately. After periods of storage of 0, 36, 60, and 168 hr at 4°C, aliquots were drawn from each of the stability samples and combined with fibrinogen at 1 mg/ml. After incubation for appropriate intervals at 37°C, aliquots of the digested fibrinogen were withdrawn and assayed by SDS-PAGE. The stability of hementin, at least in the crude mixture, was far better than had been hoped for. While it is possible for stability to decrease with increasing purification, the retention of activity over the 7-day period of the stability study implied that rapid processing might not be as critical to success as had been thought.

Surprisingly, all samples retained fibrinogenolytic activity, including both filtrate and retentate. At the time, it was thought that activity in the filtrate was due to membrane leakage. A YM 100 membrane should retain the great majority of a protein of molecular weight 120 kD. It is possible to have leakage above the nominal molecular weight cutoff of a membrane, due to a physical defect in the membrane, an unusual protein shape, or proteolysis of a high-molecular-weight form to an active low-molecular-weight form. Also, of course, there is small amount of leakage at any molecular weight. However, in retrospect, it became evident that the transmission of activity through the membrane was the first piece of evidence to indicate that the molecular weight of hementin might be less than 120 kD.

5.5.5   The second step in isolation: ion exchange chromatography

Despite a small loss of activity by leakage, ultrafiltration remained a desirable first step in isolation of hementin. Ultrafiltration permitted buffer exchange to the ion exchange starting conditions (20 mM NaHCO$_3$ and 100 μM CaCl$_2$, pH 8.0). A 600-μl (6 mg) portion of crude posterior gland extract was combined with 2 ml of starting buffer and ultrafiltered at 4°C against 9 ml of the buffer. Both the retentate and the filtrate were chromatographed on a Pharmacia Mono Q® column using a gradient to 500 mM NaHCO$_3$ and 100 μM CaCl$_2$ (pH 8.0). The retentate chromatogram is shown in Figure 26, panel A. The filtrate was enriched in peak II relative to peak I, implying that peak I contains low-molecular-weight material.

With only 100 μg total protein loaded (for method development), peaks I and II were very well resolved. When the full sample (6 mg) was injected for preparative purposes, peak II shifted to an earlier retention time. A shift to earlier retention on increased loading is a common problem in purification. If the major component can be made to elute before the minor component, the retention shift will not harm the separation as greatly as if the major component elutes after the major component.

One-ml retentate fractions were collected and assayed for fibrinogenolytic activity and for protein composition by 7.5% acrylamide SDS-PAGE. These results are shown in Figure 26, panels B and C, respectively. Activity is evident in fractions 31 to 45. In the remaining fractions, fibrinogen appears as an essentially undigested band above 200 kD. The digestion pattern has been assigned as Y$_{hem\ 1-3}$, three bands at molecular weights of 183 to 204 kD and D$_{hem\ 1-3}$, three bands of molecular weight 119 to 140 kD, D$_{hem\ 4}$, a band with molecular weight 102 kD, and E$_{hem}$ with a molecular weight of 62.5 kD.[288]

As described below, the peak labeled I (fractions 33 to 38) was ultimately determined to contain almost all of the hementin. Based on the peak area of peak I in the retentate vs. the filtrate, it was estimated that 87% of the hementin had been retained on ultrafiltration, giving a combined yield of 3.8% following buffer exchange and IEC. At this point, the peak I pool was concentrated and analyzed by reducing and nonreducing SDS-PAGE and by Edman degradation. Both disulfide-reduced and unreduced protein appeared to be nearly homogeneous on Coomassie staining. Densitometric analysis indicated that the area of the main band was about 65% of the total, with about 15 very minor bands of 1 to 5% each. Therefore, when Edman sequencing showed a sequence EVYTNYASFL, with the principal sequence being 50 times more abundant than the background, it was thought that hementin had been successfully sequenced from the peak I pool. However, the yield on Edman sequencing was about 10 times greater than had been estimated from the band area on CBB staining. This was an important clue that SDS-PAGE had failed to visualize a major contaminant.

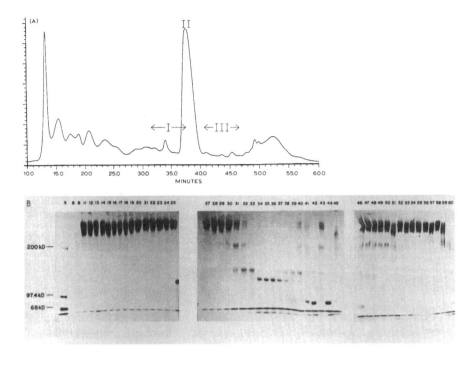

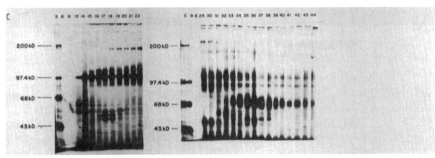

Figure 26 High performance anion exchange chromatography of *Haementeria ghilianii* salivary extract. Separation was achieved on a Mono Q® column at a flow rate of 1 ml/min, with buffers containing 20 mM NH$_4$HCO$_3$ and 100 μM CaCl$_2$, pH 8.0. The gradient was from 0–0.5 M NaCl. **(A)** Elution profile at 280 nm. **(B)** SDS-PAGE pattern of fibrinogen + column eluate showing digestion of fibrinogen in fractions 31–45. **(C)** Nonreducing SDS-PAGE (10% acrylamide, silver stain) of fractions from IEC separation. In (B) and (C), molecular weight markers of standard proteins are indicated on the vertical axes. (Reproduced with permission of Elsevier Science from Swadesh, J. K., Huang, I.-Y., and Budzynski, A. Z., *J. Chromatogr.*, 502, 359, 1990.)

The complex pattern of fibrinogenolytic activity observed in the fractions from IEC was a confounding observation. The strongest fibrinogenolysis appeared to be in fractions 38 to 45, with the activity in fractions 34 to 37 being slightly less, and traces of activity appearing even in the column wash (fractions 46 to 60). Yet, as is demonstrated in the next section, following RPLC almost all of the activity was observed in peak I. Many explanations were possible. For example, we considered the possibility that an inhibitor might coelute with peak I but not peak III; following RPLC, the inhibitor might then be removed, making the activity of peak I manifest. Another explanation would be the existence of a second protease in peak III; incapable of degrading fibrinogen to a lower molecular weight without synergy with hementin, the synergystic enzyme would be removed by RPLC. However, the simplest explanation, and the explanation that I prefer, is that either the digestion time of the peak III fractions was longer than that of the peak I fractions or that the enzyme:substrate ratio was higher, as might occur with a pipetting error.

5.5.6   *The third step in isolation: reversed phase chromatography*

Because it was not clear which of the IEC fractions 31 to 47 contained the activity, it was decided to divide the material into three fractions, labeled I, II, and III in Figure 26, and proceed with purification. Despite the tendency of RPLC to denature enzymes, sometimes irreversibly, microscale experiments on a Vydac® C-4 column indicated that activity could be recovered from a TFA-water-isopropanol mobile phase system. Therefore, fractions I and III were separately chromatographed, fractions collected, and aliquots assayed for fibrinogenolytic activity. The fibrinogenolytic activity correlated with the peak labeled IB in Figure 27, panel A. Based on the area of peak IB, it was estimated that IEC fraction III had only 10% of the activity. There were two surprises. First, RPLC peak IA (at about 34 minutes) was virtually invisible on the gel. Apparently, this was a low-molecular-weight peptide, barely visible at the dye front of the silver-stained gel. The sequence was found to be EVYTNYASFL. The second surprise was the 80-kD band in fraction 41. This was an enzymatically inactive component of sequence S(D)TGEEGA(K)RDV. This 80-kD component would coelute with hementin on SDS-PAGE or size exclusion chromatography. RPLC may be one of the few ways to separate that particular impurity.

Impurity profiling of the column eluate again provided compelling evidence that the active fraction was a protein of molecular weight about 80 kD. As shown in the figure, the activity is centered in RPLC fraction 38, as is the 80-kD protein. A densitometric scan of that band, previously unpublished, is shown in Figure 28. The densitometric scan makes clear what the photograph does not: the isolate contained essentially no 120-kD material. The sole significant contaminants of the isolate were at slightly lower molecular weight (44 kD). Based on the densitometrogram, the purity was estimated to be about 90%.

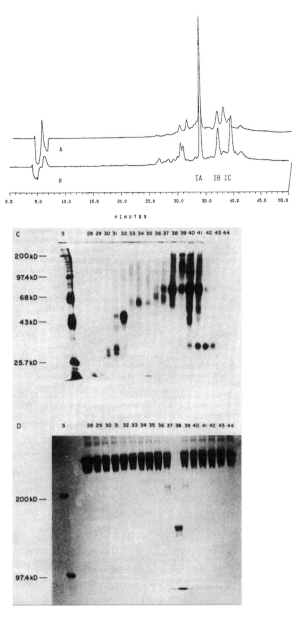

Figure 27 RPLC fractionation of IEC-purified hementin. Peaks I and III from Figure 26 were separately applied to a Vydac® C₄ reversed phase column, eluted at 0.75 ml/min from 5–70% 2-propanol with a gradient of 0.1% aqueous TFA. The peak at 34–35 min is designated IA, the peak at 38 min IB (hementin), and the peak at 40–41 min IC. Detection at 220 nm of **(A)** IEC peak III and **(B)** IEC peak I. **(C)** Nonreducing SDS-PAGE (10% acrylamide, silver stain) of fractions from RPLC separation. **(D)** SDS-PAGE pattern of fibrinogen + column eluate showing digestion of fibrinogen in fraction 38. (Reproduced with permission of Elsevier Science from Swadesh, J. K., Huang, I.-Y., and Budzynski, A. Z., *J. Chromatogr.*, 502, 359, 1990.)

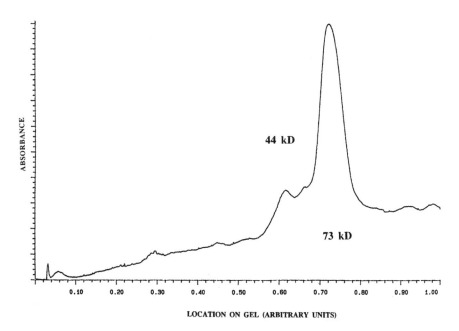

Figure 28 Densitometrogram of purified hementin. Lane 38 (active hementin) from the gel in Figure 26 (IB) was scanned by densitometry on an LKB 2202 Ultrascan laser densitometer. The large peak represents hementin. The smaller peak is an impurity of molecular weight 44 kD.

Even though the weight purity was 90%, low-molecular-weight materials were evident in the Edman sequence. The hementin sequence was TTLTE-PEPDL-D?-Y?-D?, where the question marks indicate tentative, previously unpublished assignments. The EVYTNYASFL sequence predominant after ion exchange was still seen in the RPLC isolate. Another sequence, perhaps L(W)XXPAEF, was also observed. The EVYTNYASFL sequence was identified as the main component of RPLC peak IA. The L(W)XXPAEF sequence probably represents the 44-kD contaminant in Figure 28. We also performed SDS-PAGE under reducing and nonreducing conditions of RPLC fraction IB and verified the molecular weight assignments of 73 kD (nonreducing conditions) and 82 kD (reducing conditions). A joint IEC-RPLC purification factor can be calculated from the area fraction on IEC (about 3.5%) and RPLC (about 15%) as $1/(0.035 \times 0.15) = 190$. Based on the purification factor, the concentration of hementin in the leech salivary extract can be estimated to be about 0.5%.

5.5.7   Activity testing

As shown in Figure 27, panel C, almost all of the activity was concentrated in RPLC fraction 38. The amount of hementin available was so small that the concentration could only be estimated by sequencer data. As has been pointed out, sequencer data tends to underestimate concentrations since proteins wash out or are incompleted cleaved. One may, however, upper- and lower-bound the concentration. On the ABI sequencer, correcting for volume used, about 30 pmol of hementin were observed in the RPLC IB peak; on the Beckman sequencer, about 60 pmol. Since the efficiency of the ABI sequencer is about 50% and that of the Beckman about 80%, an upper bound for the amount present in the sample would be 75 pmol (6 µg), while an absolute lower bound would be 30 pmol (2.5 µg). Using these approximations and defining activity as conversion of fragments to size D_{hem} or smaller, the activity is no less than 20 µmol fibrinogen per min per mg hementin, and no greater than 70 µmol per mg per min. The specific activity of the starting material was estimated to be no greater than 0.1 µmol fibrinogen per min per mg hementin. These values represented purification factors in the vicinity of 200- to 700-fold as calculated from the increase in specific activity. This value is close to the joint purification factor of 190 observed on IEC-RPLC.

5.5.8   The final step: sample accounting

In microscale processes, analysis consumes so much of the sample that simple calculation of yield is of little interest. Also, co-elution of impurities with the analyte of interest may make direct measurement impossible. Only after completing the multi-step purification can a chromatographic analytical method be devised. Retrospective accounting for the analyte of interest is important, helping to indicate which process steps need improvement. Even very approximate figures are useful. Because it was shown on RPLC that about 60% of the 80 kD band was an inactive species unrelated to hementin, the values given in Table 3 are corrected to compensate for the co-migrating band. A correction was also be made for known sample consumption. Table 3 shows the measured and consumption-corrected values for hementin at each stage of the process. The uncorrected values are in parentheses.

These estimates, though approximate, established that — excluding sample consumption for testing — the yield could not have been less than 20% and perhaps as much as 50%. Since another 10 µg of hementin were known to have been lost into the filtrate of the initial ultrafiltration and another 5 to 10 µg into IEC peaks II and III, no less than 40% and up to 80% of the material was accounted for.

Table 3   Hementin Recovery

	Method of Measurement				
	SDS-PAGE (dye binding)		280 nm peak area		Sequencer
Extract	—	(72 µg)	—	(84 µg)	—	—
IEC peak I	67 µg	(51 µg)	—	—	—	—
RPLC peak IB	—	—	—	—	15–37 µg	(2.5–6 µg)

5.5.9   The role of ion exchange in hementin purification

The purification of hementin was a typical, though difficult, protein purification problem. Ion exchange offered a high capacity first step which reduced the mass of contaminants to the level that a low-capacity method, RPLC, could be used. Also, because the goal of the project was to recover an active protein, IEC was ideal, since IEC conditions were nondenaturing. The principal drawback of IEC was that the major contaminant eluted after the peak of interest. Therefore, on attempting to scale up the separation, resolution was partly lost. Ion exchange also had the drawback of having relatively low resolving power. Ion exchange was unable to resolve an impurity of the same molecular weight as hementin.

5.5.10   Conclusion

The goals of the project were to confirm the molecular weight of hementin, determine the N-terminal amino acid sequence, and provide sufficient purified protein for biochemical studies of the fibrinolytic activity. These goals were all attained. Many of the issues that become important in devising a scalable process were identified, particularly

- Measuring the component of interest in a complex mixture
- Devising a quantitative and meaningful activity test
- Assessing stability
- Identifying and sequencing those contaminants that were particularly difficult to separate from the species of interest

I hope that the reader will take from this case history a sense of the philosophy behind industrial science. Industrial science is centered around generating information for the purpose of making decisions. For the industrial analyst, there are three cornerstones to participating in that process: understanding the mechanisms of separation and detection, intensive review of data to extract as much information as possible, and continual re-examination of one's working assumptions. The industrial analyst is often in the position of being required to make the correct interpretation even when he or she would wish for the opportunity to try other experiments. To find the middle road between doing insufficient experimental work and repeating experiments past the point that repetition serves to further illuminate is one of the most difficult aspects of working in industry. It is my hope that this case history will help those entering industry to understand how one may navigate that middle road.

5.6   Acknowledgments

Thanks are due to Drs. Cynthia Randall and Kálmán Benedek for assistance in preparation of this chapter, to Vivian Arrizon-Lopez and Robert Slocum of Beckman Instruments for locating the amino acid analysis on the 160B shown in Figure 7, and to S. Tinsley Preston, III, of Preston Publications for having supplied an original copy of Figure 11.

References

1. Adams, B. A. and Holmes, E. L., Absorptive properties of synthetic resins. Part I., *J. Soc. Chem. Ind. Trans. Comm.*, 54, 1, 1935.
2. Small, H., *Ion Chromatography*, Plenum Press, New York, 1989.
3. Horvath, C. G., Preiss, B. A., and Lipsky, S. R., Fast liquid chromatography: an investigation of operating parameters and the separation of nucleotides on pellicular ion exchangers, *Anal Chem.*, 39, 1422, 1967.
4. Hamilton, P. B., Ion exchange chromatography of amino acids. A single column, high resolving, fully automatic procedure, *Anal. Chem.*, 35, 2055, 1963.
5. Benson, Jr., J. V. and Patterson, J. A., Accelerated automatic chromatographic analysis of amino acids on a spherical resin, *Anal. Chem.*, 37, 1108, 1965.
6. Kirkland, J. J., Techniques for high-performance liquid-liquid and ion exchange chromatography with controlled surface porosity column packings, *J. Chromatogr. Sci.*, 7, 361, 1969.
7. Schmuckler, G., High-performance liquid ion-exchange chromatography, *J. Liq. Chromatogr.*, 10, 1887, 1987.
8. Sellergren, B. and Shea, K. J., Chiral ion-exchange chromatography. Correlation between solute retention and a theoretical ion-exchange model using imprinted polymers, *J. Chromatogr. A*, 654, 17, 1993.
9. Fritz, J. S., Principles and applications of ion-exclusion chromatography, *J. Chromatogr.*, 546, 111, 1991.
10. Yamamoto, A., Hayakawa, K., Matsunaga, A., Mizukami, E., and Miyazaki, M., Retention model of multiple eluent ion chromatography. *A priori* estimations of analyte capacity factor and peak intensity, *J. Chromatogr.*, 627, 17, 1992.
11. Law, B. and Weir, S., Quantitative structure-retention relationships for secondary interactions in cation-exchange liquid chromatography, *J. Chromatogr. A*, 657, 17, 1993.
12. Jandera, P. and Churáček, J., Gradient elution in liquid chromatography. I. The influence of the composition of the mobile phase on the capacity ratio (retention volume, band width, and resolution) in isocratic elution — theoretical considerations, *J. Chromatogr.*, 91, 207, 1974.
13. Gjerde, D. T., Schmuckler, G., and Fritz, J. S., Anion chromatography with low-conductivity eluents. II, *J. Chromatogr.*, 187, 25, 1980.
14. Renn, C.N. and Synovec, R.E., Refractive index gradient detector of biopolymers separated by high-temperature liquid chromatography, *J. Chromatogr.*, 536, 289, 1991.
15. Kissinger, P. T. and Radzik, D. M., Liquid chromatography/electrochemistry in pharmaceutical analysis, in *HPLC in the Pharmaceutical Industry*, Fong, G. W. and Lam, S. K., Eds., Marcel Dekker, New York, 1991, chap. 4.
16. Rocklin, R. D., Electrochemical detection, in *A Practical Guide to HPLC Detection*, Parriott, D., Ed., Academic Press, San Diego, 1993, chap. 6.
17. Wang, J., Electrochemical detection for liquid chromatography, in *HPLC Detection: Newer Methods*, Patonay, G., Ed.,VCH Publishers, New York, 1992, chap. 5.

18. Walton, H. F. and Rocklin, R. D., *Ion Exchange in Analytical Chemistry*, CRC Press, Boca Raton, FL, 1990, chap. 4.
19. Kissinger, P. T., Biomedical applications of liquid chromatography-electrochemistry, *J. Chromatogr.*, 488, 31, 1989.
20. Rothbart, H. L., Ion exchange separation processes, in *An Introduction to Separation Science*, Karger, B. L., Snyder, L. R., and Horváth, Cs., Eds., Wiley-Interscience, New York, 1973, chap. 12.
21. Strazhesko, D. N., Strelko, V. B., Belyakov, V. M., and Rubanik, S.C., Mechanism of cation exchange on silica gels, *J. Chromatogr.*, 102, 191, 1974.
22. Chicz, R. M., Shi, Z., and Regnier, F. E., Preparation and evaluation of inorganic anion-exchange sorbents not based on silica, *J. Chromatogr.*, 359, 121, 1986.
23. Huhn, G. and Müller, H., Polymer-coated cation exchangers in high-performance ion chromatography: preparation and application, *J. Chromatogr.*, 640, 57, 1993.
24. Kutsuna, H., Ohtsu, Y., and Yamaguchi, M., Synthesis and characterization of highly stable polymer-coated aminosilica packing material for high-performance liquid chromatography, *J. Chromatogr.*, 635, 187, 1993.
25. Small, H., Twenty years of ion chromatography, *J. Chromatogr.*, 546, 3, 1991.
26. Arshady, R., Beaded polymer supports and gels. I. Manufacturing techniques, *J. Chromatogr.*, 586, 181, 1991.
27. Kim, H.-B., Hayashi, M., Nakatani, K., Kitamura, N., Sasaki, K., Hotta, J.-I., and Masuhara, H., *In situ* measurements of ion exchange processes in single polymer particles: laser trapping microspectroscopy and confocal fluorescence microspectroscopy, *Anal. Chem.*, 68, 409, 1996.
28. Wu, S.-L., Benedek, K., and Karger, B. L., Thermal behavior of proteins in high-performance hydrophobic-interaction chromatography, *J. Chromatogr.*, 359, 3, 1986.
29. Velayudhan, A. and Horváth, Cs., On the stoichiometric model of electrostatic interaction chromatography for biopolymers, *J. Chromatogr.*, 367, 160, 1986.
30. Rounds, M. A. and Regnier, F. E., Evaluation of a retention model for high-performance ion-exchange chromatography using two different displacing salts, *J. Chromatogr.*, 283, 37, 1984.
31. Drager, R. R. and Regnier, F. E., Application of the stoichiometric displacement model of retention to anion-exchange chromatography of nucleic acids, *J. Chromatogr.*, 359, 147, 1986.
32. Ståhlberg, J., Jönsson, B., and Horváth, Cs., Theory for electrostatic interaction chromatography of proteins, *Anal. Chem.*, 63, 1867, 1991.
33. Manning, J., Limiting laws and counterion condensation in polyelectrolyte solutions I. Colligative properties, *J. Chem. Phys.*, 51, 924, 1969.
34. Mazsaroff, I., Vàrady, L., Mouchawar, G. A., and Regnier, F. E., Thermodynamic model for electrostatic-interaction chromatography of proteins, *J. Chromatogr.*, 499, 63, 1990.
35. Majors, R. E., New chromatography columns and accessories at the 1992 Pittsburgh conference, Part I, *LC-GC*, 10(3), 188, 1992.
36. Majors, R. E., New chromatography columns and accessories at the 1994 Pittsburgh conference, Part II, *LC-GC*, 12(4), 278, 1994.
37. Afeyan, N. B., Fulton, S. P., and Regnier, F. E., Perfusion chromatography packings for proteins and peptides, *J. Chromatogr.*, 544, 267, 1991.

38. Yang, M.-H., Chang, K.-C., and Lin, J.-Y., Multifunctional ion-exchange stationary phases for high-performance liquid chromatography, *J. Chromatogr. A*, 722, 87, 1996.
39. Saari-Nordhaus, R. and Anderson, Jr., J. M., Ion chromatographic analysis of anions using a solid-phase chemical suppressor, *Am. Lab.*, 26 (January), 28C, 1994.
40. Siriraks, A. and Stillian, J., Determination of anions and cations in concentrated bases and acids by ion chromatography, *J. Chromatogr.*, 640, 151, 1993.
41. Jackson, P. E., Romano, J. P., and Wildman, B. J., Studies on system performance and sensitivity in ion chromatography, *J. Chromatogr. A*, 706, 3, 1995.
42. Pohl, C. A. and Johnson, E. L., Ion chromatography — the state of the art, *J. Chromatogr. Sci.*, 18, 442, 1980.
43. Polite, L. N., McNair, H. M., and Rocklin, R. D., Linearity in ion chromatography, *J. Liq. Chromatogr.*, 10, 829, 1987.
44. Sjögren. A. and Dasgupta, P. K., Two-dimensional conductimetric detection in ion chromatography. Analyte identification, quantitation of very weak acid anions, and universal calibration, *Anal. Chem.*, 67, 2110, 1995.
45. Gunasingham, H. and Fleet, B., Hydrodynamic voltammetry in continuous-flow analysis, in *Electroanalytical Chemistry: A Series of Advances*, Vol. 16, Bard, A. J., Ed., Marcel Dekker, New York, 1989, 89.
46. Brunt, K., Electrochemical detectors for high-performance liquid chromatography and flow analysis systems, *Trace Analysis*, Vol. 1, Lawrence, J. F., Ed., Academic Press, New York, 1981, 47–120.
47. Rocklin, R. D., Detection in ion chromatography, *J. Chromatogr.*, 546, 175, 1991.
48. Han, S. H., Lee, K. S., Cha, G. S., Liu, D., and Trojanowicz, M., Potentiometric detection in ion chromatography using multi-ionophore membrane electrodes, *J. Chromatogr.*, 648, 283, 1993.
49. Trojánek, A. and De Jong, H. G., Fast scan A. C. voltammetry for better resolution of chromatographically overlapping peaks, *Anal. Chim. Acta*, 141, 115, 1982.
50. St. Claire, III, R. L. and Jorgenson, J. W., Characterization of an on-column electrochemical detector for open-tubular liquid chromatography, *J. Chromatogr. Sci.*, 23, 186, 1985.
51. Kafil, J. B., Cheng, H.-Y., and Last, T. A., Quantitation of nucleic acids at the picogram level using high-performance liquid chromatography with electrochemical detection, *Anal. Chem.*, 58, 285, 1986.
52. Bollet, C., Oliva, P., and Caude, M., Partial electrolysis electrochemical detector in high-performance liquid chromatography, *J. Chromatogr.*, 149, 625, 1977.
53. Mayer, W. J. and Greenberg, M. S., Determination of some carbamate pesticides by high-performance liquid chromatography with electrochemical detection, *J. Chromatogr.*, 208, 295, 1981.
54. Elrod Jr., L., Schmit, J. L., and Morley, J. A., Determination of isoproteronol sulfate on surfaces using high-performance liquid chromatography with electrochemical detection, *J. Chromatogr. A*, 723, 235, 1996.
55. Buchberger, W., Trends in the combination of high-performance liquid chromatography and electroanalytical methids, *Chromatographia*, 30, 577, 1990.
56. Krull, I. S., Ding, X.D., Selavka, C., and Nelson, R., Electrochemical detection in HPLC, *LC Mag.*, 2(3), 214, 1984.
57. Johnson, D. C., Dobberpuhl, D., Roberts, R., and Vandeberg, P., Pulsed amperometric detection of carbohydrates, amines and sulfur species in ion chromatography — the current state of research, *J. Chromatogr.*, 640, 79, 1993.

58. Weber, S. G., The dependence of current on flow rate in thin-layer electrochemical detectors used in liquid chromatography. A clarification, *J. Electroanal. Chem.*, 145, 1, 1983.

59. Roe, D. K., Comparison of amperometric electrochemical detectors for HPLC through a figure of merit, *Anal. Letts.*, 16, 613, 1983.

60. Kordorouba, V. and Pelletier, M., Ion chromatography using an electrochemical detector: response to non-electroactive anions, *J. Liq. Chromatogr.*, 11, 2271, 1988.

61. Fenn, R. J., Siggia, S., and Curran, D. J., Liquid chromatography detector based on single and twin electrode thin-layer electrochemistry: application to the determination of catecholamines in blood plasma, *Anal. Chem.*, 50, 1067, 1978.

62. Chen, Q., Wang, J., Rayson, G., Tian, B., and Lin, Y., Sensor array for carbohydrates and amino acids based on electrocatalytic modified electrodes, *Anal. Chem.*, 65, 251, 1993.

63. Hoogvliet, J. C., Reijn, J. M., and van Bennekom, W. P., Multichannel amperometric detection system for liquid chromatography and flow inhection analysis, *Anal. Chem.*, 63, 2418, 1991.

64. Thannhauser, T. W., McWherter, C. A., and Scheraga, H. A., Peptide mapping of bovine pancreatic ribonuclease A by reverse-phase high-performance liquid chromatography. II. A two-dimensional technique for determination of disulfide pairings using a continuous-flow disulfide detection system, *Anal. Biochem.*, 149, 322, 1985.

65. Henderson, S. K. and Henderson, D. E., Enhanced UV detection on sugar phosphates by addition of a metal complex to the HPLC mobile phase, *J. Chromatogr. Sci.*, 23, 222 (1985)

66. Yamauchi, S., Nakai, C., Nimura, N., Kinoshita, T., and Hanai, T., Highly sensitive detection of nonreducing carbohydrates by liquid chromatography, *Analyst*, 118, 769, 1993.

67. Yamauchi, S., Nakai, C., Nimura, N., Kinoshita, T., and Hanai, T., Development of a highly sensitive fluorescence reaction detection system for liquid chromatographic analysis of reducing carbohydrates, *Analyst*, 118, 773, 1993.

68. Blau, K. and Dodge, C.S., Continuous nephelometric monitoring of glycosaminoglycans eluted from chromatographic columns, *Anal. Biochem.*, 58, 650, 1974.

69. Dollinger, G., Cunico, B., Kunitani, M., Johnson, D., and Jones, R., Practical on-line determination of biopolymer molecular weights by high-performance liquid chromatography with classical light-scattering detection, *J. Chromatogr.*, 592, 215, 1992.

70. Hagel, L., Interferometric concentration determination of dextran after gel chromatography, *Anal. Chem.*, 50, 569, 1978.

71. Petrie, L. M., Jakel, M. E., Brandvig, R. L., and Kroenig, J. G., Ion chromatography of sulfur dioxide, sulfate ion, and dithionate ion in aqueous mineral leachates, *Anal. Chem.*, 65, 952, 1993.

72. Buchholz, A. E., Verplough, C. I., and Smith, J. L., A method for the simultaneous measurement of less than a part-per-million of chloride, nitrate, and sulfate in aqueous samples by nonsuppressed ion chromatography, *J. Chromatogr. Sci.*, 20, 499, 1982.

73. Kadnar, R. and Rieder, J., Determination of anions in oilfield waters by ion chromatography, *J. Chromatogr. A*, 706, 301, 1995.

74. Kadnar, R. and Rieder, J., Determination of anions in amine solutions for sour gas treatment, *J. Chromatogr. A*, 706, 339, 1995

75. Hu, W. and Haraguchi, H., New approach for controlling the partitioning of analyte ions in ion chromatography witha water mobile phase, *J. Chromatogr. A*, 723, 251, 1996.

76. Rocklin, R. D., Pohl, C. A., and Schibler, J. A., Gradient elution in ion chromatography, *J. Chromatogr.*, 411, 107, 1987.

77. Dabek-Zlotorzynska, E. and Dlouhy, J. F., Simultaneous determination of alkali, alkaline-earth metal cations and ammonium ion environmental samples by gradient ion chromatography, *J. Chromatogr.*, 638, 35, 1993.

78. Klancke, J. W., Determination of monovalent and divalent cations and chloride in the carbacephalosporin loracarbef by ion chromatography, *J. Chromatogr.*, 637, 63, 1993.

79. Smith, R. E., Yourtee, D., Bean, T., and McQuarrie, R. A., Ion chromatography on a new moderate capacity anion exchange column, *J. Chromatogr. Sci.*, 31, 366, 1993.

80. Amey, M. D. H. and Bridle, D. H., Application and development of ion chromatography for the analysis of transition metal cations in the primary coolants of light water reactors, *J. Chromatogr.*, 640, 323, 1993.

81. Jensen, D., Weiss, J., Rey, M. A., and Pohl, C. A., Novel weak acid cation-exchange column, *J. Chromatogr.*, 640, 65, 1993.

82. Nair, L. M., Saari-Nordhaus, R., and Anderson, Jr., J. M., Simultaneous separation of alkali and alkaline-earth cations on polybudaiene-maleic acid-coated stationary phase by mineral acid eluents, *J. Chromatogr.*, 640, 41, 1993.

83. Lee, D. P., A new anion exchange phase for ion chromatography, *J. Chromatogr. Sci.*, 22, 327, 1984.

84. Xianren, Q., Chong-yu, X., and Baeyens, W., Computer-assisted predictions of resolution, peak height and retention time for the separation of inorganic anions by ion chromatography, *J. Chromatogr.*, 640, 3, 1993.

85. Benson, J. R. and Woo, D. J., Polymeric columns for liquid chromatography, *J. Chromatogr. Sci.*, 22, 386, 1984.

86. Brenman, L. and Schmuckler, G., Quantitative determination of anions in fertilizers using ion chromatography, *LC-GC*, 11(4), 298, 1993.

87. Fujiwara, M., Matsushita, T., Kobayashi, T., Yamashoji, Y., and Tanaka, M., Preparation of an anion-exchange resin with quaternary phosphonium chloride and its adsorption behavior for noble metal ions, *Anal. Chim. Acta*, 274, 293, 1993.

88. Michigami, Y., Kuroda, Y., Ueda, K., and Yamamoto, Y., Determination of urinary fluoride by ion chromatography, *Anal. Chim. Acta*, 274, 299, 1993.

89. Cassidy, R. M. and Elchuk, S., Dynamic and fixed-site ion-exchange columns with conductimetric detection for the separation of inorganic ions, *J. Chrom. Sci.*, 21, 454, 1983.

90. Cassidy, R. M. and Sun, L., Optimization of the anion-exchange separation of metal-oxalate complexes, *J. Chromatogr. A*, 654, 105, 1993.

91. Janoš, P., Separation of some metals as their anionic oxalate complexes by reversed-phase ion-interaction chromatography, *J. Chromatogr.*, 635, 257, 1993.

92. Buldini, P. L., Sharma, J. L., and Mevoli, A., Determination of inorganic ions in carboxylic acids by ion chromatography, *J. Chromatogr. A*, 654, 123, 1993.

93. Buldini, P. L., Sharma, J. L., and Sharma, S., Ion-chromatographic determination of inorganic anions and cations in some reagents used in the electronics industry, *J. Chromatogr. A*, 654, 113, 1993.

94. Smith, R. E., Ion chromatography in the manufacture of multilayer circuit boards, *J. Chromatogr.*, 546, 369, 1991.

95. Pereira, C. F., Application of ion chromatography to the determination of inorganic ions in foodstuffs, *J. Chromatogr.*, 624, 457, 1992.

96. Frenzel, W., Schepers, D., and Schulze, G., Simultaneous ion chromatographic determination of anions and cations by series conductivity and flame photometric detection, *Anal. Chim. Acta*, 277, 103, 1993.

97. Krokhin, O. V., Smolenkov, A. D., Svintsova, N. V., Obrezkov, O. N., and Shpigun, O. A., Modified silica as a stationary phase for ion chromatography, *J. Chromatogr. A*, 706, 93, 1995.

98. Läubli, M. W. and Kampus, B., Cation analysis on a new poly(butadiene-maleic acid)-based column, *J. Chromatogr. A*, 706, 99, 1995.

99. Bettmer, J., Cammann, K., and Robecke, M., Determination of organic ionic lead and mercury species with high-performance liquid chromatography using sulphur reagents, *J. Chromatogr.*, 654, 177, 1993.

100. Janoš, P., Determination of stability constants of metal complexes from ion chromatographic measurements, *J. Chromatogr.*, 641, 229, 1993.

101. Dasgupta, P. K., Ion chromatography: the state of the art, *Anal. Chem.*, 64, 775A, 1992.

102. Marchetto, A., Mosello, R., Tartari, G. A., Muntau, H., Bianchi, M., Geiss, H., Serrini, G. and Lanza, G. S., Precision of ion chromatographic analyses compared with that of other analytical techniques through intercomparison exercises, *J. Chromatogr. A*, 706, 13, 1995.

103. Rowland, A. P., Woods, C., and Kennedy, V. H., Control of errors in anion chromatography applied to environmental research, *J. Chromatogr. A*, 706, 229, 1995.

104. Debowski, J. K. and Wilde, N. D., Determination of isocyanuric acid, ammeline, and melamine in crude isocyanuric acid by ion chromatography, *J. Chromatogr.*, 639, 338, 1993.

105. Cox, G. B., Loscombe, C. R., Slucutt, M. J., Sugden, K., and Upfield, J. A., The preparation, properties and some applications of bonded ion-exchange packings based on microparticulate silica gel for high-performance liquid chromatography, *J. Chromatogr.*, 117, 269, 1976.

106. Green, J. C. and Donaldson, R. M., The role played by ion chromatography in the assessment of amines for two-phase erosion corrosion control in nuclear electric's steam-water circuits, *J. Chromatogr.*, 640, 303, 1993.

107. Krol, J., Alden, P. G., Morawski, J., and Jackson, P. E., Ion chromatography of alkylamines and alkanolamines using conductivity detection, *J. Chromatogr.*, 626, 165, 1992.

108. Perry, T. L. and Schroeder, W. A., The occurrence of amines in human urine: determination by combined ion exchange and paper chromatography, *J. Chromatogr.*, 12, 358, 1963.

109. Tanaka, K., Ishizuka, T., and Sunahara, H., Chromatography of aliphatic amines on an anion-exchange resin, *J. Chromatogr.*, 172, 484, 1979.

110. Dumont, P. J., Fritz, J. S., and Schmidt, L. W., Cation exchange chromatography in nonaqueous solvents, *J. Chromatogr. A*, 706, 109, 1995.

111. Adler, H., Margoshes, M., Snyder, L. R., and Spitzer, C., Rapid chromatographic method to determine polyamines in urine and whole blood, *J. Chromatogr.*, 143, 125, 1977.

112. Villanueva, V. R. and Adlakha, R., Automated analysis of common basic amino acids, mono-, di-, and polyamine phenolic amines, and indoleamines in crude biological samples, *Anal. Biochem.*, 91, 264, 1978.

113. Gehrke, C. W., Kuo, K. C., Ellis, R. L., and Waalkes, T.P., Polyamines — an improved automated ion-exchange method, *J. Chromatogr.*, 143, 345, 1977.

114. Watanabe, S., Sato, S., Nagase, S., Tomita, M., Saito, T., and Ishizu, H., Automated quantitation of polyamines by improved cation-exchange high-performance liquid chromatography using a pump equipped with a plunger washing system, *J. Liq. Chromatogr.*, 16, 619, 1993.

115. Moore, S. and Stein, W. H., Photometric ninhydrin method for use in the chromatography of amino acids, *J. Biol. Chem.*, 176, 367, 1954.

116. Hirs, C. H. W., Stein, W. H., and Moore, S., The amino acid composition of ribonuclease, *J. Biol. Chem.*, 211, 941, 1954.

117. Benedek, K. and Swadesh, J. K., HPLC of proteins and peptides in the pharmaceutical industry, in *HPLC in the Pharmaceutical Industry*, Vol. 47, Fong, G. W. and Lam, S. K., Eds., Marcel Dekker, New York, 1991, chap. 11.

118. Stone, K. L. and Williams, K. R., High-performance liquid chromatographic peptide mapping and amino acid analysis in the subnanomole range, *J. Chromatogr.*, 359, 203, 1986.

119. Beecher, G. R., Design and assembly of an inexpensive, automated, microbore amino acid analyzer: separation and quantitation of amino acids in physiological fluids, *Adv. Exp. Med. Biol.*, 105, 827, 1978.

120. Udenfriend, S., Stein, S., Böhlen, P., Dairman, W., Leimgruber, W., and Weigele, M., Fluorescamine: a reagent for assay of amino acids, peptides, proteins, and primary amines in the picomole range, *Science*, 178, 871, 1972.

121. Benson, J. R. and Hare, P. E., *o*-Phthalaldehyde: fluorogenic detection of primary amines in the picomole range: comparison with fluorescamine and ninhydrin, *Proc. Natl. Acad. Sci. U.S.A.*, 72, 619, 1975.

122. Dong, M. W., Gant, J. R., and Benson, J. R., Characterization and performance of a high speed amino acid analysis column, *Am. Biotech. Lab.*, 3, 34, 1985.

123. Hadžija, O. and Keglević, D., Simple method for the separation of amino acids, amino sugars, and amino alcohols related to the peptidoglycan components on a standard amino acid analyzer, *J. Chromatogr.*, 138, 458, 1977.

124. Gorbics, L., Urge, L., Otvos-Papp, E., and Otvos, Jr., L., Determination of amino sugars in synthetic glycopeptides during the conditions of amino acid analysis utilizing precolumn derivatization and high-performance liquid chromatographic analysis, *J. Chromatogr.*, 637, 43, 1993.

125. Yamada, H., Moriya, H., and Tsugita, A., Development of an acid hydrolysis method with high recoveries of tryptophan and cysteine for microquantities of protein, *Anal. Biochem.*, 198, 1, 1991.

126. Hartkopf, J. and Erbersdobler, H. F., Stability of furosine during ion-exchange chromatography in comparison with reversed-phase high-performance liquid chromatography, *J. Chromatogr.*, 635, 151, 1993.

127. Kedenburg, C.-P., A lithium buffer system for accelerated single-column amino acid analysis in physiological fluids, *Anal. Biochem.*, 40, 35, 1971.

128. Murayama, K. and Shindo, N., Recommended method for the analysis of amino acids in biological materials, *J. Chromatogr.*, 143, 137, 1977.

129. Ishimitsu, S., Fujimoto, S., and Ohara, A., Quantitative analysis of the isomers of hydroxyphenylalanine by ion-exchange chromatography, *Chem. Pharm Bull.*, 24, 2556, 1976.

130. Houpert, Y., Tarallo, P., and Siest, G., Amino acid analysis by ion-exchange chromatography using a lithium elution gradient. Influence of methanol concentration and sample pH, *J. Chromatogr.*, 115, 33, 1975.
131. Chin, C. C. Q., Ion-exchange chromatography of some amino acid derivatives found in proteins, *Meth. Enzymol.*, 106, 17, 1984.
132. Atkin, G. E., and Ferdinand, W., Accelerated amino acid analysis: studies on the use of lithium citrate buffers and the effect of *n*-propanol, in the analysis of physiological fluids and protein hydrolyzates, *Anal. Biochem.*, 38, 313, 1970.
133. Murren, C., Stelling, D., and Felstead, G., An improved buffer system for use in single-column gradient-elution ion-exchange chromatography of amino acids, *J. Chromatogr.*, 115, 236, 1975.
134. Lou, M. F., A split stream ion exchange chromatographic method for isolating amino acids and peptides, *Anal. Biochem.*, 55, 51, 1973.
135. Rahm, J., Analytical separation of amino acids on a cation-exchange resin cross-linked with *m*-divinylbenzene, *J. Chromatogr.*, 115, 455, 1975.
136. Vandeberg, P. J. and Johnson, D. C., Pulsed electrochemical detection of cysteine, cystine, methionine, and glutathione at gold electrodes following their separation by liquid chromatography, *Anal. Chem.*, 65, 2713, 1993.
137. Fung, Y.-S. and Mo, S.-Y., Determination of amino acids and proteins by dual-electrode detection in a flow system, *Anal. Chem.*, 67, 1121, 1995.
138. Skelly, N. E., Gradient elution in the separation of chlorophenols by ion exchange, *Anal. Chem.*, 33, 271, 1961.
139. Small, H., Stevens, T. S., and Bauman, W. C., Novel ion exchange chromatographic method using conductimetric detection, *Anal. Chem.*, 47, 1801, 1975.
140. Mayer, W. J. and Greenberg, M. S., A comparison of differential pulse and D. C. amperometric detection modes for the liquid chromatographic determination of oxalic acid, *J. Chromatogr. Sci.*, 17, 614, 1979.
141. Kasai, Y., Tanimura, T., Tamura, Z., Tanimura, T., and Ozawa, Y., Automated determination of carboxylic acids by anion-exchange chromatography with specific color reaction, *Anal. Chem.*, 49, 655, 1977.
142. Goudie, A. J., and Rieman, III, W., Chromatographic separation and determination of fruit acids, *Anal. Chim. Acta*, 26, 419, 1962.
143. Hirayama, N., Maruo, M., and Kuwamoto, T., Determination of dissociation constants of aromatic carboxylic acids by ion chromatography, *J. Chromatogr.*, 639, 333, 1993.
144. Leubolt, R. and Klein, H., Determination of sulphite and ascorbic acid by high-performance liquid chromatography withe electrochemical detection, *J. Chromatogr.*, 640, 271, 1993.
145. Chiu, G., Analysis of mixtures of cyanovaleric and other carboxylic acids by liquid chromatography, *J. HRC&CC*, 9, 57, 1986.
146. Chiu, G., Separation of carboxylic acids by temperature programmed liquid chromatography, *J. HRC&CC*, 9, 410, 1986.
147. Grosjean, D. J., Van Neste, A., and Parmar, S. S., Analysis of atmospheric carboxylic acids using single column ion exclusion chromatography with ultraviolet detection, *J. Liq. Chromatogr.*, 12, 3007, 1989.
148. Stinson, E. A., Subers, M. H., Petty, J., and White, Jr., J. W., The composition of honey. V. Separation and identification of the organic acids, *Arch. Biochem. Biophys.*, 89, 6, 1960.

149. Jahangir, L. M. and Samuelson, O., Cation-exchange resins and non-ionic cross-linked polymers in analysis of aqueous solutions for organic contaminants, *Anal. Chim. Acta*, 100, 53, 1978.

150. Lange, H.W. and Hempel, K., Automated chromatography of aromatic acids, aldehydes, and alcohols with an amino acid analyzer, *J. Chromatogr.*, 59, 53, 1971.

151. Peldszus, S., Huck, P. M., and Andrews, S. A., Determination of short-chain aliphatic, oxo- and hydroxy acids in drinking water at low microgram per liter concentrations, *J. Chromatogr. A*, 723, 27, 1996.

152. Lee, H. K. and Hoffman, N. E., Effect of temperature on the retention of simple organic compounds in ion-exchange HPLC, *J. Chromatogr. Sci.*, 32, 97, 1994.

153. Feibush, B. and Santasania, C. T., Hydrophilic shielding of hydrophobic, cation- and anion-exchange phases for separation of small analytes: direct injection of biological fluids onto high performance chromatographic columns, *J. Chromatogr.*, 544, 41, 1991.

154. Vreeken, R. J., Ghijsen, R. T., Frei, R. W., de Jong, G. J., and Brinkman, U. A. Th., Coupling of ion-pair liquid chromatography and thermospray mass spectrometry via phase-system switching with a polymeric trapping column, *J. Chromatogr. A*, 654, 65, 1993.

155. Freifelder, D., *Molecular Biology*, 2nd ed., Jones and Bartlett, Boston, 1987, 60–64.

156. Floyd, T. R., Yu, L. W., and Hartwick, R. A., Use of diluted anion-exchange and hydrophobic properties in separating single-stranded oligodeoxyribonucleotides on mixed-ligand stationary phases, *Chromatographia*, 21, 402, 1986.

157. Floyd, T. R., Cicero, S.E., Fazio, S. D., Raglione, T. V., Hsu, S.-H., Winkle, S. A., and Hartwick, R. A., Mixed-mode hydrophobic ion exchange for the separation of oligonucleotides and DNA fragments using HPLC, *Anal. Biochem.*, 154, 570, 1986.

158. Basiuk, V. A. and Chuko, A. A., Selectivity of bonded stationary phases containing uracil derivatives for liquid chromatography of nucleic acid components, *J. Chromatogr. Sci.*, 31, 120, 1993.

159. Cohn, W. E., The anion-exchange separation of ribonucleotides, *J. Am. Chem. Soc.*, 72, 1471, 1950.

160. Wade, H. E., The fractionation of phosphate esters on ion-exchange resin by a new system of pH-gradient elution, *Biochem. J.*, 77, 534, 1960.

161. Pal, B. C., Novel application of a sugar-borate complexation for separation of ribo-, 2'-deoxyribo-, and arabinonucleosides on cation exchange resin, *J. Chromatogr.*, 148, 545, 1978.

162. Bates, M. and Avdalović, N., The value of Dowex®-50 in fractionation of nucleotides from acid-soluble pool. A caution, *Anal. Biochem.*, 61, 508, 1974.

163. Breter, H.-J., Seibert, G., and Zahn, R. K., Single-step separation of major and rare ribonucleosides and deoxyribonucleosides by high-performance liquid cation-exchange chromatography for the determination of the purity of nucleic acid preparations, *J. Chromatogr.*, 140, 251, 1977.

164. Floridi, A., Palmerini, C. A., and Fini, C., Simultaneous analysis of bases, nucleosides, and nucleoside mono- and polyphosphates by high-performance liquid chromatography, *J. Chromatogr.*, 138, 203, 1977.

165. Alonso, J., Nogues, M. V., and Cuchillo, C. M. Separation of RNA derivatives by high-performance anion-exchange liquid chromatography, *J. Liq. Chromatogr.*, 8, 299, 1985.

166. Alpert, A. J. and Regnier, F. E., Preparation of a porous microparticulate anion-exchange chromatography support for proteins, *J. Chromatogr.*, 185, 375, 1979.

167. Frenkel, K. and Klein, C. B., Methods used for analyses of "environmentally" damaged nucleic acids, *J. Chromatogr.*, 618, 289, 1993.

168. Köster, H. and Frank, R., Automated chromatography of protected oligonucleotides with high separation efficiency, *Chromatographia*, 9, 497, 1976.

169. Kothari, R. M., Some aspects of fractionation of DNA on an IR-120 Al³⁺. II. Effect of the physical state of DNA on chromatographic profiles, *J. Chromatogr.*, 53, 580, 1970.

170. Kothari, R. M., Some aspects of fractionation of DNA on an IR-120 Al³⁺. IV. Effect of protein association on the chromatographic profiles of DNA, *J. Chromatogr.*, 56, 151, 1971.

171. Kothari, R.M., Some aspects of fractionation of DNA on an IR-120 Al³⁺. VI. The effect of pH and temperature variation on the chromatographic profiles of DNA, *J. Chromatogr.*, 59, 194, 1971.

172. Shankar, V. and Joshi, P. N., Fractionation of RNA on a metal ion equilibrated cation exchanger. I. Chromatographic profiles of RNA on an Amberlite® IR-120 (Al⁺³) column, *J. Chromatogr.*, 90, 99, 1974.

173. Wulfson, A. N. and Yakimov, S. A., HPLC of nucleotides. II. General methods and their development for analysis and preparative separation: an approach to selectivity control, *J. HRC&CC*, 7, 442, 1984.

174. Singhal, R. P., Griffin, G. D., and Novelli, G. D., Separation of transfer ribonucleic acids on polystyrene anion exchangers, *Biochemistry*, 23, 5083, 1976.

175. Müller, W., Fractionation of DNA restriction fragments with ion-exchangers for high-performance liquid chromatography, *Eur. J. Biochem.*, 155, 203, 1986.

176. Ozaki, H., Wada, H., Takeuchi, T., Makino, K., Fukui, T., and Kato, Y., Behaviour of single-stranded oligodeoxyribonucleotides on a DEAE-5PW anion-exchange column, *J. Chromatogr.*, 322, 243, 1985.

177. McLaughlin, L. W. and Bischoff, R., Resolution of RNA using high-performance liquid chromatography, *J. Chromatogr.*, 418, 51, 1987.

178. Colpan, M. and Riesner, D., High-performance liquid chromatography of high-molecular-weight nucleic acids on the macroporous ion exchanger, Nucleogen®, *J. Chromatogr.*, 296, 339, 1984.

179. Coppella, S. J., Acheson, C. M., and Dhurjati, P., Isolation of high-molecular-weight nucleic acids for copy number analysis using high-performance liquid chromatography, *J. Chromatogr.*, 402, 189, 1987.

180. Drager, R. R. and Regnier, F. E., High-performance anion-exchange chromatography of oligonucleotides, *Anal. Biochem.*, 145, 47, 1985.

181. Garon, C. F. and Peterson, L. L., An improved method for the isolation of supercoiled DNA molecules using ion-exchange column chromatography, *Gene Anal. Technol.*, 4, 5, 1987.

182. Sawai, H., Analysis and purification of synthetic large oligodeoxyribonucleotides by HPLC on RPC-5 like resin, *Nucleic Acids Res.*, 17, 113, 1986.

183. Baba, Y., Prediction of the behavior of oligonucleotides in high-performance liquid chromatography and capillary electrophoresis, *J. Chromatogr.*, 618, 41, 1993.

184. Baba, Y., Fukuda, M., and Yoza, N., Computer-assisted retention prediction system for oligonucleotides in gradient anion-exchange chromatography, *J. Chromatogr.*, 458, 385, 1988.

185. Bourque, A. J. and Cohen, A. S., Quantitative analysis of phosphorothioate oligonucleotides in biological fluids using fast anion-exchange chromatography, *J. Chromatogr.*, 617, 43, 1993.

186. Murray, Jr., E. D. and Clarke, S., Synthetic peptide substrates for the erythrocyte protein carboxyl methyltransferase, *J. Biol. Chem.*, 259, 10722, 1984.

187. Bonnerjea, J., Oh, S., Hoare, M., and Dunnill, P., Protein purification: the right step at the right time, *BioTechnology*, 4, 954, 1986.

188. Siezen, R. J., Kaplan, E. D., and Anello, R. D., Superior resolution of γ-crystallins from microdissected eye lens by cation-exchange high-performance liquid chromatography, *Biochem. Biophys. Res. Comm.*, 127, 153, 1985.

189. Hatch, R. G., Very-high-speed anion-exchange chromatography of proteins using monodisperse nonporous polymer particles, *J. Chromatogr. Sci.*, 31, 469 1993.

190. Mikeš, O., Štrop, P., Hostomská, Z., Smrž, M., Slováková, S., and Čoupek, J., Ion-exchange derivatives of Spheron. V. Sulphate and sulpho derivatives, *J. Chromatogr.*, 301, 93, 1984.

191. Atassi, M. Z., Perlstein, M. T., Rosenblatt, M. C., and Rocek, P., Fully automated simple analytical peptide chromatography on the amino acid analyzer at the 10^{-8} mole level, *Anal. Biochem.*, 49, 164, 1972.

192. Hartley, R. W., Peptide analysis and preparation with an amino acid analyzer, *Anal. Biochem.*, 46, 676, 1972.

193. Dizdaroglu, M., Krutzsch, H. C., and Simic, M. G., Separation of angiotensins by high-performance liquid chromatography on a weak anion-exchange bonded phase, *Anal. Biochem.*, 123, 190, 1982.

194. Crimmins, D. L., Gorka, J., Thoma, R. S., and Schwartz, B. D., Peptide characterization with a sulfoethyl aspartamide column, *J. Chromatogr.*, 443, 63, 1988.

195. Bennett, H. P. J., Use of ion-exchange Sep-Pak® cartridges in the batch fractionation of pituitary peptides, *J. Chromatogr.*, 359, 383, 1986.

196. Weber, S. G., Tsai, H., and Sandberg, M., Electrochemical detection of dipeptides with selectivity against amino acids, *J. Chromatogr.*, 638, 1, 1993.

197. Takahashi, N., Takahashi, Y., Ishioka, N., Blumberg, B., and Putnam, F. W., Application of an automated tandem high-performance liquid chromatographic system to peptide mapping of genetic variants of human serum albumin, *J. Chromatogr.*, 359, 181, 1986.

198. Randall, C. S., Malefyt, T. R., and Sternson, L. A., Approaches in the analysis of peptides, in *Peptide and Protein Drug Delivery*, Lee, V. H. L., Ed., Marcel Dekker, New York, 1991, chap. 5.

199. Vanacek, G. and Regnier, F. E., Macroporous high-performance anion-exchange supports for proteins, *Anal. Biochem.*, 121, 156, 1982.

200. Kato, Y., Nakamura, K., and Hashimoto, T., New high-performance cation exchanger for the separation of proteins, *J. Chromatogr.*, 294, 207, 1984.

201. Weinbrenner, W. F. and Etzel, M. R., Competitive adsorption of α-lactalbumin and bovine serum albumin to a sulfopropyl ion-exchange membrane, *J. Chromatogr. A*, 662, 414, 1994.

202. Kato, Y., Nakamura, K., Yamazaki, Y., and Hashimoto, T., Comparison of high-performance ion-exchange chromatography and gel electrophoresis in protein separations, *J. Chromatogr.*, 318, 358, 1985.

203. Boulias, C., Pang, H., Mastronardi, F., and Moscarello, M. A., The isolation and characterization of four myelin basic proteins from the unbound fraction during CM52 chromatography, *Arch. Biochem. Biophys.*, 322, 174, 1995.

204. Huisman, T. H. J. and Dozy, A. M., Studies on the heterogeneity of hemoglobin. IV. Chromatographic behavior of different human hemoglobins on anion-exchange cellulose (DEAE-cellulose), *J. Chromatogr.*, 7, 180, 1962.

205. Gooding, K. M., Lu, K.-C., and Regnier, F. E., High-performance liquid chromatography of hemoglobins. I. Determination of hemoglobin A_2, *J. Chromatogr.*, 164, 506, 1979.

206. Alpert, A. J., Cation-exchange high-performance liquid chromatography of proteins on poly(aspartic acid)-silica, *J. Chromatogr.*, 266, 23, 1983.

207. Burke, D. J., Duncan, J. K., Siebert, C., and Ott, G. S., Rapid cation-exchange chromatography of hemoglobins and other proteins, *J. Chromatogr.*, 359, 533, 1986.

208. Narum, D. L., Welling, G. W., and Thomas, A. W., Ion-exchange-immuno-affinity purification of a recombinant baculovirus Plasmodium falciparum apical membrane antigen, PF83/AMA-1, *J. Chromatogr. A*, 657, 357, 1993.

209. Yang, Y.-B., Harrison, K., and Kindsvater, J., Characterization of a novel stationary phase derived from a hydrophilic polystyrene-based resin for protein cation-exchange high-performance liquid chromatography, *J. Chromatogr. A*, 723, 1, 1996.

210. Kaltenbrunner, O., Tauer, C., Brunner, J., and Jungbauer, A., Isoprotein analysis by ion-exchange chromatography using a linear pH gradient combined with a salt gradient, *J. Chromatogr.*, 639, 41, 1993.

211. Greve, K. F., Hughes, D. E., Richberg, P., Kats, M., and Karger, B. L., Liquid chromatographic and capillary electrophoretic examination of intact and degraded fusion protein CTLA4Ig and kinetics of conformational transition, *J. Chromatogr. A*, 723, 273, 1996.

212. Mhatre, R., Nashabeh, W., Schmalzing, D., Yao, X., Fuchs, M., Whitney, D., and Regnier, F., Purification of antibody Fab fragments by cation-exchange chromatography and pH gradient elution, *J. Chromatogr. A*, 707, 225, 1995.

213. Josić, Dj., Hofmann, W., Wieland, B., Nuck, R., and Reutter, W., Anion-exchange high-performance liquid chromatography of membrane proteins from liver and Morris hepatomas, *J. Chromatogr.*, 359, 315, 1986.

214. Imaoka, S. and Funae, Y., Ion-exchange high-performance liquid chromatography of membrane-bound protein cytochrome P-450, *J. Chromatogr.*, 375, 83, 1986.

215. Cacia, J., Quan, C. P., Vasser, M., Sliwkowski, M. B., and Frenz, J., Protein sorting by high-performance liquid chromatography I. Biomimetic interaction chromatography of recombinant human deoxyribonuclease I on polyionic stationary phases, *J. Chromatogr.*, 634, 229, 1993.

216. Kopaciewicz, W. and Regnier, F. E., Mobile phase selection for the high-performance ion-exchange chromatography of proteins, *Anal. Biochem.*, 133, 251, 1983.

217. Roush, D. J., Gill, D. S., and Willson, R. C., Anion-exchange chromatographic behavior of recombinant rat cytochrome b_5. Thermodynamic driving forces and temperature dependence of the stoichiometric displacement parameter Z, *J. Chromatogr.*, 653, 207, 1993.

218. Jungbauer, A., Uhl, K., Schulz, P., Tauer, C., Gruber, G., Steindl, F., Buchacher, A., Schoenhofer, W., and Unterluggauer, F., Displacement effects in large-scale chromatography?, *Biotechnol. Bioeng.*, 39, 579, 1992.

219. Herzig, C. M., Schoeppe, W., and Scherberich, J., Angiotensinase A (aminopeptidase A): properties of chromatographically purified isoforms from human kidney, *J. Chromatogr.*, 625, 73, 1992.

220. Richey, J. S., Optimal pH conditions for ion exchangers on macroporous supports, *Meth. Enzymol.*, 104, 223, 1984.

221. Mazsaroff, I., Bischoff, R., Tice, P. A., and Regnier, F. E., Influence of mobile phase pH on high-performance liquid chromatographic column loading capacity, *J. Chromatogr.*, 437, 429, 1988.

222. Malmquist, G. and Lundell, N., Characterization of the influence of displacing salts on retention in gradient-elution ion-exchange chromatography of peptides and proteins, *J. Chromatogr.*, 627, 107, 1992.

223. Peelen, G. O. H., de Jong, J. G. N., and Wevers, R. A., High-performance liquid chromatography of monosaccharides and oligosaccharides in a complex biological matrix, *Anal. Biochem.*, 198, 334, 1991.

224. Hotchkiss, Jr., A. T., Hicks, K. B., Doner, L. W., and Irwin, P. L., Isolation of oligogalacturonic acids in gram quantities by preparative H.P.L.C., *Carb. Res.*, 215, 81, 1991.

225. Yamauchi, S., Nimura, N., and Kinoshita, T., Protamine-coated silica gel as packing material for high performance liquid chromatography of carbohydrates, *Analyst*, 118, 161, 1993.

226. Kragten, E. A., Kamerling, J. P., and Vliegenthart, J. F. G., Composition analysis of carboxymethlcellulose by high-pH anion-exchange chromatography with pulsed amperometric detection, *J. Chromatogr.*, 623, 49, 1992.

227. De Ruiter, G. A., Schols, H. A., Voragen, A. G. J., and Rombouts, F. M., Carbohydrate analysis of water-soluble uronic acid-containing polysaccharides with high-performance anion-exchange chromatography using methanolysis combined with TFA hydrolysis is superior to four other methods, *Anal. Biochem.*, 207, 176, 1992.

228. Martens, D.A. and Frankenberger, Jr., W. T., Determination of glycuronic acids by high-performance anion chromatography with pulsed amperometric detection, *Chromatographia*, 30, 651, 1990.

229. Martens, D.A. and Frankenberger, Jr., W. T., Determination of saccharides by high-performance anion-exchange chromatography with pulsed amperometric detection, *Chromatographia*, 29, 7, 1990.

230. Reddy, G. P., Hayat, U., Abeygunawardana, C., Fox, C, Wright, A. C., Maneval, D., Bush, C. A., and Morris, Jr., J. G., Purification and determination of the structure of capsular polysaccharide of *Vibrio vulnificus* M06-24, *J. Bacteriol.*, 174, 2620, 1992.

231. Paskach, T. J., Lieker, H.-P., Reilly, P. J., and Thielecke, K., High-performance anion-exchange chromatography of sugars and sugar alcohols on quaternary ammonium resins under alkaline conditions, *Carb. Res.*, 215, 1, 1991.

232. Kynaston, J. A., Fleming, S. C., Laker, M. F. and Pearson, A. D. J., Simultaneous quantification of mannitol, 3-O-methyl glucose and lactulose in urine by HPLC with pulsed electrochemical detection, for use in studies of intestinal permeability, *Clin. Chem.*, 39, 453, 1993.

233. Pastore, P., Lavagnini, I., and Versini, G., Ion chromatographic determination of monosaccharides from trace amounts of glycosides isolated from grape musts, *J. Chromatogr.*, 634, 47, 1993.

234. van Riel, J. and Olieman, C., Selectivity control in the anion-exchange chromatographic determination of saccharides in dairy products using pulsed amperometric detection, *Carb Res.*, 215, 39, 1991.

235. Tabata, S. and Dohi, Y., An assay for oligo-(1→4) →(1→4)-glucantransferase activity in the glycogen debranching enzyme system by using HPLC with a pulsed amperometric detector, *Carb. Res.*, 230,179, 1992.

236. LaCourse, W. R. and Johnson, D. C., Optimization of waveforms for pulsed amperometric detection (P.A.D.) of carbohydrates following separation by liquid chromatography, *Carb. Res.*, 215, 159, 1991.

237. Palu, M. and Samuelson, O., Influence of aromatic compounds in sugar chromatography on anion-exchange resins, *Talanta*, 24, 264, 1977.

238. Tomiya, N., Suzuki, T., Awaya, J., Mizuno, K., Matsubara, A., Nakano, K., and Kurono, M., Determination of monosaccharides and sugar alcohols in tissues from diabetic rats by high-performance liquid chromatography with pulsed amperometric detection, *Anal. Biochem.*, 206, 98, 1992.

239. Togami, D. W., Poulsen, B. J., Batalao, C. W., and Rolls, W. A., Separation of carbohydrates and carbohydrate derivatives by HPLC with cation-exchange columns at high pH, *BioTechniques*, 10, 650, 1991.

240. Scobell, H. D. and Brobst, K. M., Rapid high-resolution separation of oligosaccharides on silver form cation-exchange resins, *J. Chromatogr.*, 212, 51, 1981.

241. Blanken, W. M., Bergh, M. L. E., Koppen, P. L., and van den Eijnden, D.H., High-pressure liquid chromatography of neutral oligosaccharides: effects of structural parameters, *Anal. Biochem.*, 145, 322, 1985.

242. Kunitani, M. and Kresin, L., High-performance liquid chromatographic analysis of carbohydrate mass composition in glycoproteins, *J. Chromatogr.*, 632, 19 1993.

243. Lampio, A. and Finne, J., Sugar analysis of glycoproteins and glycolipids after methanolysis by high-performance liquid chromatography with pulsed amperometric detection, *Anal. Biochem.*, 197, 132, 1991.

244. Stefansson, M. and Westerlund, D., Ion-pair LC of carbohydrates at alkaline pH. Separation and retention behavior of glycoconjugates, *Chromatographia*, 35, 55, 1993.

245. Ogawa, H. K., Takeuchi, Y., Uchibori, H., Matsumoto, I., and Seno, N., Determination of N-acetylneuraminic acid and N-glycolylneuraminic acid in glycoproteins by high-performance liquid chromatography without derivatization, *J. Chromatogr.*, 612, 145, 1993.

246. Mrochek, J. E., Dinsmore, S. R., Tormey, D. C., and Waalkes, T. P., Protein-bound carbohydrates in breast cancer, liquid-chromatographic analysis for mannose, galactose, fucose, and sialic acid in serum, *Clin. Chem.*, 22, 1516, 1976.

247. Ishihara, K., Kameyama, J. -I., and Hotta, K., Development of an HPLC method to estimate hexosamines and its application to determine mucin content in rat and human gastric mucosa, *Comp. Biochem. Physiol.*, 104B, 781, 1993.

248. Hicks, K. B., *Advances in Carbohydrate Chemistry and Biochemistry*, Vol. 46, Tipson, R. S. and Horton, D., Eds., Academic Press, New York, 1988, 17.

249. Ammeraal, R. N., Delgado, G. A. Tenbarge, F. L., and Friedman, R. B., High-performance anion-exchange chromatography with pulsed amperometric detection of linear and branched glucose oligosaccharides, *Carb. Res.*, 215, 179, 1991.

250. van der Hoeven, R. A. M., Niessen, W. M. A., Schols, H. A., Bruggink, C., Voragen, A. G. J., and van der Greef, J., Characterization of sugar oligomers by on-line high-performance anion-exchange chromatography-thermospray mass spectrometry, *J. Chromatogr.*, 627, 63, 1992.

251. Townsend, R. R., Atkinson, P. H., and Trimble, R. B., Separation of high-mannose isomers from yeast and mammalian sources using high pH anion-exchange chromatography, *Carb. Res.*, 215, 211. 1991.

252. Hardy, M. R. and Townsend, R. R., Separation of positional isomers of oligosaccharides and glycopeptides by high-performance anion-exchange chromatography with pulsed amperometric detection, *Proc. Natl. Acad. Sci. U.S.A.*, 85, 3289, 1988.

253. Kamerling, J. P., Gruter, M., Hoffman, R. A., Kragten, E. A., and Vliegenthart, J. F. G., A useful fractionation method in the characterization of bacterial exopolysaccharides, arabinoxylans, and substituted celluloses, *Carb. Neth.*, 9, 27, 1993.

254. Niessen, W. M. A., van der Hoeven, R. A. M., van der Greef, J., Schols, H. A, Voragen, A. G. J., and Bruggink, C., Recent progress in high-performance anion-exchange chromatography-thermospray mass spectrometry of oligosaccharides, *J. Chromatogr.*, 647, 319, 1993.

255. Holmbeck, S. and Lerner, L., Separation of hyaluronan oligosaccharides by the use of anion-exchange HPLC, *Carb. Res.*, 239, 239, 1993.

256. Watson, E., Bhide, A., Kenney, W. C., and Lin, F.-K., High-performance anion-exchange chromatography of asparagine-linked oligosaccharides, *Anal. Biochem.*, 205, 90, 1992.

257. Reddy, G. P. and Bush, C. A., High-performance anion exchange chromatography of neutral milk oligosaccharides and oligosaccharide alditols derived from mucin glycoprotein, *Anal. Biochem.*, 198, 278, 1991.

258. Hernandez, L. M., Ballou, L., and Ballou, C. E., Separation of yeast asparagine-linked oligosaccharides by high-performance anion-exchange chromatography, *Carb. Res.*, 203, 1, 1990.

259. Linhardt, R. J., Gu, K. N, Loganathan, D., and Carter, S. R., Analysis of glycosaminoglycan-derived oligosaccharides using reversed-phase ion-pairing and ion-exchange chromatography with suppressed conductivity detection, *Anal. Biochem.*, 181, 288, 1989.

260. Somlyo, B., Kovats, E., Keler, T., and Nowotny, A., Column liquid chromatography of endotoxins, *J. Chromatogr.*, 525, 329, 1990.

261. Thurl, S., Offermanns, J., Müller-Werner, B., and Sawatzki, G., Determination of neutral oligosaccharide fractions from human milk by gel permeation chromatography, *J. Chromatogr.*, 568, 291, 1991.

262. Ohta, M., Matsuura, F., Kobayashi, Y., Shigeta, S., Ono, K., and Oka, S., Further characterization of allergenically active oligosaccharitols isolated from a sea squirt H-antigen, *Arch. Biochem. Biophys.*, 290, 474, 1991.

263. Heyraud, A. and Leonard, C., Analysis of oligouronates on reversed-phase ion-pair H.P.L.C.: role of the mobile phase, *Carb. Res.*, 215, 105, 1991.

264. Koizumi, K., Okada, Y., and Fukuda, M., High-performance liquid chromatography of mono- and oligo-saccharides on a graphitized carbon column, *Carb. Res.*, 215, 67, 1991.

265. Suzuki-Sawada, J., Umeda, Y., Kondo, A., and Kato, I., Analysis of oligosaccharides by on-line high-performance liquid chromatography and ion-spray spectrometry, *Anal. Biochem.*, 207, 203, 1992.

266. Soga, T., Inoue, Y., and Yamaguchi, K., Determination of carbohydrates by hydrophilic interaction chromatography with pulsed amperometric detection using postcolumn pH adjustment, *J. Chromatogr.*, 625, 151, 1992.

267. Green, E. D. and Baenziger, J. U., Separation of anionic oligosaccharides by high-performance liquid chromatography, *Anal. Biochem.*, 158, 42, 1986.

268. Hotchkiss, Jr., A. T., Haines, R. M., and Hicks, K. B., Improved gram-quantity isolation of malto-oligosaccharides by preparative HPLC, *Carb. Res.*, 242, 1, 1993.

269. Feste, A. S. and Khan, I., Separation of glucooligosaccharides and polysaccharide hydrolysates by gradient elution electrophilic interaction chromatography with pulsed amperometric detection, *J. Chromatogr.*, 630,129, 1993.

270. Takahashi, N., Wada, Y., Awaya, J., Kurono, M., and Tomiya, N., Two-dimensional elution map of GalNAc-containing N-linked oligosaccharides, *Anal. Biochem.*, 208, 96, 1993.

271. Femia, R. E. and Weinberger, R., Determination of reducing and non-reducing carbohydrates in food products by liquid chromatography with post-column catalytic hydrolysis and derivatization: comparison with refractive index detection, *J. Chromatogr.*, 402, 127, 1987.

272. Christ, W., Arndt, G., and Schulze, G., Separation behavior of nicotinamide and thionicotinamide derivatives on Dowex® 1-X8, *J. Chromatogr.*, 106, 418, 1975.

273. Illingworth, D. R. and Portman, O. W., An improved method for separating the products of lecithin and lysolecithin catabolism, *J. Chromatogr.*, 73, 262, 1972.

274. Rindi, G. and De Giuseppe, L., A new chromatographic method for the determination of thiamine and its mono-, di-, and tri-phosphates in animal, *Biochem. J.*, 78, 602, 1961.

275. Fahey, R. C., Newton, G. L., Dorian, R., and Kosower, E. M., Analysis of biological thiols: quantitative determination of thiols at the picomole level based upon derivatization with monobromobimanes and separation by cation-exchange chromatography, *Anal. Biochem.*, 111, 357, 1981.

276. Callmer, K. and Davies, L., Separation and determination of vitamin B_1, B_2, B_6, and nicotinamide in commercial vitamin preparations using high-performance cation-exchange chromatography, *Chromatographia*, 7, 644, 1974.

277. Virtanen, V. and Lajunen, L. H. J., High-performance liquid chromatographic method for simultaneous determination of clodronate and some clodronate esters, *J. Chromatogr.*, 617, 291, 1993.

278. Shamsi, S. A. and Danielson, N. D., Ion chromatography of polyphosphates and polycarboxylates using a naphthalenetrisulfonate eluent with indirect photometric and conductivity detection, *J. Chromatogr. A*, 653, 153, 1993.

279. Yokoyama, Y., Kondo, M. and Sato, H., Determination of alkylbenzenesulphonates in environmental water by anion-exchange chromatography, *J. Chromatogr.*, 643, 169, 1993.

280. Pan, N. and Pietrzyk, D. J. Separation of anionic surfactants on anion exchangers, *J. Chromatogr.*, 706, 327, 1995

281. Klein, H. and Leubolt, R., Ion-exchange high-performance liquid chromatography in the brewing industry, *J. Chromatogr.*, 640, 259, 1993.

282. Massardier, V. and Vialle, J., Investigation of liquid chromatographic systems for the separation of sulphonium salts, *Anal. Chim. Acta*, 281, 391, 1993.

283. Rounds, M.A. and Nielsen, S. S., Anion-exchange high-performance liquid chromatography with post-column detection for the analysis of phytic acid and other inositol phosphates, *J. Chromatogr. A*, 653, 148, 1993.

284. Jones, N. R. and Burt, J. R., The separation and determination of sugar phosphates, with particular reference to extracts of fish tissue, *Analyst*, 85, 810, 1960.

285. Rotman, B., Uses of ion exchange resins in microbiology, *Bacteriol. Rev.*, 24, 251, 1960.

286. Budzynski, A. Z., Olexa, S. A., and Sawyer, R. T., Composition of salivary gland extracts from the leech Haementeria ghilianii, *Proc. Soc. Exp. Biol. Med.*, 168, 259, 1981.

287. Budzynski, A. Z., Olexa, S. A., Brizuela, B. S., Sawyer, R. T., and Stent, G. S., Anticoagulant and fibrinolytic properties of salivary proteins from the leech *Haementeria ghilianii*, *Proc. Soc. Exp. Biol. Med.*, 168, 266, 1981.

288. Kirschbaum, N. E. and Budzynski, A. Z., A unique proteolytic fragment of human fibrinogen containing the Aα COOH-terminal domain of the native molecule, *J. Biol. Chem.*, 265, 13669, 1990.

289. Sawyer, R. T., Munro, R., and Jones, C. P., A novel thrombolytic directed against platelet-rich clots, *Thromb. Haem.*, 65(6), 1191, 1991.

290. Malinconico, S. M., Katz, J. B., and Budzynski, A. Z., Hementin: anticoagulant protease from the salivary gland of the leech *Haementeria ghilianii*, *J. Lab. Clin. Med.*, 103, 44, 1984.

291. Calbiochem catalog, San Diego, CA, 1994/5.

292. Electricwala, A., Sawyer, R. T., and Atkinson, T., Hementin: a novel anticoagulant from leech, *Thromb. Haem.*, 65(6), 1210, 1991.

293. Swadesh, J. K., Huang, I.-Y., and Budzynski, A. Z., Purification and characterization of hementin, a fibrinogenolytic protease from the leech *Haementeria ghilianii*, *J. Chromatogr.*, 502, 359, 1990.

chapter six

Gel permeation
chromatography

Rajesh G. Beri, Laurel S. Hacche, and Carl. F. Martin

0-8493-2682-6/97/$0.00+$.50

6.1   Introduction

6.1.1   General

Gel permeation chromatography (GPC) first established itself as the method of choice for the analysis of synthetic polymers and, to a far lesser extent, of high-molecular-weight biopolymers. That situation has changed substantially in the recent past, as the speed and efficiency of gel permeation columns has permitted the fractionation of compounds of lower molecular weights. Still, no discussion of GPC would be complete without a thorough discussion of the nature of polymers and of the information about the composition of polymeric analytes that can be extracted in a few minutes by GPC. This chapter will cover all aspects of gel permeation chromatography, beginning with a full description of the nature of polymers.

Polymers are high-molecular-weight compounds derived from smaller monomeric units. The simplest kind of polymer, the linear homopolymer, is conceptualized as a repeating sequence of n monomeric units, A, joined head-to-tail, written A_n. Many homopolymers are formed using a chain initiator I and a chain terminator T, so that the structure of a linear homopolymer is more correctly written I-A_n-T. For small homopolymers (termed oligomers), the contribution of the end groups, I and T, to the physical characteristics of the polymer is not negligible, but as n increases, the relative effects of the end groups decline to the point that they may be neglected.

If two homopolymers, A_n and B_m, are joined end to end, the resulting polymer is known as a diblock copolymer, while if monomeric units are randomly incorporated into the chain, the result is termed a random copolymer. Polymers or copolymers may, of course, polymerize in other than head-to-tail fashion, introducing branching. When the branching forms regular side-chains from a definable main chain, the polymer is known as a side-chain polymer. Cyclic polymers, in which the monomer end units are joined, are known particularly in biological systems. Proteins are examples of linear heteropolymers consisting of the 20 normal amino acids which often exhibit cross-links from one section of the main chain to another.[1] Examples are given in Figure 1.

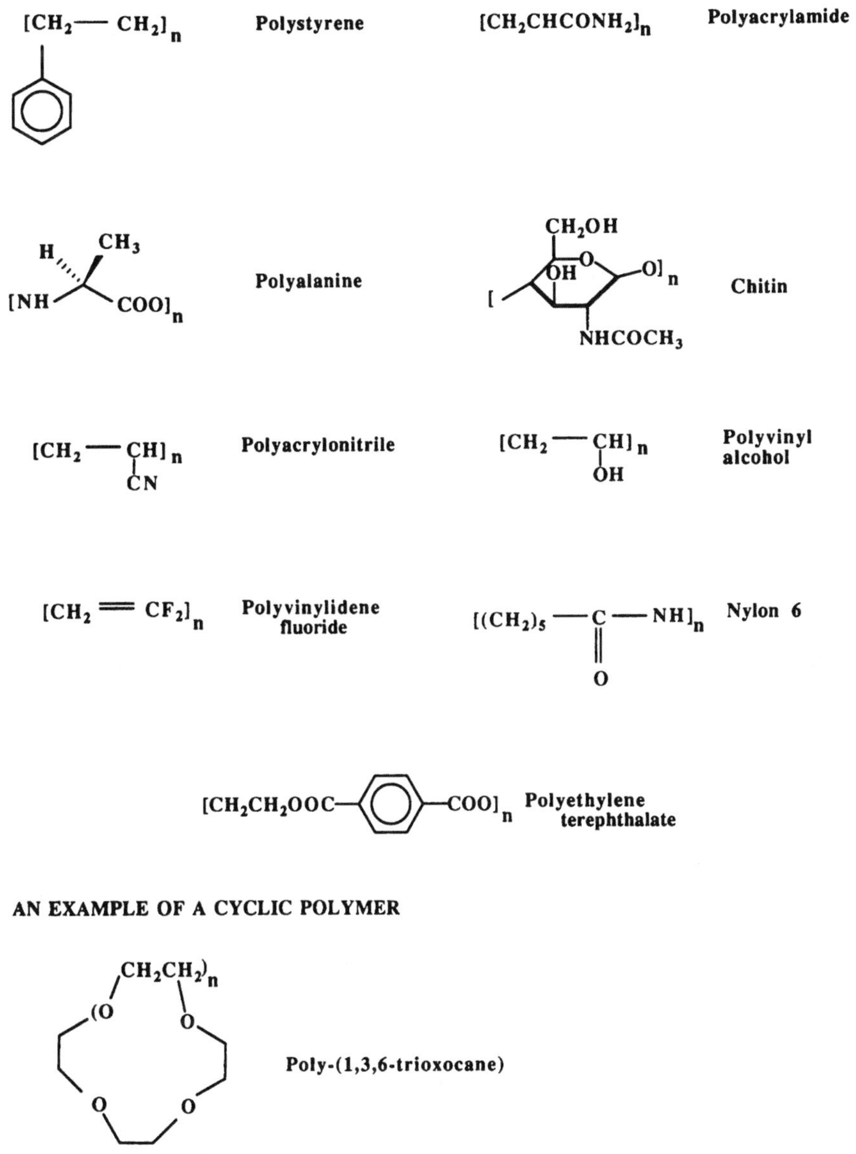

$[CH_2 — CH_2]_n$ **Polystyrene** $[CH_2CHCONH_2]_n$ **Polyacrylamide**

Polyalanine

Chitin

$[CH_2 — CH]_n$ **Polyacrylonitrile** $[CH_2 — CH]_n$ **Polyvinyl alcohol**

$[CH_2 = CF_2]_n$ **Polyvinylidene fluoride** $[(CH_2)_5 — C — NH]_n$ **Nylon 6**

$[CH_2CH_2OOC—\bigcirc—COO]_n$ **Polyethylene terephthalate**

AN EXAMPLE OF A CYCLIC POLYMER

Poly-(1,3,6-trioxocane)

Figure 1 Diagram of structures of simple polymers.

The dissolution of a polymer in a solvent is a remarkable process, the solid first swelling and then individual chains separating and disentangling;[2,3] intertangled chains can also pin one another against a solid surface to slow dissolution.[4] Although a low-molecular-weight substance typically reaches equilibrium composition after a few seconds of mixing, the dissolution of a polymer can take days to reach equilibrium. The presence of one polymer can decrease the solubility of another through steric exclusion;[5,6]

i.e., one polymer may preferentially interact with the solvent. A small molecule has a limited number of ways in which it may interact with itself or its neighbors, so in the presence of a competing interaction with a solvent molecule, clusters of solute are quickly dispersed. In polymers, however, a great multiplicity of interactions between a molecule and itself or a molecule and its neighbors is possible. While individually weak, the cumulative statistical effect is such that polymers often dissolve slowly and, even after dissolution, display structural properties rarely observed in solutes of low-molecular-weight. These are termed secondary, tertiary, and quaternary conformational effects in order of the increasing distance of the interaction required to maintain intra- or intermolecular associations.[1] The term "distance" has a special meaning, described below. Self-association or aggregation phenomena are an example of quaternary structure observed not only in biological chemistry, but also in synthetic polymers.[7]

At equilibrium in solution, main chain bond angles vary in a statistical mechanically random fashion such that the polymeric backbone traces a seemingly random path through space. The path is not, however, truly random, since the distribution of bond angles that are observed are determined from statistical mechanical considerations of intramolecular and solute-solvent interactions.[8,9] Absent any strong constraints on intra- or intermolecular organization, a polymer is said to exhibit the statistical coil conformation. If, in the course of polymerization, a stereoregular conformation is generated and if the energetic barriers to conformational interconversion are high, the resulting polymer may not exhibit the statistical coil conformation, but rather may be confined to defined conformations, such as the isotactic or syndiotactic conformations of polypropylene shown in Figure 2. Even if conformational equilibration in solution is rapid, regular structures, such as the α-helix and β-sheet observed in proteins, may be thermodynamically favored (see Figure 3). As is seen in Figure 3, the NH–OC hydrogen bonds of the α-helix are between residues i and i + 3, i.e., between neighbors at close distance in the chain. The interactions of the β-sheet are likewise at relatively close distance in the chain. In Figure 3 is shown an example of a polypeptide turn similar to that found in some β-sheets. Therefore, α-helices and β-sheets are termed secondary structural features to indicate that the distance along the chain is small for these interactions. Regular secondary structures have been classified in considerable detail.[10,11] By contrast, tertiary structure involves interactions between much more chain-distant residues, and quaternary structure involves interactions between distinct chains. The formation of tertiary structure is often the result of interactions of elements of secondary structure. Omega-loops, which require the interaction of sequence-distant residues without an intervening segment of regular structure, might also be classified as tertiary structural elements.[12]

Quaternary structure is a special case of aggregation between distinct chains. In some aggregates, the interactions involved in holding molecules together do not involve a definable group of residues. In a number of structures, however, the interactions are extremely regular. In DNA, hemoglobin,

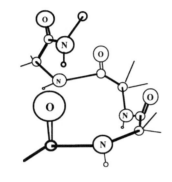

ISOTACTIC **SYNDIOTACTIC**

Figure 2 Isotactic and syndiotactic polypropylene.

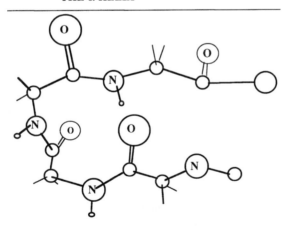

THE α-HELIX

A β-BEND AND β- SHEET

Figure 3 Examples of higher-order structure in proteins: the α-helix and β-sheet.

collagen, and poly(β-glucose), for example, interstrand interactions are well defined.

Cross-links, which impose strong conformational constraints on the intervening segment of the chain, generally are not classified as elements of secondary or tertiary structure. Disulfide cross-links in protein may certainly stabilize both secondary and tertiary structure, and such cross-links have the

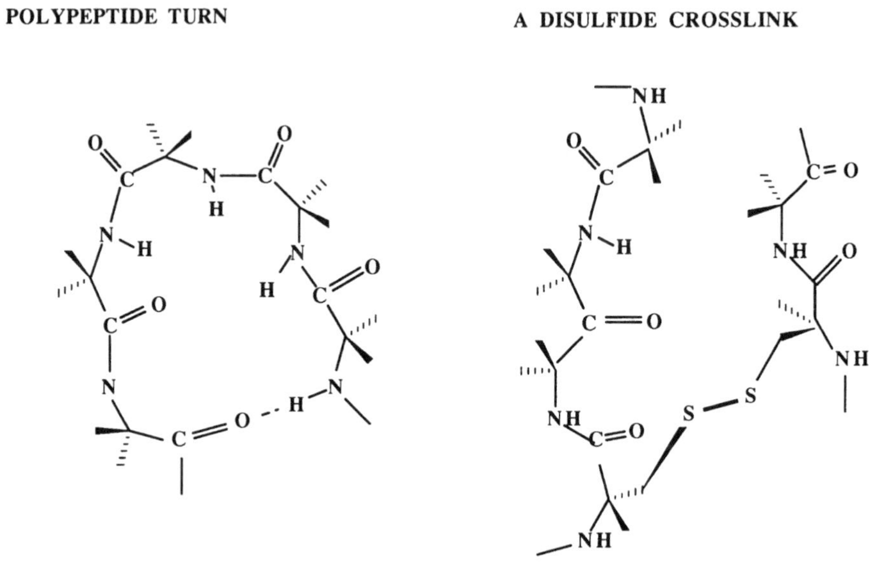

POLYPEPTIDE TURN   **A DISULFIDE CROSSLINK**

Figure 4 Examples of long-range interactions in protein: a disulfide cross-link and a polypeptide turn.

important property that they are reversibly formed. Metals are often involved in stabilizing certain conformations by forming cross-links within or between chains. The roles of metals — most importantly calcium, but also iron, copper, zinc, and others — in maintaining the structure of metalloproteins are well established. Zinc, which can coordinate with oxygen, nitrogen, and sulfur, is particularly important in enzymes such as alcohol dehydrogenase and regulatory proteins such as transcription factors.[13]

The types of interactions that determine polymer structure are relatively few. Hydrogen bonding, solvophobicity, and ionic interactions predominate, but several unanticipated nonbonded interactions, such as the interaction of disulfide bonds with aromatic rings have been proposed to be important in some polymeric systems.[14-17] Some typical interactions are depicted in Figure 4. One concept central to an understanding of the nature of these interactions in determining conformation is that of competition among solute-solvent, solute-solute, and solvent-solvent interactions. In other words, defining a structure for a given polymer implies that the solvent and the temperature are also defined. Above or below the given temperature, or under conditions of different pH or ionic strength, one equilibrium structure may deform or convert to another equilibrium structure. Statistical coil polymers, though lacking a regular structure, undergo expansion or contraction in molar volume as a function of temperature and solution conditions.

A biologically formed protein or polynucleotide, while exhibiting some limited heterogeneity in length or microheterogeneity in the main chain sequence or in pendant prosthetic groups, has an essentially unique sequence. Different forms of a protein or polynucleotide often can be resolved chromatographically or at least described by degradative analysis. Synthetic polymers, on the other hand, typically exhibit much higher degrees of heterogeneity in size. For this reason, synthetic polymers are usually described in terms of distributions around some mean value of chain length and branching, while proteins and polynucleotides are described as individual species of which variants occur. Of course, if a protein or polynucleotide is formed by *in vitro* synthesis or by degradation, it may exhibit such a high degree of heterogeneity that it may best be described by a distribution. At present, polysaccharides are generally treated as distributions, but the possibility of synthesizing them by stepwise or template synthesis may alter the situation, as it may for synthetic polymers. Similarly, as the resolution of chromatographic methods rises, it becomes increasingly possible to characterize a distribution as populations of individual species.

A comparison of the terminology used to describe essentially homogeneous polymers and heterogeneous polymers is presented in Table 1. The molecular parameters used to characterize both classes of polymer are molecular weight, molar volume, radius of gyration, and length of principal axes. Higher-order moments of the molecular weight average are used to characterize the shape of the distribution of synthetic polymers. These parameters are not useful in describing defined polymers such as proteins.

Table 1 Terminology Used To Describe Polymers

Topological	Linear	Branched	Cyclic	Cross-linked
Compositional	Homopolymer		Block or random copolymer	
Copolymer types	Statistical	Random	Block or graft	Alternating
Conformational	Atactic	Isotactic	Syndiotactic	

6.1.2   Conformation of a polymer in solution

The conformation of a polymer in solution is the consequence of a competition between solute intra- and intermolecular forces, solvent intramolecular forces, and solute-solvent intermolecular forces. Addition of a good solvent to a dry polymer causes polymer swelling and disaggregation as solvent molecules adsorb to sites which had previously been occupied by polymer intra- and intermolecular interaction. As swelling proceeds, individual chains are brought into bulk solution until an equilibrium solubility is attained.

The nature of the solvent has a considerable effect on polymer solution properties, particularly if the polymer is a polyelectrolyte. Ionic interactions are long-range interactions; therefore, a polyelectrolyte in water, particularly if homopolymeric, tends to expand to a size much greater than the random coil. At low polymer concentration and low solution ionic strength, the charge density on the polymer is greatest and swelling is most extreme. Swelling is reduced if low-molecular-weight salts are included in solution, since these mask charge. Ionization may be reduced by control of pH, reducing intramolecular repulsion. Because most organic molecules have some degree of hydrophobicity, the result of reducing the charge on a polymer may be decreased solubility. Intramolecular attractions counterbalance repulsions in polymeric ampholytes, such as proteins. This effect is seen when ampholytes (amphoteric polyelectrolytes) are added to protein solutions. Protein solubility may be increased by many orders of magnitude by the addition of ampholytes. In some cases, polymer-solvent interactions noncovalently cross-link chains and inhibit dissolution, as when aqueous solutions of amines are added to polycarboxylate polymers.

Chromatographic surfaces add another layer of complexity in polymer solution behavior beyond that introduced by the statistical chain and solvent effects. The materials used in chromatography invariably contain some degree of hydrophobicity, some capacity for hydrogen bonding, or some ionic character and therefore tend to adsorb or repel analytes. In gel permeation chromatography, polymers may be partially excluded from pores by both steric considerations and ionic interactions. A gradient of chemical potential is created between the stagnant and mobile phases, an example of a Donnan effect.[18] Therefore, if a solution contains both nonexcluded and excluded ions, the concentration of the smaller ions may rise above that expected in the absence of the polymeric ion, an effect that may be called ion inclusion. Examples of the Donnan effects have been observed in the chromatography of lignosulfates, which are anionic and tend to be excluded from Sephadex™ (Pharmacia; Piscataway, NJ) by the negative charge on that material.[19,20] Addition of a high-molecular-weight lignosulfate suppresses ion inclusion of a lower-molecular-weight component, while addition of salt to the mobile phase suppresses both ion inclusion and ion exclusion. Similarly, in chromatographing dextrans, monocarboxylation and dicarboxylation were sufficient to generate additional peaks on a diol-modified Lichrosphere® (Merck; Darmstadt, Germany) packing for the excluded species.[21] A theoretical treatment, using a model that incorporates polyelectrolyte charge and the charge on the chromatographic packing was developed and applied to poly-L-glutamic acid and polystyrene sulfonate.[22]

A similar effect may exist for hydrophobic interaction between solute and stationary phase, as one solute may adsorb more strongly to the stationary phase than another. It has also been remarked that a flexible polymer confined to a pore should be at a lower entropy than one in bulk solution, leading to exclusion in excess of that expected for a simple geometric solid.[23] Even in the absence of interactions, a high concentration of a small component can lead to an excluded volume effect, since the effective volume inside the pore is reduced.

Gel permeation chromatography (GPC), sedimentation velocity, viscosimetry, osmometry, electrophoretic mobility, nuclear magnetic resonance (NMR), and light-scattering are techniques used to characterize polymers; of these, GPC is perhaps the most widely used. GPC is often called gel filtration chromatography (GFC) or size-exclusion chromatography (SEC). These terms will be used interchangeably below. The following sections will describe instrumentation used in GPC, viscosimetry, and light-scattering and applications of GPC in polymer chemistry. While it is not possible to explore in great detail the effects of temperature and solution conditions on the results obtained by these techniques, a few examples of the critical role of an understanding of the physical chemistry of macromolecules in predicting chromatographic behavior are presented.

6.1.3 Column materials

Materials that have been used for gel permeation chromatography include carbohydrates, methacrylates, silicas, and polystyrene.[24-34] Even the vesicles of plant cell walls have been used.[35] The chromatographic particle contains pores or interstices into which molecules may enter. One may conceptualize the pores according to a network model.[36] Usually, solvent is freely permeable, as are molecules of about the size of the solvent, while high-molecular-weight solutes are partially excluded by steric reasons.[23,36-42] Solvent in the pores or interstices is often called the "stagnant phase", while solvent outside is called the "mobile phase". The two solvent phases are in dynamic exchange. A molecule that is freely permeable, therefore, experiences a larger solvent volume than one that is not freely permeable. The volume of the phase outside the pores is the excluded or void volume (V_o), while the combined stagnant and mobile phase volume is the total column or permeation volume (V_t). The total volume less the excluded volume is the pore volume. The pore volume is one of the most important characteristics defining a GPC column, since a high porosity permits the greatest resolution of small molecules from large. If a solute elutes at elution volume V_e, the distribution constant can be expressed as:[43]

$$K' = (V_e - V_o)/(V_t - V_o)$$

which is simply the ratio of the pore volume experienced by the solute relative to that experienced by solvent. To maximize the resolving power of a column, the porosity should be maximal, the pore size just slightly larger than the size of the analyte and narrowly distributed, and the excluded volume as small as possible, a condition that is satisfied by small particles. High porosity requires that pores be numerous, a condition that may be met only at the cost of reducing the structural integrity of the particle. For large analytes, meeting the simultaneous requirements of large and abundant pores, while maintaining a capacity sufficient for detection, is particularly challenging.

One defining characteristic of GPC materials is rigidity. Silica, zirconium-stabilized silica, and controlled pore glass are extremely rigid materials that can readily tolerate the relatively high pressures used for high performance liquid chromatography (HPLC). At the other extreme of rigidity are soft gels formed from natural polymers, such as agarose and dextran. Soft gels are unable to tolerate high pressure and are unsuitable for HPLC. Synthetic polymers, such as acrylamide, polystyrene, and poly(methylmethacrylate), are in between these extremes. Both natural and synthetic polymers have been modified by cross-linking to form a much more rigid matrix than those observed for materials that have not been cross-linked. Agarose, which is commercially available as Sepharose™ (Pharmacia), is cross-linked with epichlorohydrin or other difunctional reagents and sold commercially as Sepharose™ CL-B.[26] Cross-linking is used to reduce the water-solubility of dextran. Cross-linked dextran is sold as Sephadex™ (Pharmacia). By copolymerizing allyldextran with bisacrylamide, a material sold as Sephacryl™ (Pharmacia) is produced.[26] Polystyrene is typically copolymerized with divinyl benzene to produce a cross-linked copolymer, poly(styrene-divinyl benzene) or PS-DVB. The properties of PS-DVB materials vary widely depending on such factors as:

- Ratio of divinyl benzene to styrene used in polymerization
- Distribution of ortho-, meta-, and para-isomers in the divinyl benzene copolymer
- Level of contamination of divinyl benzene with ethylvinyl benzene
- Conditions used for polymerization

Typically an oil-water suspension polymerization process is used. The characteristics of the resultant product are dependent on the initiator and the surfactant used.[26] Following polymerization, copious washing is required to remove the initiator, surfactant, monomer, and co-solvents. One can also perform the polymerization using a suspension of tiny, monodisperse latex spheres as seeds.[26,34,44] Acrylamide and copolymers of acrylamide with bisacrylamide as a cross-linking agent are moderately rigid, but extremely difficult to produce with reproducible porosity in the small particle sizes required for HPLC. In recent years, polyhydroxymethacrylate phases produced by Showa Denko (Tokyo, Japan) have proven themselves to be moderately rigid and with reproducible porosity. The hydroxylated polyether phase produced by ToyoSoda (Tokyo, Japan) as type PW, of which the major constituent is a polyglycerol,[33] is moderately rigid. The ToyoSoda type SW is silica based.[31] Other materials of moderate rigidity that have shown promise include polyvinyl alcohol and hydroxyethylmethacrylate.

The surface characteristics and solvent compatibility of GPC materials are characteristics crucial to their chromatographic functionality. Ion exclusion due to charge repulsion between ionized negative groups and analytes is also a potential problem,[45] as are ion inclusion[19] and ion exchange. Silica-based phases, which are very stable in nonaqueous media, are less stable in water.[46] Forms stable to aqueous solutions of pH 2 to 7 are generally available, and surface modification has produced phases stable in aqueous solution to about pH 9. Silica-based phases without surface modification are prone to hydrogen-bonding and ionic or hydrogen-bonding interactions with solutes, so modification with hydrophilic groups such as glyceropropylsilane has been a common tactic in passivating silica for aqueous solution.[29,47,48] In nonaqueous solvents at high temperature, silica-based materials are often the most stable phase possible. PS-DVB phases have been steadily improving, such that most of the commercially available phases are capable of handling the majority of organic solvents. Highly polar solvents such as DMSO or DMF may damage PS-DVB, particularly at high temperature. Macromolecular adsorption onto polymeric phases in nonaqueous solvents is a common observation. For example, poly(methyl vinyl ether co-maleic anhydride) adsorbs onto a cross-linked, hydroxylated polymethacrylate gel in nonaqueous solvents or in an aqueous solvent at low pH,[49] while exhibiting little or no adsorption at high pH.[50] Polyvinyl alcohol gels were found to exhibit adsorptive effects superimposed on size exclusion effects for a number of solutes.[51] Surface modification of PS-DVB to eliminate adsorption of aqueous solutes has been attempted, but remains a difficult goal. In aqueous solvents, unmodified PS-DVB is commonly used as a reversed phase material.[52,53] The Showa Denko hydroxymethacrylate series has a limited tolerance for organic solvents, but it is perhaps the least adsorptive material available for aqueous GPC. The TSK-Gel™ PW-type material is frequently cited as exhibiting severe adsorption in aqueous media due to hydrophobic interactions (Tosoh Corp.; Tokyo, Japan).[54] The surface of Sephacryl™ in aqueous media has been noted to exhibit hydrophobic adsorption at low pH and ion exchange behavior at high pH.[55] Carbohydrate-based polymers are unsuitable for use in organic solvents. Polyacrylamide, perhaps the least adsorptive material for aqueous chromatography, is also intolerant of organic solvents. Polymeric phases such as PS-DVB, polyacrylamide, and PMMA tend to be susceptible to chemical degradation such as slow oxidation or hydrolysis, generating negatively charged groups capable of adsorption by ion exchange interactions.

Another important aspect of characterization of GPC materials is the porosity, defined above as the ratio of the volume within pores to that of the mobile phase outside the pores. The greater the porosity, the greater the resolving power of the column. The narrower the distribution in pore size, the greater the resolution of the column within a narrower range. In other words, if the pore size varies widely within a material, the analysis can cover a broad molecular weight range at low resolution. A narrow distribution of pore size increases resolution over a particular range at the expense of lower and higher molecular weights. A ratio of internal to external pore volume of about 1 is observed in well packed columns.[29] Raising the porosity, while perhaps technically feasible, increases the fragility of the particles. Reduction of the particle size, by permitting more efficient column packing, can serve to improve the ratio of internal pore volume to mobile phase pore volume. Below a certain limit, however, the channel between particles becomes a size on the order of that of typical polymers; therefore, particles of diameter 10 to 20 μm are preferred for high-molecular-weight polymers, and particles of diameter about 3 to 7 μm for smaller polymers. Smaller particles offer much greater resistance to flow than large particles and are substantially more fragile. Thus, there are unavoidable tradeoffs among speed of analysis, resolution, and column durability. PS-DVB is preferred for high-molecular-weight organic polymers, since the pore size can be made much larger than that of silica.

Separation by size can occur even on a nonporous material, as the flow in the center of a flow channel is faster than that near the walls. Since large molecules are excluded from the wall regions, they tend to travel in the faster flow down the flow channel center, a phenomenon known as hydrodynamic chromatography (HDC).[56-58] As discussed in Chapter 1, a mixed-mode form of HDC called "slalom chromatography" has found application in DNA analysis. HDC, however, is far less efficient as a separation process than GPC.

The presence of pores or microreticulae in a particle permits a macromolecule to partition between the stagnant and the mobile phase, leading to the slower elution of small molecules. GPC materials are generally classified as macroporous or microporous, although the distinction may be ambiguous for some materials. Macroporous particles can be conceptualized to be composed of small beads; the pores are the voids between the beads. Microporous materials can be conceptualized as being composed of mesh, like a natural sponge. Carbohydrates and some polystyrenes are considered to be microporous, while silicas and some polystyrenes are considered to be macroporous. More recently, PS-DVB has been recognized to contain both macropores and smaller micropores.[59] There seems to be no fundamental difference between materials of different pore structures, with convection having an important role in all materials.[60] A significant problem with surface modification of polymeric phases to reduce analyte adsorption is that the pore size may be altered.

6.1.4   Choice of solvent conditions for GPC

Understanding the physical chemistry of polymers facilitates doing a separation by GPC. As has been described, solution conditions affect polymer shape and conformation and interactions between the stationary phase and the analyte. In nonaqueous GPC, devising a mobile phase is relatively straightforward. Assuming that a good solvent for the analyte is known, problems may arise because of adsorption of the analyte to the stationary phase, incompatibility of the solvent with the stationary phase, or limitations of detection. A different solvent may be chosen by consideration of the solvent polarity parameter.[2] If, for example, adsorption to the stationary phase is observed, a solvent with a slightly lower or higher polarity parameter may be chosen. If, as is often the case, the refractive index of the solvated solute is too near that of the solvent, a solvent of similar polarity but different refractive index may be chosen. If solvent choices are limited, but solubility is inadequate, the temperature of the mobile phase may be raised to increase solubility. Therefore, the most difficult step in nonaqueous GPC is identifying those solvents in which the sample is freely soluble. The bulk of nonaqueous GPC is performed in a few solvents, such as chloroform, THF, and DMF. In determining sample solubility, one should permit gentle overnight mixing, since dissolution of polymers in nonaqueous solvents is often slow. Rapid mixing can lead to shear degradation,[61] particularly for viscous solutions. High-molecular-weight polymers (strongly interacting polymers present at high concentration) are the most susceptible to shear degradation. Ultrasound may also cause degradation.[62]

The situation is much more complex in aqueous GPC, since both the pH and the ionic strength may require adjustment. The principal difficulty in protein GPC is usually adsorption to the stationary phase. Conformational equilibria between the folded and unfolded states are another significant complication. Although conformational equilibria have the greatest effect when the solvent or the stationary phase promotes unfolding, as in hydrophobic interaction or reversed phase chromatographies,[63,64] retention shifts or multiple peaks due to pH denaturation[65] and thermal unfolding in GPC are observed. With charged polymers, such as ionic homopolymers and DNA, adsorption and ion exclusion or inclusion are both significant problems. Polymerization initiators or terminators may be unanticipated sources of hydrophobicity in an otherwise hydrophilic synthetic polymer. Variable swelling of a charged polymer depending on the pH and ionic strength may lead to poor reproducibility in chromatography. In general, pH 7 and a solution of 0.1 M ionic strength comprise a good starting point for developing a chromatographic method using aqueous GPC. Halide salts, particularly NaCl, are of particular utility, although precautions to prevent corrosion of the system are advisable. Sodium nitrate and phosphate salts are also in common use in aqueous GPC. If adsorption of the polymer to the chromatographic phase is observed, the mobile phase may be modified by increasing the ionic strength, raising or lowering the pH, or raising the temperature. Chaotropes, such as urea, promote dissolution and unfolding of some biopolymers and may be added to the mobile phase. Since proteins and DNA are readily detected with UV, loading can be varied over a wide range to minimize inclusion or exclusion effects and optimize separation.

6.1.5   GPC of small molecules

It may be easiest to understand GPC on a practical level by considering the separation of small molecules. It has long been recognized that small molecules — such as salts,[45,66] dialkyl phthalates,[67] oligostyrenes,[67] and epoxy oligomers[68] — can be separated by GPC. As a nonadsorbtive, isocratic method, the selectivity of GPC relative to that of other high performance chromatographies is poor. For example, separating dimethyl and diethyl phthalate involves a difference in molecular weight of 15% and in carbon number of 20%. Since separation of homologs differing by two carbons is usually easily accomplished by reversed phase or normal phase chromatography, GPC has not been seen to be advantageous. Also, the compounds of synthetic interest tend to be concentrated in the molecular weight range 200 to 800, and the chromatographic resolution requirement is to separate compounds differing by a single methylene carbon, i.e., a molecular weight difference of 2 to 7%.

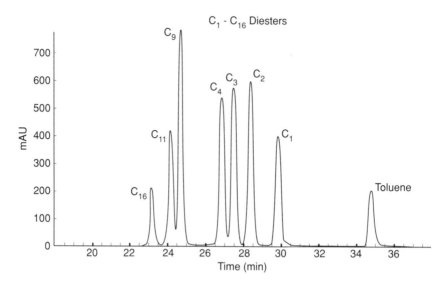

Figure 5 GPC of *n*-alkyl diesters of biphenyl-4,4′-dicarboxylic acid. Separation was accomplished on four 30 × 7.5-mm, 50-μm columns in tetrahydrofuran at 1 ml/min. (From Swadesh, J. K., Stewart, Jr., C. W., and Uden, P. C., *Analyst*, 118, 1123, 1993. With permission.)

Recently, a detailed comparison of GPC with reversed phase and normal phase separations has been conducted in the context of determination of the purity of a number of homologs of mono- and diesters of biphenyl-4,4′-dicarboxylic acid (BDCA).[69] The diesters were the synthetic precursors of the liquid crystalline monoesters. The separation of the diesters by GPC is shown in Figure 5. While resolution was, indeed, lost as the carbon number increased, near-baseline separation was obtained with the C1, C2, C3, and C4 homologs and simultaneously with the C9, C11, and C16 homologs. For the monoesters, C3, C4, C8, C11, and C15 were separated simultaneously; presumably, the C1 and C2 were also simultaneously separable. Similar separations of the monoesters were performed on reversed phase liquid chromatography (RPLC), and of the diesters on normal phase liquid chromatography (NPLC), but GPC was the only chromatographic method capable of analyzing the precursor diester and the product monoester simultaneously. Both NPLC and RPLC required extensive method development to select the mobile phase. The resolution on NPLC or RPLC was better than that on GPC, but not remarkably better. The peak shape symmetry on NPLC and RPLC was poor, while it was excellent on GPC. Peak shape symmetry is an important feature in evaluating the purity of a compound. This confluence of favorable factors — high resolution in a reasonable analysis time, the capability of simultaneous analysis of synthetic precursors with products, good peak symmetry, and the ease of method development — combined to make GPC attractive in the analysis of the esters of biphenyl-4,4′-dicarboxylic acid.

The separation of the esters of biphenyl-4,4′-dicarboxylic acid shown above is by no means the first nor the last word on small molecule GPC. A number of excellent separations were achieved many years ago,[67,68,70] but progress toward improving resolution has been slow. Of the factors that affect resolution, pore size can be optimized to enhance resolution in the key region of molecular weight. Faster flow may help to enhance pore utilization. Smaller particle size and reduced extracolumn effects may be useful in decreasing diffusive broadening. Smaller particle size and faster flow imply higher pressure unless temperature is raised. Examples of GPC separations approaching the resolution available on NPLC or RPLC are appearing in the literature,[71-74] indicating a growing interest in the technique. One example is the separation of the buckminsterfullerenes C_{60} and C_{70}.[75] Two 60 × 2-cm, 10-μ particle size, 500-Å pore size PS-DVB columns (Showa Denko; Tokyo, Japan) were operated in toluene.

Even with optimized GPC columns, high-molecular-weight polymers cannot be separated into their individual components. The best separation seen for the BDCA esters above involved a molecular weight difference of about 5%. A small polymer has a degree of polymerization of 100, so adjacent homologs differ by about 1%. High-molecular-weight polymers may have a degree of polymerization of 100,000 or more, so adjacent homologs differ by only 0.001% in molecular weight. Since there is no possibility of resolving individual homologs, synthetic polymers are commonly analyzed using the statistics of distribution.

6.1.6   Polymer molecular weight distribution statistics

The weight-average molecular weight (M_w), number-average (M_n) molecular weight, and the polydispersity ratio (P) can be calculated as:[76,77]

$$M_w = \sum_{i=1}^{n} (h_i M_i) \bigg/ \sum_{i=1}^{n} (h_i)$$

$$M_n = \sum_{i=1}^{n} (h_i) \bigg/ \sum_{i=1}^{n} (h_i M_i)$$

$$P = M_w / M_n$$

where n is the number of chromatographic slices, h_i is the SEC curve height at the ith volume increment, M_i is the molecular weight of the species eluted in the ith retention volume increment, and P is the polydispersity. For a monodisperse sample such as a protein, $M_w = M_n$, so P = 1. Other quantities that are used in polymer analysis are the z-average molecular weight, M_z:

$$M_z = \sum_{i=1}^{n} (h_i) M_i^2 \bigg/ \sum_{i=1}^{n} h_i M_i$$

the viscosity average molecular weight:[77]

$$M_v = \left[\sum_{i=1}^{n} \left(h_i M_i^{1+a} \right) \Big/ \sum_{i=1}^{n} \left(h_i M_i \right) \right]^{1/a}$$

where a is the Mark-Houwink exponent as defined in Section 6.2.7 and the peak maximum molecular weight (M_p). In practice, then, one needs to know the molecular weight corresponding to each retention time. One approach to calculating M_n, M_w, M_p, and M_z from GPC retention time is to use multiple narrow-molecular-weight standards to construct a calibration curve of molecular weight vs. retention time. Narrow standards of a few common polymers, such as polystyrene, polyethylene oxide, sodium polystyrene sulfonate, and pullulan, are commercially available. Molecular weight ranges, solvents, and suppliers are listed in Table 2. As an example of homemade standards, glycosaminoglycans were fractionated from bovine tracheal chondroitin sulfate A on Sephadex™ G200, then characterized by molecular weight on Pharmacia Superose® 6 and 12 HR 10/30 columns.[78] Alternatively, one may use a single, extremely well characterized broad standard.[79] It is generally agreed that a cubic fit of the logarithm of molecular weight of standards vs. retention is preferable to a linear fit.

Several problems arise in practice in determining a polymer distribution from GPC retention alone. The greatest difficulty arises from the assumptions involved in relating molecular weight to size. A number of parameters have been proposed to relate retention to molecular weight, including projection length, radius of gyration, contour length, viscosity radius, and diffusion coefficient or Stokes radius.[28-40,80,81] Polymer size and retention on GPC are structure dependent. In other words, one cannot directly calibrate a column for one polymer using another as a standard. The universal calibration approach[81] has been used as an empirical correction to convert the molecular weights obtained from one set of standards for use with a sample of a different composition. To perform universal calibration requires careful measurement of the viscosity of well characterized samples. The approach has been automated, using an in-line viscosimetric detector.[82] A significant advantage of in-line determination of viscosity is the elimination of effects from low-molecular-weight contaminants. Universal calibration is described below.

Branching in the polymer chain affects the relationship between retention and molecular weight.[83] Universal calibration has been used with some success in branched polymers, but there are also pitfalls. Viscosimetry[84-91] and other instrumental methods have proved to be useful. A computer simulation of the effects of branching on hydrodynamic volume and the detailed effects observable in GPC is available in the literature.[92,93] In copolymer analysis, retention may be different for block and random copolymers, so universal calibration may be difficult. However, a UV-VIS detector, followed by a low-angle light-scattering (LALLS) detector and a differential

Table 2 Molecular Weight Standards

Polymer	Typical solvents	Typical molecular weight	Polydispersity	Vendors
Polystyrene	THF, toluene	$162–2 \times 10^7$	Narrow	Phenomenex™, Waters, Polymer Laboratories
	CHCl$_3$	$345–235 \times 10^3$	Broad	Phenomenex™
PMMA	HFIP, CHCl$_3$,	$1000–150 \times 10^3$	Narrow	Phenomenex™, Waters, Polymer Laboratories
	Acetic acid	$17,600–513 \times 10^3$	Broad	Phenomenex™
Polyisoprene	THF, hydrocarbons	$1000–300 \times 10^3$	Narrow	Waters, Polymer Laboratories
Polybutadiene	THF, hydrocarbons	$1000–100 \times 10^3$	Narrow	Waters, Polymer Laboratories
Polypropylene	Hydrocarbons	$1220–348 \times 10^3$	Broad	Phenomenex™
PolyTHF		$1000–500 \times 10^3$	Narrow	Waters, Polymer Laboratories
Pullulan	Aqueous	$5800–1.6 \times 10^6$	Narrow	Phenomenex™, Waters
Dextran	Aqueous	$505–4.9 \times 10^6$	Broad	Phenomenex™
PEO	Aqueous, alcohols, DMF	$26,000–860 \times 10^3$	Narrow	Phenomenex™, Waters, Polymer Laboratories
		$122,700–700,500$	Broad	Phenomenex™
PEG	Aqueous, alcohols, DMF	$62–22,000$	Narrow	Phenomenex™, Waters, Polymer Laboratories
PAA	Aqueous, alcohols, DMF	$1930–958 \times 10^3$	Narrow	Phenomenex™, Polymer Laboratories
Polyacrylamide	Aqueous, morpholine	$11,530–1.14 \times 10^6$	Narrow	Phenomenex™
PSS/NaPSS	Aqueous	$4800–708 \times 10^3$	Narrow	Pressure Chemical

refractive index (DRI) detector, has been used for accurate characterization of a polystyrene-polyethylene oxide co-polymer, and this approach contrasted with a dual detector method requiring compositional information.[94] Dimers occurring on termination of polymerization were noted.

Fluctuations in retention due to changes in flow rate lead to exponential changes in calculated molecular weight. Even with the excellent flowrate control of modern pumps, calculated molecular weights vary several percent from run-to-run based on flowrate correction. Flowrate fluctuations interfere with the use of a broad standard for calibration. Inhomogeneity in packing pore size distribution has been implicated as a significant concern in obtaining column-to-column reproducibility.[89,95,96] In blends of packing materials, the calibration graph may be linear piecewise but with discontinuities. Although the effect is difficult to characterize with narrow standards, the result is that the error in associating a molecular weight with a retention time fluctuates about a mean.

High concentration can lead to distortion of the distribution as observed on GPC. This may occur in several ways. At high concentration, high-molecular-weight polymers may not dissolve completely in the time allotted. High-molecular-weight polymers can also shear as they pass through the column frits.[97,98] More typically, viscous fingering causes inhomogeneities in flow.[99,100] Also, there may be inadequate pore volume within a molecular weight regime. Partial sample exclusion, particularly of the higher-molecular-weight standards, is a common problem, leading to early elution of samples. The activity of the solute may change as it is diluted during chromatography, causing steric or ion exclusion. For these reasons, it is preferable to use the minimum sample concentration possible. The sensitivity of the detector, however, may lower-bound the minimum concentration.

Analyte adsorption can be a serious problem. Typically, adsorbtion leads to late elution of standards and samples. Although more common in aqueous than in nonaqueous GPC, adsorption should always be suspected if standards show distorted peak shape. Some nonaqueous solvents, notably THF, absorb water from the atmosphere, so retention may vary with the age of the solvent. As part of developing a nonaqueous GPC method, it is wise to test the effects on retention of dry and wet solvent, since adsorption can distort the molecular weight distribution. With some solvents, such as N,N,-dimethylformamide (DMF), one may wish to test the effect of the addition of salts such as LiBr. In developing an aqueous GPC method, varying the salt content and pH are important steps in obtaining reliable, rugged methods.

6.1.7   Nonaqueous GPC

The single most important consideration in nonaqueous GPC is sample solubility. Although adsorption is not an infrequent problem, finding a solvent for a polymer is usually the hard step in analysis. The most common solvents for nonaqueous GPC are toluene, tetrahydrofuran, chloroform, and DMF. A number of potentially useful solvents are toxic, corrosive, or expensive, considerations which may weigh heavily in the choice of a solvent.[101] Examples of applications of solvents used in GPC include the use of toluene for asphalt[102] and vinyl chloride oligomers;[103] tetrahydrofuran for alkyd resins[104] and poly(styrene-co-butadiene);[105] chloroform for polyethylene and polypropylene glycol;[106] trifluoroethanol and LiBr for nylon 6, nylon 4,6, and nylon 6,6;[107] pyridine and N-methylpyrrolidone for poly(2-vinylpyridine);[108] trichlorobenzene at elevated temperature for polypropylene,[109] polyethylene,[110] and poly(ethylene-co-vinyl acetate);[91] decalin for polypropylene;[111] *o*-chlorophenol at elevated temperature for nylon 6;[112] benzyl alcohol for polyamides;[101] and a mixture of phenol-trichlorobenzene at elevated temperature for poly(ether ether ketone).[113] The Shodex™ OHPak SB-800 HQ series columns, which are based on a polyhydroxymethacrylate packing, were found to be superior to silica or PS-DVB packings in the separation of polycarbonate, phenol-formaldehyde resols, and polydimethylaminostyrene in DMF.[114] The incorporation of an amine into the OH-Pak SB800 causes it to behave as a base in DMF, reducing ionic and hydrophobic adsorption. Extensive compilations of suitable solvents are available in the literature.[115]

Detection is also frequently a key issue in polymer analysis, so much so that a section below is devoted to detectors. Only two detectors, the ultraviolet-visible spectrophotometer (UV-VIS) and the differential refractive index (DRI), are commonly in use as concentration-sensitive detectors in GPC. Many of the common polymer solvents absorb in the UV, so UV detection is the exception rather than the rule. Refractive index detectors have improved markedly in the last decade, but the limit of detection remains a common problem. Also, it is quite common that one component may have a positive RI response, while a second has a zero or negative response. This can be particularly problematic in co-polymer analysis. Although such problems can often be solved by changing or blending solvents, a third detector, the evaporative light-scattering detector, has found some favor.

6.1.8   GPC of proteins, polynucleotides, and carbohydrates

Proteins, polynucleotides, and carbohydrates normally are analyzed in aqueous systems. It has been recently recognized that proteins can be chromatographed in nonaqueous systems.[116] Proteins are monodisperse, so obtaining the optimum resolution is often an issue. Even at maximal resolution, however, GPC is not competitive with electrophoretic means for purity determination. Also, hydrophobic interactions between analyte and stationary phase are common. A typical application was the measurement of the strength of complexation of erythropoeitin complexed to the monomer and the dimer of the extracellular portion of the erythropoeitin receptor.[117] A Superdex® 200 column (Pharmacia) was coupled in series to a Knauer UV-VIS detector, a Wyatt Mini-Dawn® light-scattering detector, and a Polymer Laboratories PL-RI refractive index detector. The monomer-dimer equilibrium of human growth hormone was studied using a 6 × 40-mm TSK-Gel™ G3000SWXL guard column with 7-μ particle size packing, with the short residence time enabling the detection of aggregates with half-lives as short as 5 seconds.[118] The molecular weights of the myofibrillar proteins, titin (formerly known as connectin; $M_w = 2 \times 10^6$) and nebulin ($M_w = 560 \times 10^3$), were determined in the denaturant, guanidinium hydrochloride, on a 60-cm TSK6000-PW and a 30-cm column packed with Superose®-6 cross-linked agarose.[119] Particularly on the agarose support, the retention volume in guanidine was not identical to that in a nondenaturing solvent. The effect could not be explained by a change in viscosity radius, suggesting that the porosity of the supports is altered by the denaturant. Ion exclusion was intensified by the addition of a zwitterionic detergent in a pI-dependent manner, making it possible to separate isozymes.[120] A glycerolpropylsilane modified silica column, the Waters™ Protein-Pak™ 125 (Waters Corp.; Milford, MA), was used to separate potato invertase inhibitor and potato hemaglutinin in the presence of glycine derivatives.

Polynucleotides tend to be of such high-molecular-weight that, even if monodisperse, additional resolution is of little help in many analytical problems. Electrophoresis has tended to predominate in polynucleotide analysis, although ion exchange has proven increasingly useful. The substances used in anti-sense DNA technology, described in Chapter 3, are of such a molecular weight that GPC could be used to advantage. These synthetic substances, typically of the size of 10 to 30 base pairs, probably could be analyzed for deletion variants by GPC.

Polymeric carbohydrates are usually encountered as distributions, so high resolution is rarely important. Of all biological macromolecules, carbohydrates are particularly amenable to analysis by GPC because hydrophobic interactions are typically weak. A section below is devoted to the analyses of carboxymethylcellulose and xanthan. Other examples of polysaccharides of interest are hyaluronic acid,[62] polymers of β-glucose,[121-125] heparin,[126,127] cellulose and chitin,[128] and Mucorales extracellular polysaccharides.[129]

6.1.9   GPC of synthetic water-soluble polymers

Many synthetic water-soluble polymers are easily analyzed by GPC. These include polyacrylamide,[130] sodium poly(styrenesulfonate),[131] and poly (2-vinyl pyridine).[132] An important issue in aqueous GPC of synthetic polymers is the effect of solvent conditions on hydrodynamic volume and therefore retention. Ion inclusion and ion exclusion effects may also be important. In one interesting case, samples of polyacrylamide in which the amide side chain was partially hydrolyzed to generate a random copolymer of acrylic acid and acrylamide exhibited pH-dependent GPC fractionation.[130] At a pH so low that the side chain would be expected to be protonated, hydrolyzed samples eluted later than untreated samples, perhaps suggesting intramolecular hydrogen bonding. At neutral pH, the hydrolyzed samples eluted earlier than untreated samples, an effect that was ascribed to enlargement of the radius of gyration. Porous glass columns exhibited ion exclusion of sodium polystyrene sulfonate, but hydrophobic adsorption to polymeric columns.[131] Poly(allyl amine) was fractionated on a cationic-sized exclusion column using viscosimetric detection.[132]

6.1.10 GPC of copolymers

Copolymer analysis presents a special challenge. First, the solubility properties of the monomeric units may differ substantially, complicating the process of selecting a suitable solvent. Also, conformation and therefore retention may be dependent on the linear sequence. The copolymer A_nB_n might exhibit different retention than $(AB)_n$. If the spectroscopic properties of the monomeric units differ, coupled detectors can be used to estimate the composition.[133] For example, the UV absorbance of poly(styrene-methyl methacrylate) would be directly dependent on the styrene content, while a concentration-sensitive detector, such as the evaporative light-scattering detector described below, could be used to estimate the total amount of analyte at a given retention time. For copolymer analysis, viscosimetric detection may also be of use.[133] On the other hand, if the monomeric units are spectroscopically similar, as in poly (methyl methacrylate-decyl methacrylate), analysis may be more difficult. Several investigators have demonstrated the utility of adsorptive chromatography in copolymer analysis.[134-136]

6.1.11 Universal calibration

As was mentioned above, there is no direct relationship between molecular weight and GPC retention. Instead, the GPC retention is determined by a complex set of factors, including molecular shape and chain branching. The many factors proposed to determine GPC retention include the molecular weight, the second virial coefficient (Stokes radius), viscosity radius, radius of gyration, mean linear projected length, and contour length. Of these, the viscosity radius R_η is the most central in determining GPC retention.[40] This was first demonstrated in a study of linear polystyrene, poly(methylmethacrylate), polybutadiene, and polyvinyl chloride; branched polymers such as star polystyrene and poly(methylmethacrylate) and comb polystyrene; and heterograft copolymers of polystyrene on poly(methylmethacrylate).[81] It was found that a plot of log ([η] M) vs. elution volume gave a linear relation.

For a statistical coil, the product of polymer intrinsic viscosity and molecular weight is directly proportional to the cube of the root-mean-square radius of gyration R_G:[77,137]

$$[\eta] \, M \, \alpha \, R_G^3$$

The radius of gyration may be related to an equivalent hydrodynamic sphere, which is conceptualized as a solid sphere of radius R_e.[138] This sphere is often considered to represent a solvated polymer molecule. If so, the relation above can be written as:

$$[\eta]\, M \propto R_e^3$$

such that the product of intrinsic viscosity and molecular weight is proportional to the equivalent hydrodynamic volume.

Then, from the empirical Mark-Houwink relationship,[77,139-141] the viscosity can be related to the viscosity-average molecular weight as $[\eta] = KM_v^a$. It should be noted that alternative forms for this empirical equation have been presented. For example, because the Mark-Houwink relationship does not fit well in the low-molecular-weight region, an equation of the form $1/[\eta] = a + bM^{-1/2}$ was proposed.[141] Confining the discussion to the Mark-Houwink relationship, it can be seen that for monodisperse statistical coil polymers, $KM^{a+1} \propto R_e^3$. In other words, given specified solution conditions, the molecular weight is directly related to the hydrodynamic volume. Since the hydrodynamic volume is the variable that controls the retention, a plot of the product of $\log([\eta]M)$ vs. the retention volume should give a single curve for all polymers. Then, from knowledge of the retention volume and viscosity, the molecular weight of an unknown sample can be determined. If the relationship between molecular weight and viscosity has been established for polymers of the type of the unknown sample, the viscosity can be calculated, rather than measured. For example, if the viscosities and molecular weights of a series of monodisperse polystyrene and polybutadiene standards have been measured, polystyrene standards can be used to calibrate a column for polydisperse butadiene samples.

Several significant difficulties arise in practice in performing universal calibration. First, obtaining truly monodisperse samples in sufficient quantities for measurement of intrinsic viscosity may be a significant undertaking. For polydisperse samples, the viscosity molecular weight may be far different than the average molecular weight. Second, because the hydrodynamic volume is dependent on solution conditions, the values of the Mark-Houwink constants, K and a, usually depend on solution conditions. Standards and samples must be evaluated in the same solution. Third, due to imprecision in measurement, literature values for Mark-Houwink constants vary widely. Finally, many polymers differ significantly from the statistical coil conformation. Given well characterized, monodisperse standards, this is not an impediment to applied chromatography. Since there are significant difficulties to obtaining such standards, however, there is a temptation to calculate the molecular weights necessary for calibration by combining theory with a molecular size derived by another technique or, equivalently, to calculate size from molecular weight data. There are many pitfalls to such calculations. A solvated macromolecule has around it a layer of solvation, so the partial specific volume (estimated by measuring the volume change on dissolving a known weight of polymer in a known volume of solvent) may differ greatly from the volume observed crystallographically. The effective or equivalent hydrodynamic volume, which includes the shell of tightly bound solvent molecules, is greater than the volume of the macromolecule less its solvation sphere. The viscosity, which is not directly dependent on molecular weight, is highly dependent on the molecular shape of the solvated molecule.[142] The hydrodynamic volume depends on the Stokes radius and a shape factor,[40] both of which are dependent on solution conditions. Empirically, the viscosity seems to follow the same functional form as the shape factor.[40] The Scheraga-Mandelkern equation has proven useful to estimate the shape of certain polymers.[143]

Recent advances have greatly decreased the difficulties raised by these cautionary points. In particular, on-line viscosimetry and multi-angle light-scattering make it possible to determine the molecular weight and viscosity of samples as a function of elution volume. With such detectors, effects of solute-column packing interactions become unimportant, since the properties of narrow fractions can be measured. These detectors will be discussed in greater detail below.

6.1.12 Corrections for flow phenomena

A number of corrections may have to be made for instrumental factors. This is particularly true when detectors are used in series. There may be significant band broadening in the upstream detector flow cell. Also, except for monodisperse samples, one cannot assume that the maximum in the signal of one detector corresponds to the maximum of another detector, since the sensitivity of each detector to different molecular weights may not be equivalent. The simplest correction is for interdetector volume. It has been recommended that the interdetector volume be measured by correlating the response to a monodisperse standard in each of the two detectors.[96,144,145] It is also recommended that a multiple detector system be built up sequentially, verifying the function of each detector. An iterative, sample-independent method of estimating interdetector lag has also been proposed.[146] When flow splitting is used to distribute sample to two detectors, the relative flows through the detectors can be affected by solution viscosity, which changes with the elution of a polymer.[147] The correct M_p for each detector, however, can be found by scaling the data by a proportionality factor.

Axial flow correction is the correction for peak broadening due to diffusion or extracolumn effects.[144-149] Most theoretical treatments begin from Tung's equation.[150,151] One can correct for effects of axial dispersion by use of monodisperse standards, fitting the peak shape to a band-spreading function. Lacking truly monodisperse standards, one must be careful that the response of the detectors does not represent the peak differently. The relationship between peak distortion and molecular moments, demonstrating the effects of axial distortion, has been explored.[152] An application with the polypropylenes Daplen PT55 and KS-10 indicated that peak broadening is not a universal function but is specific to a given polymer.[153]

6.1.13   The critical role of the pump in gel permeation chromatography

An extremely useful analysis of the effects of flowrate variation on the values calculated for molecular weight from columns calibrated by standards has been compiled.[154] Except for very high-molecular-weight polymers, column performance is relatively independent of average flow. Flow, however, can vary over the course of a day or during a run. Day-to-day variation can be minimized by timing the pump. Maintaining check valve function is critical in obtaining reproducible flow. Variation in flow over the course of a day can arise from a number of factors, the most serious being dissolution of atmospheric gas in the mobile phase. Continuous degassing of solvents with helium (which has a low solubility) or maintaining a degassed solvent under a helium atmosphere helps to minimize the associated decrease in pump rate. Slow leaks on the inlet side of the pump are another source of flowrate error. Fluctuations in flow, which can arise from solute viscosity effects, can greatly affect calculated values of the molecular weight. To obtain reliable values of molecular weight, one may, of course, run numerous replicates of samples and standards and average results. It is much less wasteful to use pumps that have extremely good flowrate stability. As little as 1% variation in flow will cause about 1% variation in molecular weight if the variation is random or increasing or decreasing across the peak, but about 10 to 20% if the 1% deviation is run-to-run or continuously increasing or decreasing from the beginning of the run. On-line flowrate correction, while certainly technically feasible, adds complexity to an already complex system. Some modern pumps are capable of extremely good flowrate reproducibility, and it may be easier to install these for the analysis rather than use an inferior pump and try to correct the complications introduced by it. The Hewlett-Packard® diaphragm pump and the Applied Biosystems syringe pump are examples of systems with the extraordinarily high flowrate precision required for GPC. In a recent work,[69] the claims of flowrate stability by Hewlett-Packard® for its Model 1090 were found to be fully warranted, the total range of flowrate variation being about 0.2% within a series of runs and less than 1% between two sets of runs. Even extremely well maintained piston pumps tend to have a far lower degree of flow rate reproducibility.

6.2   Detectors for gel permeation chromatography

6.2.1   General

In the absence of adsorption, inclusion, or exclusion, a polymer is fractionated on a GPC column according to the hydrodynamic volume.[40,138] The hydrodynamic volume is a function of monomer identity, as well as polymer molecular weight, branching, and cross-linking. The polymer chains in any given chromatographic fraction have roughly the same hydrodynamic volume.

In the absence of any nonsize exclusion effects or branching effects, the concentration distribution within the chromatographic peak describes the polymer molecular weight distribution. The concentration distribution is deduced from the response of a suitable detector. Detectors which respond to physicochemical properties other than concentration may also be of use in GPC. Over the past 25 years, a variety of detectors have been developed. This section reviews detectors available to the chromatographer. The detectors used in GPC can be grouped as

- Concentration-sensitive detectors, such as the refractive index detector or UV-VIS spectrophotometer
- Molecular-weight-sensitive detectors, such as a laser light-scattering photometer

Figure 6 summarizes the various detectors that can be used in conjunction with a GPC system. It should be noted that direct determination of polymer molecular weight by mass spectrometry is increasingly popular. For example, in the field of automotive coatings, acrylic macromonomers of molecular weight <3000 were separated by GPC and directly inserted into an electrospray mass spectrometer.[155] By this means, cross-linking agents, additives, stabilizers, and coalescing solvents could be differentiated from a complex formulation.

The primary purpose in performing GPC is to determine the molecular weight and molecular weight distribution (MWD) of the chromatographed polymer. The use of concentration-sensitive detectors dictates the need for column calibration with a series of monodisperse or a few broadly dispersed polymer standards of known molecular weight. A calibration curve that relates the known molecular weight of each polymer standard to its elution volume is generated. Using this calibration curve, the unknown molecular weight of a polymer can be obtained by summing over the molecular weights of peak slices, as has been described above. The various calibration methods have been reviewed elsewhere.[79] On the other hand, a molecular-weight-sensitive detector responds to both the molecular weight as well as the concentration of the eluting polymer. Thus, they must be used in tandem with concentration-sensitive detectors. Column calibration, however, is not necessary with the use of molecular weight detectors.

6.2.2   Concentration-sensitive detectors

The output signal of concentration-sensitive detectors is proportional to the concentration or weight of polymer in the column eluent. Examples of this type include the differential refractometer and the ultraviolet-visible spectrophotometer. Infrared and fluorescence detectors are used infrequently. None of the detectors described above is truly universal; i.e., the response of these detectors varies with the chemical species, and, in case of the DRI, response is also a function of the chromatographic eluent.[156] Recently, an

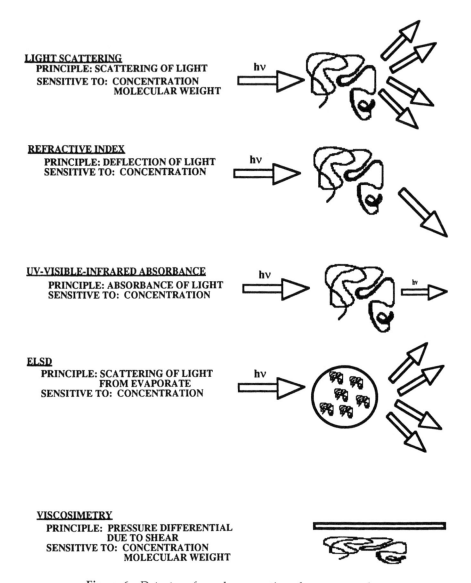

LIGHT SCATTERING
    PRINCIPLE: SCATTERING OF LIGHT
    SENSITIVE TO: CONCENTRATION
                  MOLECULAR WEIGHT

REFRACTIVE INDEX
    PRINCIPLE: DEFLECTION OF LIGHT
    SENSITIVE TO: CONCENTRATION

UV-VISIBLE-INFRARED ABSORBANCE
    PRINCIPLE: ABSORBANCE OF LIGHT
    SENSITIVE TO: CONCENTRATION

ELSD
    PRINCIPLE: SCATTERING OF LIGHT
                FROM EVAPORATE
    SENSITIVE TO: CONCENTRATION

VISCOSIMETRY
    PRINCIPLE: PRESSURE DIFFERENTIAL
                DUE TO SHEAR
    SENSITIVE TO: CONCENTRATION
                  MOLECULAR WEIGHT

Figure 6   Detectors for gel permeation chromatography.

evaporative light-scattering detector (ELSD) has been introduced.[157-166] Its manufacturers claim that instrumental response is a function of number and size of the polymer molecules and is invariant with respect to the nature of the chemical species and the eluent.[164] This claim has been firmly rebutted.[160] Linearity in instrumental response, however, is generally an idealization of real detector behavior. Although the ELSD response is markedly sigmoidal, the instrument has considerable utility.

6.2.3   Differential refractometers

Differential refractive index detectors measure the difference in refractive index between the solvent and the eluting polymer solution. The DRI can be used with almost any polymer-solvent combination, as long as the incremental increase in refractive index of the polymer solution with increase in polymer concentration is sufficient. Although a large number of differential refractometers are available in the market,[167,168] they usually conform to three main types, namely

- Deviation, also known as deflection
- Fresnel
- Interferometric

Deviation refractometers are the most commonly used. This version of the DRI measures the deflection in the location of a light beam on the surface of a photodiode by the difference in refractive index between the polymer solution and pure solvent. The Fresnel-type refractometers operate on the principle that the intensity of light reflected from a glass-liquid interface is dependent on the incident angle and the RI difference between the two phases. The deviation and Fresnel detectors typically have cell volumes of 5 to 10 µl, detection limits of about 5×10^{-6} refractive index units (RIU), and a range of 10^{-7} to 10^{-3} RIU.[156] The deflection-type DRI is relatively insensitive to the buildup of contaminants on the sample cell and is therefore of special utility in laboratories that process large numbers of samples, such as industrial laboratories.

The recently introduced Wyatt/Optilab® DSP DRI (Wyatt Tech.; Santa Barbara, CA) is based on the interferometric principle. As shown in Figure 7, it consists of a collimated light beam, plane polarized at 45° to the horizontal. The first Wollaston prism splits the beam into two components, a vertically polarized beam passing through the sample cell and a horizontally polarized beam passing through the reference cell. The two are recombined in the second prism, quarterwave plate, and analyzer to yield a plane-polarized beam rotated at an angle $\varphi/2$ with respect to the original beam. The phase difference φ is directly proportional to the difference in refractive index between the sample and reference. The sensitivity of this instrument is a full order of magnitude greater (about 10^{-7} RIU) than conventional DRI detectors. The response of this instrument is independent of solvent; unlike the first two types, it does not require recalibration for different solvents. It is also claimed to be insensitive to temperature induced changes in solvent RI, although thermal equilibration is still required.

6.2.4   UV-VIS detectors

These detectors are based on selective absorption of ultraviolet or visible light by the desired solute. A large number of single or multiple wavelength

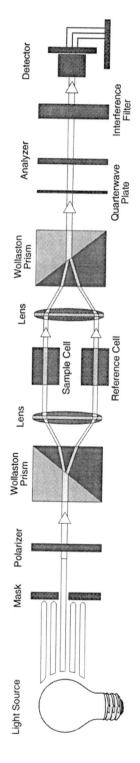

Figure 7 Schematic of the Wyatt/OptiLab® Model DSP refractive index detector. (Reprinted with permission of Wyatt Technology Corp; Santa Barbara, CA.)

detectors are available in the market.[168,169] UV/VIS detectors are insensitive to solvent flow rate and temperature changes and have exceptional baseline stability. These detectors are very reliable and easy to operate, exhibiting linearity over many decades of response. Their sensitivity is typically almost two orders of magnitude greater than differential refractometers, permitting nanogram detection of samples.[156] Despite these many advantages, the use of UV-VIS in GPC is rather limited. Unlike the DRI, the UV-VIS cannot be used with all solvent-polymer pairs. Often a UV-transparent solvent that solubilizes the polymer cannot be found. In this context, it should be noted that tetrahydrofuran, the most commonly used polymer solvent, can be used at wavelengths greater than 240 nm if unstabilized solvent is used.

6.2.5 IR and fluorescence detectors

The infrared (IR) spectrophotometer has sensitivity comparable to a differential refractometer. Both conventional and Fourier transform (FT) spectrophotometers have been used.[170-175] The acquisition speed of the FT spectrophotometer has been shown to be useful for on-line analysis.[172-174] The IR spectrometer is particularly useful when the relative amount of a functional group incorporated into a polymer must be quantitated. This function makes it particularly useful in co-polymer analysis. The principal limitations of IR are that solvents may absorb strongly in a region of interest and IR absorbances, particularly those of polymers, tend to be broad bands not uniquely characteristic of a single functional group. Solvent absorption has been overcome by nebulization and deposition onto a germanium disk for later analysis, an approach that was illustrated with copoly(ethylene-vinylacetate) and high density polyethylene chromatographed in trichlorobenzene and jet oil lubricant in unstabilized tetrahydrofuran.[176]

Fluorescence detectors, discussed in Chapter 1, are extremely sensitive; picogram quantities of sample can sometimes be detected. However, most polymers (with the exception of certain proteins) are not fluorescent and thus these detectors are rarely used in GPC. Proteins, particularly those containing tryptophan, fluoresce intensely and are readily detected. Because both the IR and the fluorimetric detector are selective for certain functional groups, rather than being sensitive to analyte mass, there are many pitfalls in quantitation. These and other detectors have been reviewed.[177,178]

6.2.6 Evaporative light-scattering detector

As mentioned previously, this detector has been claimed to be universal, i.e., it has been reported that the response of the ELSD is not a function of the nature of the solute. Although this is not, strictly speaking, true, the detector is almost universal. The detector functions by nebulizing column effluent into droplets, which are evaporated in a heated gas stream. A droplet of evaporate, containing some solvent, remains. Light of a wavelength considerably smaller than the evaporate particles, which are 5 to 15 μm in size,[161]

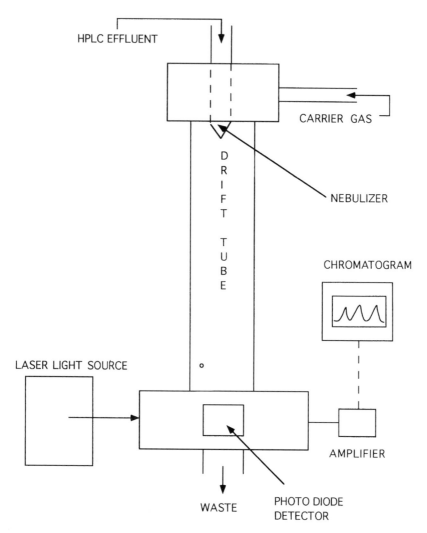

Figure 8 Construction of the evaporative light-scattering detector. (From Stockwell, P. B. and King, B. W., *Am. Lab.*, 23, 19, 1991. With permission.)

is scattered by a phenomenon known as Mie scattering. The intensity of scattered light is detected by a photomultiplier tube. As shown in Figure 8, the GPC eluent is passed to a nebulizer tube where it is dispersed to droplets by a high velocity stream of carrier gas. The nebulized gas is then evaporated in a temperature-controlled drift tube to form small particles of nonvolatile solute. These particles are carried to a light-scattering photometer in the base of the instrument. The scattered light intensity is measured at a 60° forward angle. The magnitude of light scattered is related to the number and size of the particles. A commercially available evaporative light-scattering detector is the Varex Universal HPLC Detector (Varex; Burtonsville, MD). The response of this detector has been shown to be invariant to a first approximation with respect to:[157,163]

- Solvent type and solvent composition
- Polymer type and polymer molecular weight
- Temperature
- Flow rate

The limit of detection for this instrument is about 10 µg/ml for polystyrene in 2-butanone,[163] which is close to two orders of magnitude higher than that of the deflection-type DRI. Moreover, the response of the ELSD is linear over only two decades in concentration.[163] The ELSD is a useful backup detector when the DRI or UV detectors are not appropriate, e.g., when the UV absorbance or RI change is a function of copolymer composition as well as concentration or in gradient elution systems where changes in solvent composition cause drift in baselines of the UV and DRI detectors. Compounds about as volatile as the solvent are poorly detected by ELSD.

6.2.7   Molecular-weight-sensitive detectors

As their name implies, these detectors are sensitive to both molecular weight and concentration of the polymer. Although a variety of techniques (e.g., colligative property determination, scattering, and sedimentation) can provide us with the molecular weight of polymers, not all are amenable for real time on-line measurement. Currently, only laser light-scattering and viscosimetry are useful as methods of chromatographic detection. Laser light-scattering is an absolute measurement technique, so polymer molecular weights are determined without recourse to prior column calibration with molecular weight standards. The response of an on-line viscosimeter, by contrast, is dependent on the intrinsic viscosity of the polymer solution. The intrinsic viscosity of the polymer solution is related to its molecular weight through the Mark-Houwink-Sakurada equation described below:[79,179]

$$[\eta] = K \, M^a$$

where $[\eta]$ is the viscosity, M is the molecular weight, and K and a are the Mark-Houwink constants. Using an on-line viscosimeter requires either prior knowledge of the Mark-Houwink constants or determination of a universal calibration curve with molecular weight standards. The on-line viscosimeter is thus not a true absolute molecular weight detector. It is, however, classified as a molecular-weight-sensitive detector. Another detector that could also be included in this type is the OROS 801 molecular size detector (Oros Instruments Ltd.; Slough, Berkshire, England).[180] This detector is based on the principle of photon correlation spectroscopy, also known as dynamic light-scattering or quasielastic light-scattering. Using the Stokes equation, or Stokes radius, it measures the diffusion coefficient (D_T) of the eluting species from which molecular size and molecular weight can be determined. The manufacturer has specified that certain assumptions regarding molecular shape should be made in order to determine molecular weight,[180] so the applicability of this detector is limited to materials of known shape, such as spherically shaped globular proteins and polymer latex suspensions. Photon correlation spectroscopy has been used for characterization of poly(lactic acid) nanoparticles of radius 50 to 200 nm[181] and gelatin.[182]

6.2.8   Light-scattering photometer

Light-scattering photometers work on the principle of Rayleigh scattering. The Rayleigh principle states that the excess light scattered by a polymer solution over that scattered by the pure solvent is directly proportional to the polymer molecular weight. The theory of light-scattering and its application to on-line molecular weight measurement in GPC has been reviewed.[183-185] There are two types of light-scattering instruments currently available in the market: LALLS, a low angle (about 5 to 6°) scattering method,[185] and MALLS, a multiple angle[186] scattering method (15 angles from 5 to 175°, depending upon solvent/glass refractive index). A more recent design by Precision Detectors (Amherst, MA) uses just 15° and 90° collection to obtain accurate data for Gaussian coils at any molecular weight and most other molecules with a R_g <100 nm.[187] The technique, TALLS, was used for high-molecular-weight polystyrenes and long chain branched polyesters.[188] The characteristics of commercial light-scattering photometers are shown in Table 3. The low angle, laser light-scattering photometer can only provide molecular weight information. The advantage in using a multiple angle, light-scattering photometer is that the size (radius of gyration, R_G) of the polymer is additionally obtained.[185] Using both radius of gyration and molecular weight information, scaling relationships (Figure 9) that allow determination of polymer conformation in dilute solution can be established.[189] In addition, the MALLS instrument has no moving parts, no lenses, mirrors or prisms that could become misaligned, damaged or dirty. It is expected that MALLS will provide more reliable and reproducible measurements and prove to be a more versatile instrument than the LALLS.

Both LALLS and MALLS can also be used to estimate polymer branching. The degree of polymer branching can be determined using the Zimm-Stockmeyer branching index, which is given by:[190]

$$g_m = (M_{lin} \backslash M_{br})_{v_i}^{a+1}$$

where M_{lin} and M_{br} are the molecular weights of the linear and branched polymers, respectively, at the same elution volume v_i, and a is the Mark-Houwink-Sakurada exponent of the linear polymer. This equation can be applied only if universal calibration is valid. Quantitative branching information obtained using GPC-LALLS is shown in Figure 10.[191,192] Branching in

Table 3 Characteristics of Commercial GPC Light-Scattering Detectors

Model	Cell volume (μl)	Scattering volume (nl)	Angles (degrees)	Laser
KMX-6	10	35	2–3, 3–4, 3–7, 4.5–5.5, 6–7, 90, 175	2-mW He Ne 633 nm
CMX-100[a]	10.5	40	5.1–6.1	2-mW He Ne 633 nm
L-8[b]	30	100	5	—
MiniDawn®[c]	67	500	18 angles in equidistant cot θ from –1.6 to 2.0, i.e, 5–175°, depending on solvent and cell refractive indices	2-mW He Ne 633 nm
Mini-Dawn®[c]	—	20	45, 90, 135	20-mW 690-nm semiconductor
RALLS[d]	—	—	90	670-nm laser
T60[d]	—	—	30, 90	950-nm IR laser
PD2000W[e]	10	—	15 or 30 (optional)	2-mW 670-nm semiconductor

[a] LDC Analytical; Riviera Beach, FL.

[b] Tosoh Corporation; Tokyo, Japan.

[c] Wyatt Technology; Santa Barbara, CA.

[d] Viscotek Corp.; Houston, TX.

[e] Precision Detectors; Amherst, MA.

Source: Stuting, H. H., Krull, I. S., Mhatre, R., Krzysko, S. C., and Barth, H. G., *LC-GC*, 7, 402, 1989. With permission.

poly(vinyl galactose ketone), poly(3-O-methacryloly gluconic acid), and poly(2-(D-glucopyranosyl-3-oxo-methyl)-acrylic acid ethyl ester) has been determined using GPC-MALLS.[193]

The development of light-scattering detectors has substantially improved the reliability of the molecular weight data obtained via GPC. Errors associated with the use of calibration methods and the determination of polymer elution volume are overcome by the use of GPC-LS.[194] The use of light-scattering detectors, however, requires accurate determination of the specific refractive index increment (SRI, mathematically defined as dn/dc) of the polymer of interest in the solvent to be applied for light-scattering experiments. The second virial coefficient, A_2, for the polymer in solution must also be determined.[195] Interdetector volumes must be accurately determined, and, even so, the low and high ends of the molecular weight distribution may not be adequately determined because the sensitivity of, respectively, LS and DRI are poorest at the extrema.[196]

The (dn/dc) values are tabulated extensively in the literature.[197,198] They can also be measured offline, using a refractometer.[198] Differential refractometers commercially available for this application include the Brice-Phoenix[198]

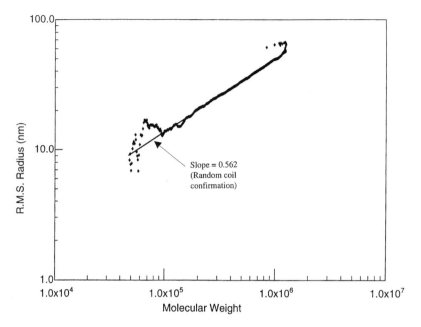

Figure 9 Scaling relationship between root-mean-square radius and molecular weight for polystyrene using GPC/MALLS. (From Wyatt, P. J., Hicks, D. L., Jackson, C., and Wyatt, G. K., *Am. Lab.*, 20, 108, 1988. With permission.)

and the Wyatt/OptiLab® DSP described above. The second virial coefficient, which arises from polymer-polymer interactions is more difficult to determine. This parameter is determined by plotting light-scattering results as a function of scattering angle and polymer concentration.[197] Fortunately, at concentrations typically employed in GPC, the contribution of the second virial coefficient can often be neglected. Exceptions occur when there are significant activity effects due to polymer-polymer or polymer-solvent interaction.

With a three-component system, such as a polymer in an aqueous salt solution, preferential adsorption of one component to the polymer can affect the analysis of light-scattering data.[199] Such interactions can affect the SRI. Therefore, measurements of the SRI must be made at constant chemical potential. Constant chemical potential is achieved experimentally by dialyzing the solvent and polymer solution to equilibrium through a membrane permeable to the solvent but impermeable to the polymer.[199]

6.2.9   On-line viscosimetry

The need for a viscosimeter capable of online measurement of the intrinsic viscosity of a polymer sample arose when it was shown that polymer molecules elute in order of decreasing hydrodynamic volume.[81] The hydrodynamic volume can be related to the product of intrinsic viscosity and molecular weight.

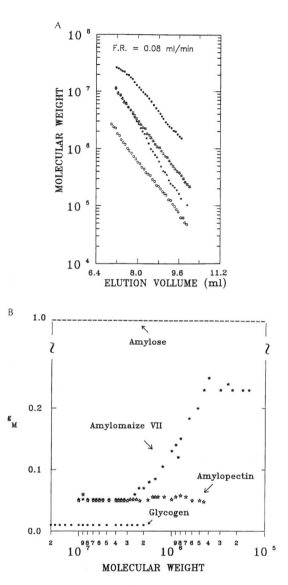

Figure 10 On-line determination of molecular weight and branching by light scattering. **(A)** Molecular weight as a function of elution volume. **(B)** Dependence of branching index, g_M, on molecular weight. (From Yu, L. P. and Rollings, J. E., *J. Appl. Polym. Sci.*, 33, 1909, 1987. With permission.)

It was shown that the logarithm of the product of intrinsic viscosity and molecular weight of polymers of different chemical and stereochemical composition, configuration, or molecular weight is linearly related to the elution volume.[81] A plot of log (η M) vs. elution volume provides a single calibration curve, useful for samples of similar composition, and is termed a universal calibration plot. However, the utility of this plot for determining the unknown molecular weight of a polymer sample is limited by the difficulty of accurate determination of Mark-Houwink constants.[85] The development of on-line viscosimetry has been most helpful in this regard.

The on-line viscosimeters currently available are adaptations of the classical dilute solution capillary viscosimeters. They work on the principle of measuring the pressure drop across a capillary with a differential pressure transducer. The pressure drop can be related to the reduced or inherent viscosity of the sample via Poiseuille's law.[84] Intrinsic viscosity is determined using the equation:

$$[\eta] = \lim_{c \to 0} \eta_{red} = \lim_{c \to 0} \eta_{sp}/c$$

where η_{red} is the reduced viscosity and η_{sp} is the specific viscosity, i.e., the difference between solution viscosity and solvent viscosity divided by the solvent viscosity:[77]

$$(\eta_{solution} - \eta_{solvent})/\eta_{solvent}$$

Since the concentration of polymer eluting from the GPC column is dilute (about 10 µg/ml), the approximation involved in taking the limit is valid and the viscosimeter provides accurate measurement of sample intrinsic viscosity. Using the intrinsic viscosity value at each elution volume and a column calibration curve generated by chromatographing polymer standards of known molecular weights, the molecular weight and weight distribution of any polymer are determined.

There are several commercial viscosimeters available. The Waters™ viscosimeter is a single capillary unit available as an integral part of their 150CV GPC system. The Du Pont viscosimeter (E.I. Du Pont de Nemours & Co.; Wilmington, DE) consists of two capillaries in series. The Viscotek viscosimeter (Viscotek; Houston, TX) consists of four capillaries with a Wheatstone bridge design. The salient features of each detector are shown in Table 4. It should be noted that the Waters™ design incorporates an extensive pulse dampening system to achieve constant flow rate, which may compensate for small pressure changes caused by differences in viscosity between the polymer solution and pure solvent.[85] An analysis of the sources of error in calculating M_n from the use of viscosimetry with a Model 502A differential viscosimeter showed the potential of large error from even small errors in the estimate of interdetector volume, the sensitivity of the viscosimeter to low-molecular-weight species, and the need to correct for axial dispersion.[200]

The three on-line viscosimeters are capable of providing molecular weight, weight distribution, radius of gyration, polymer branching, and the absolute number average molecular weight of copolymers.[85] However, all of these quantities can be determined only after establishing a universal calibration curve or with knowledge of the Mark-Houwink constants, K and a,

Table 4   Characteristics of Commercial On-Line Viscometers

	Waters™ 150 CV	Viscotek 200	Viscotek 500Y[a]
Design	Single capillary	4-Capillary bridge	2-Capillary series
Cell volume (µl)	≈18	≈50	≈10
Output sensitivity to flow rate	Sensitive	Sensitive	Insensitive
Sensitivity to flow fluctuations	Sensitive	Insensitive	Insensitive
Calibration/accuracy	Depends on pump	Depends on matched capillaries	Depends on electronic adjustment
Sensitivity at S/N = 4	—	≈4×10^{-5} η_{sp}	<4×10^{-5} η_{sp}
Delay volume	No	No	Yes

[a] Original design by E. I. Du Pont de Nemours, Wilmington, DE.

Source: Yau, W. W., Barth, H. G., and Jackson, C., *Proc. Am. Chem. Soc. Div. Polym. Mat. Sci. Eng.*, 65, 146, 1991. With permission.

for the polymer. The K and a values are strongly dependent on the solvent and temperature. Tabulated values, therefore, should be used with caution.[179] Also, it is difficult to determine the two constants without knowledge of the molecular weight of the polymer, making the Mark-Houwink approach circular for an unknown polymer. Instead, a universal calibration curve is established, using molecular weight standards.

The reliability of data obtained using GPC with concentration-sensitive detection is dependent on universal calibration. There are notable exceptions where universal calibration is not applicable, particularly when polymers differ in conformation.[201] The radius of gyration is determined by the use of various correlation methods.[85] The applicability of correlation methods such as the Flory-Fox[202] or Ptitsyn-Eizner[203] correlation methods, requires prior knowledge of branching and conformation. The proportionality constants employed by the correlation methods are dependent on the nature of the polymer-solvent combination.[202] Also, it has been shown that the elution volume of polymers is a function of the column load.[194] A difference of about 30% in the molecular weights determined for the highest and lowest amount injected was observed.[204] In the absence of second virial coefficient effects, light-scattering has been shown to be unaffected by varying polymer injection mass,[194] since molecular weights and radius of gyration are determined with no recourse to external standards.

A few studies have reportedly used a combination of both light-scattering and viscosity detectors.[205,206] Such a combination has improved immensely polymer conformation[206] and long chain branching data[85] obtained with a single detector. It has been demonstrated that a combination of GPC-MALLS-VIS determined the exponent α of the scaling relationship ($R_G \sim M^\alpha$) to be 0.576 ± 0.002.[205] GPC-MALLS predicted $\alpha = 0.52 \pm 0.14$. The former value compares favorably with the predicted value of 0.588 for random coils in a good solvent.[156]

6.2.10   Viscosimetry-right angle light-scattering detector

Light scattering has an inherent limitation in determining the size of very small polymer molecules in solution.[85,183,184] It has been shown that for a standard helium-neon laser ($\lambda = 632.8$ nm), the minimum radius of gyration of a flexible chain molecule that can be determined is equal to 10 nm.[85] This limitation arises due to the fact that the angular dissymmetry in light scattered by small polymer molecules in solution is negligibly small.[183,184] However, taking advantage of this phenomenon, the molecular weight of a small polymer molecule can be estimated simply by measuring the scattered-light intensity at 90° to the incident angle. Simultaneous measurement of the polymer intrinsic viscosity permits determination of the radius of gyration with the well known Flory-Fox correlation.[202,207-209]

For large polymer molecules ($\langle R_G \rangle^{1/2} > 10$ nm), the angular dissymmetry in scattered light is markedly different at 0° and 90° to the incident angle. In such situations, the intrinsic viscosity data is utilized to account for this difference.[210] However, such a correction is strictly valid only for linear flexible random coil-like or spherical molecules,[202] and the data bears this out. This technique, though, has severe limitations for rod-like or branched polymers,[210] e.g., errors in molecular weight estimation are as high as 15 to 40% for rod-like molecules. Further improvements in correcting the angular dissymmetry of various shaped polymer molecules are necessary before right-angle laser light-scattering (RALLS) detectors can achieve widespread applicability.

6.2.11   Summary of detectors

The current trend in GPC is toward the use of molecular-weight-sensitive detectors, circumventing the need for calibration methods. It is preferable to use absolute methods of molecular weight determination such as light-scattering over on-line viscosimetric methods; however, in most cases the choice of a suitable detector is dependent on the specific polymer being characterized. There may be instances where it would be advantageous to employ on-line viscosimeters rather than light-scattering detectors. The future will see the introduction of new on-line detectors as well as refinements in the existing ones geared towards achieving the highest precision in molecular properties obtainable via gel permeation chromatography.

6.3   Application of aqueous size-exclusion chromatography to carbohydrates: a case history

6.3.1   Background

As has been discussed, size-exclusion chromatography (SEC) is a powerful tool for macromolecular separation and characterization. This technique separates macromolecules which vary in effective molecular size, utilizing columns that are packed with a stationary phase that has a specific pore diameter. The retention of macromolecules in the stationary phase is dependent on the hydrodynamic volume, S*. A macromolecule with an S* value much less than the stationary phase pore diameter will diffuse in and out of the pores and therefore be retained on the column for a longer period of time than a macromolecule with a greater S* value. Macromolecules are excluded from the column and show no retention when their S* value is much greater than the stationary phase pore size.[211] Johnson[212] describes the SEC chromatogram as a molecular size distribution curve, that may be expressed as size distribution in weight concentration with the application of a concentration-sensitive detector. For determination of molecular weight, SEC columns are calibrated with polymer standards, or a light-scattering detector is used to determine absolute molecular weights. Because there is a wide variety of SEC columns available which operate over a molecular weight range of 10^2 to 10^7, stationary phases must be selected to accommodate a specific molecular weight and size range.[213,214] Careful column selection can yield a chromatogram that separates each macromolecule of interest, providing sufficient information for characterization of molecular weight averages and distributions. SEC may also be used to separate macromolecules from small molecules (formulation components) or for preparative techniques such as polymer fractionation. For some systems, complete resolution of each component may not be possible. Alternate techniques which involve degradation of a specific component may be applied to increase resolution and separation efficiency.

Apart from the effect of pore size on column retention, a number of other factors may be considered in aqueous SEC.[215,216] These factors, referred to collectively as non-size-exclusion effects may be controlled by mobile phase selection. Anionic macromolecules may be repelled and excluded from a stationary phase that also carries a negative charge, whereas cationic macromolecules may show increased or permanent retention. The ionic interactions described above may be overcome by addition of salt to the mobile phase or pH adjustment. Other types of non-size-exclusion effects may occur through hydrogen bonding or hydrophobic interaction. This section will describe the analysis of two polyelectrolytes using aqueous SEC. Column, mobile phase, and detector selection will be discussed, as well as analysis of average molecular weights and polymer characterization parameters.

6.3.2 Xanthan

Xanthan (Figure 11) is a commercially important polysaccharide produced by the bacterium *Xanthomonas campestris*.[187,188] The xanthan backbone consists of a β(1-4)-linked D-glucopyranose chain with a trisaccharide side chain attached at C3 to alternate glucose residues. These side chains consist of an acetylated mannose residue, a glucuronic acid residue, and a pyruvate ketal linked to a terminal mannose residue. The acetate and pyruvate content depend on the fermentation and isolation conditions used by the supplier.

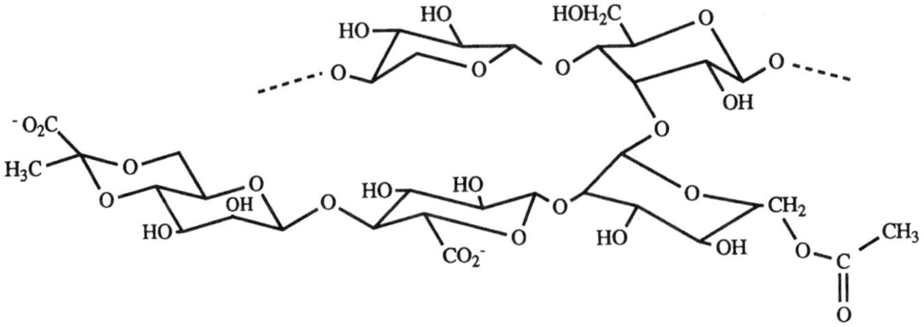

Figure 11   Structure of xanthan.

Aqueous xanthan undergoes a diffuse conformational transition that can be driven by changes in temperature, ionic strength, and the degree of ionization of the carboxyl groups. The conformational change is often described as an order-to-disorder transition. The ordered state of xanthan is a helix that may be single or double stranded. Much evidence has been published in favor a single helix, and also in favor of a double helix. The disordered state of xanthan is believed by most investigators to be a random coil in which the side-chains are not aligned with the backbone of the polymer.[219-236] This section will describe the use of SEC for analysis of xanthan from alternate suppliers. It will also describe the preparation of xanthan fractions for analysis of the xanthan ordered state.

6.3.3   Carboxymethylcellulose

Carboxymethylcellulose (CMC; see Figure 12) is a cellulose ether, produced by reacting alkali cellulose with sodium monochloracetate under rigidly controlled conditions.[237] CMC is widely used in cosmetics, foods, and pharmaceuticals as a preferred thickener, suspending aid, stabilizer, binder, and film-former. The performance of various grades of CMC is dependent on the following physical and chemical properties: the degree of polymerization (DP), the degree of substitution (DS), and the substituent distribution pattern.[238] This section will describe the use of SEC to examine CMC enzymatic degradation products. It will also present a method for analysis of CMC in an ophthalmic formulation.

6.3.4   Conditions for analysis of xanthan

Xanthan molecular weight fractions have been prepared by SEC with post-column collection using a controlled pore glass stationary phase[230] of pore size 324 and 1038 Å, a 0.05 M Na_2SO_4 mobile phase, and ultraviolet absorbance detection (λ = 206 nm, Table 5, XCPS-1). A saline mobile phase is used to reduce ionic interactions which may occur between the negatively charged xanthan side-chains and the stationary phase. It also reduces the hydrodynamic volume of the polymer, thereby increasing chromatographic resolution and

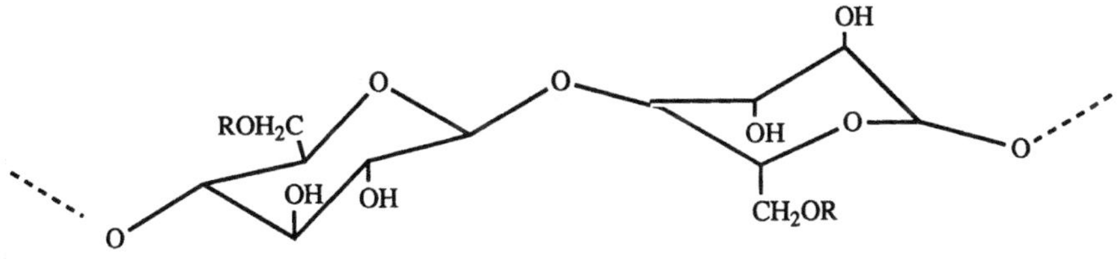

Where R = H or CH$_2$COO$^-$

Figure 12 Stucture of carboxymethylcellulose.

separation efficiency.[239] Alternate SEC methods for xanthan (Table 5, XCPS-2 and XCPS-3) employ a stationary phase consisting of a hydrophilic polymer matrix (TSK PW) in selected pore sizes of 500 Å (4000 PW), 1000 Å (5000 PW), and >1000 Å (6000 PW).[240,241] Detection is performed by coupling a differential refractometer on-line with low angle, laser light-scattering. Application of LALLS in series with the refractive index detector yields direct detection of the weight-average molecular weight (M_w). An on-line, multiangle, laser light-scattering detector is also available[186,189] which may be used for determination of M_w, A_2, and also R_G. A thorough review of light-scattering parameters is published.[195] An on-line capillary viscosimeter has been coupled with a differential refractometer and LALLS for analysis of xanthan.[242] The addition of the capillary viscosimeter provides polymer intrinsic viscosity at every SEC retention volume, leading to absolute molecular weights through universal calibration[79,81] and information on long chain branching.[86]

6.3.5 Conditions for analysis of carboxymethylcellulose

Initial analysis of carboxymethylcellulose by aqueous SEC applied a stationary phase consisting of a glycerylpropylsilyl layer covalently bonded to 10-μm Lichrosphere® silica particles (Table 5, CMC-1).[239] Nominal pore sizes used in this study were 100, 500, 1000, and 4000 Å. A number of mobile phases were examined at varying ionic strengths with refractive index detection. CMC enzymatic degradation products have been chromatographed with a Merck Fractogel® TSK column (Table 5, CMC-2) employing a pH 5 sodium sulfate/sodium acetate mobile phase.[238] Carboxymethylcellulose fractions were detected by post-column reaction with orcinol/sulfuric acid and visible detection (λ = 420 nm). Finally, CMC has been analyzed in an ophthalmic formulation by application of an Ultrahydrogel® Linear column (mixed pore size: 120 to 2000 Å) and a Ultrahydrogel® 120-Å column (Table 5, CMC-3) applying a 0.15 *M* sodium phosphate buffer (pH 7.2) and a flow rate of 0.3 ml/min. Analysis of CMC using the Ultrahydrogel® Linear column alone does not completely resolve the polymer from low-molecular-weight formulation components. The additional Ultrahydrogel® 120 column increases the retention of these components, affording complete resolution

Table 5 Chromatographic Media and Detection for Selected Hydrophilic Polymers

Polymer	Stationary phase	Supplier	Mobile phase	Detection
XCPS-1	Controlled pore Glass, CPG-10 (324 and 1038 Å)	Electronucleonics	0.05 M Na$_2$SO$_4$	Ultraviolet absorbance (λ = 206 nm)
XCPS-2	TSK 4000 PW and 5000 PW/6000 PW	ToyoSoda	0.1 M LiCl	DRI coupled with LALLS
XCPS-3	4000 and 6000 PW	ToyoSoda	0.1 M NaNO$_3$, 0.02% NaN$_3$, 0.1% ethylene glycol	DRI coupled with LALLS and capillary viscometer
CMC-1, 100, 500, 1000, and 4000	Synchropak® GPC	Synchrom	Sodium phosphate	DRI
CMC-2, HW-55(S)	Fractogel® TSK	E. Merck, Darmstadt, pH 5, 0.02% NaN$_3$	0.05 M Na$_2$SO$_4$, 0.05 M NaAcetate	Orcinol-H$_2$SO$_4$ post-column with colorimetric detection (λ = 420 nm)
CMC-3	Ultrahydrogel® Linear and Ultrahydrogel® 120	Waters™, 0.15 M sodium phosphate buffer, pH 7.2		Differential refractometer

of CMC. Detection is accomplished using a Waters™ 410 refractive index detector connected to a PE Nelson Series 900 A/D converter interfaced to a VAX 6210 computer (Digital Equipment Corporation) running PE Nelson Access*Chrom software, version 1.6 (PE Nelson, Inc.). Detector and columns are equilibrated at 30°C. Apparent carboxymethylcellulose molecular weights are determined by application of pullulan standards (Polymer Laboratories; Amherst, MA) with molecular weights of 2.37×10^4, 1×10^5, and 3.8×10^5. A calibration line is generated by plotting log(molecular weight) vs. retention time on the column. This calibration is then applied to the determination of weight-average and number-average molecular weights for CMC samples as described above. Although molecular weights determined by this method are not absolute, the values determined via pullulan calibration are sufficient for quality control and analysis of long-term polymer formulation stability. This can be accomplished by setting significant parameters for weight-average and number-average molecular weights and monitoring these carefully against appropriate standards and controls. For ophthalmic formulations where viscosity is the primary measure of product efficacy, the formulator must determine the acceptable molecular weight range that will correlate with the optimal viscosity for the product. When the polymer is chosen as a drug delivery vehicle, the formulator must determine the optimal molecular weight for sustained drug delivery. Appropriate specifications are then applied to the analysis of polymer stability in the vehicle.

6.3.6   Results: xanthan

Lecacheux et al.[240] have examined xanthan samples from various suppliers using SEC and refractive index detection coupled with on-line LALLS (Table 5, XCPS-2). Figure 13 shows LALLS and refractive index chromatograms for two xanthan samples supplied by Shell (———) and Satia (— — —). The third chromatogram (— - —) is for the polyelectrolyte scleroglucan. The Shell sample, which elutes from the column first, is assigned a M_w of 7×10^6. The Satia sample is assigned a M_w of 2.2×10^6. These results demonstrate that the fermentation and isolation conditions employed by alternate suppliers can affect the molecular weight of the native polymer.

To reduce the number of factors influencing the xanthan ordered state and eliminate molecular weight differences observed between native xanthan samples, Paradossi and Brant[230] have prepared narrow molecular weight xanthan fractions by application of SEC (Table 5, XCPS-1) to ultrasonically degraded xanthan.[241] These samples were than analyzed by light-scattering in a static or zero flow mode by application of a Sofica Model 42000 multiangle instrument. Zimm plots[243] yielded M_w values ranging from 0.8×10^5 to 15×10^5, A_2 values ranging from 13.0×10^4 to 5.35×10^4 ml-mol/g^2, and R_G values ranging from 23 to 137 nm. Xanthan fractions were also analyzed by proton nuclear magnetic resonance, optical rotatory dispersion, and osmotic pressure. Paradossi and Brant's experimental results supported the proposal that the xanthan aqueous ordered state is a double-stranded helical chain.[244]

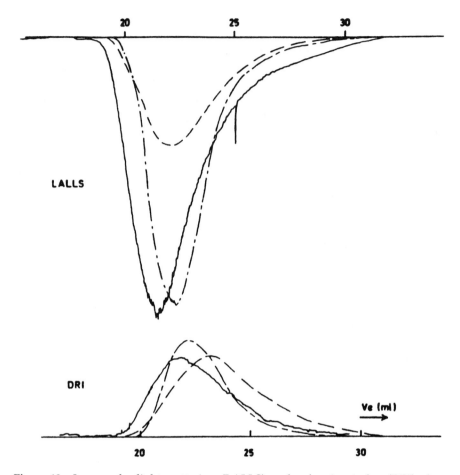

Figure 13 Low-angle, light-scattering (LALLS) and refractive index (DRI) chromatograms for two xanthan samples. The xanthan samples were supplied by Shell (———) and Satia (— — —). The third chromatogram (— · —) is for the polyelectrolyte scleroglucan. (From Lecacheux, D., Mustiere, Y., Panaars, R., and Brigand, G., *Carb. Polym.*, 6, 477, 1986. With permission.)

6.3.7   Results: carboxymethylcellulose

Hamacher and Sahm[238] (Table 5, CMC-2) have developed a SEC method for analysis of enzymatically degraded carboxymethylcellulose employing on-line carbohydrate detection using post-column reaction with orcinol/sulfuric acid and visible detection at 420 nm. Chromatograms of degraded CMC samples showed that the proportion of oligomeric products (low-molecular-weight CMC) essentially decreased on increasing the degree of substitution. This novel approach to carbohydrate detection may be employed in a mixed polymer matrix, i.e., CMC and a nonreacting synthetic, although the authors reported that there was an uncertainty in the detection of substituted glucose units. The peak area below the extinction curve decreased with increasing degree of substitution (DS), in spite of a constant number of monomeric units applied to the column. Apparently, glucose detection with orcinol/sulfuric acid is dependent on the DS. The authors were not able to determine the influence of the distribution pattern of carboxymethyl groups on the orcinol/sulfuric acid reaction.

SEC has been applied to the analysis of the stability of carboxymethylcellulose in an ophthalmic formulation (Table 5, CMC-3). Stability is evaluated by analysis of polymer concentration and apparent molecular weight. The concentration is calculated against an external standard that is run concurrently. Apparent molecular weights are determined by application of pullulan standards as described under materials and methods. A chromatogram for CMC is shown in Figure 14. Criteria for stability are established by treating a CMC standard, the CMC ophthalmic formulation, and a placebo (the formulation without CMC) with heat, acid, base, and hydrogen peroxide (Table 6). Results for these experiments are shown in Table 7. Of particular interest are the results for acid- and peroxide-stressed samples (Figures 15 and 16). Under acidic conditions, CMC may be forming some type of high molecular aggregate, whereas under peroxide treatment the polymer is degraded. Under acidic conditions, the M_w of carboxymethylcellulose in the formulation is reduced by 76.7% with an 11.8% reduction in polymer concentration by peak area analysis. Under basic conditions, the M_w of CMC in the formulation is reduced by 12.6% with a 1.8% reduction in polymer concentration by peak area analysis. With heat, the M_w of CMC in the formulation is virtually unchanged with a 9% increase (water loss) in polymer concentration by peak area analysis. Because the method is able to detect major and minor changes in polymer concentration and molecular weight, it can be used for stability analysis of carboxymethylcellulose in the ophthalmic formulation by setting appropriate parameters for M_w and M_n.

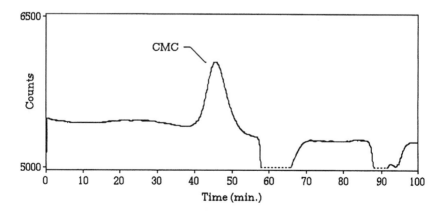

Figure 14   SEC chromatogram of carboxymethylcellulose in an ophthalmic formulation. The horizontal axis is elution time in minutes and the vertical axis is DRI response in arbitrary units.

Table 6   Criteria of Stability Conditions for
Carboxymethylcellulose Standard and Formulation

Treatment	Temperature (°C)	Time (days)	Other conditions
Heat	45	35	—
Acid	45	35	Adjusted to pH 1.5 with 1.0 N HCl
Base	45	14	Adjusted to pH 1.2 with 1.0 N NaOH
H_2O_2	45	3	3% hydrogen peroxide

Table 7   Criteria of Stability Results for Carboxymethylcellulose
Standard and Formulation

Treatment	Sample	Percent label claim by peak area (w/v)	Percent label claim by peak height (w/v)	M_w (1×10^4)	M_n (1×10^4)	P
None	Standard	100	100	9.9	7.8	1.3
	Formulation	100	100	10.3	8.1	1.3
Heat	Standard	115	119	9.8	8.0	1.2
	Formulation	109	113	10.5	8.2	1.3
Acid	Standard	—	—	570	550	1.0
		80.9	105	2.4	2.0	1.2
	Formulation	—	—	540	530	1.0
		88.2	109	2.4	2.2	1.2
Base	Standard	95.8	105	8.0	6.5	1.2
	Formulation	98.2	104	9.0	7.4	1.2
H_2O_2	Standard	35.8	44.4	—	—	—
	Formulation	84.2	84.2	—	—	—

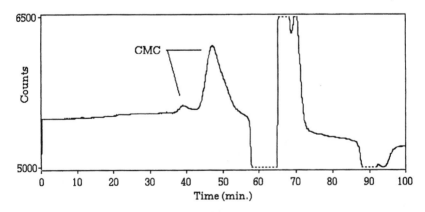

Figure 15   SEC chromatogram for carboxymethylcellulose stressed with acid. The
horizontal axis is elution time in minutes and the vertical axis is DRI response in
arbitrary units.

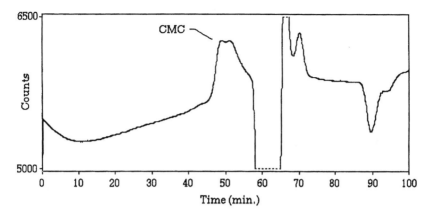

Figure 16 SEC chromatogram for carboxymethylcellulose stressed with hydrogen peroxide. The horizontal axis is elution time in minutes and the vertical axis is DRI response in arbitrary units.

References

1. Cantor, C. R. and Schimmel, P. R., *Biophysical Chemistry*. Part I. *The Conformation of Biological Macromolecules*, W. H. Freeman, San Francisco, 1980, chap. 2.
2. Yau, W. W., Kirkland, J. J., and Bly, D. D., *Modern Size Exclusion Chromatography*, John Wiley & Sons, New York, 1979, 209–222.
3. Haddon, M. R. and Hay, J. N., High-temperature size exclusion chromatography, in *Size Exclusion Chromatography*, Hunt, B. J. and Holding, S. R., Eds., Blackie, Glasgow, 1989, chap. 4.
4. Johnson, H. E. and Granick, S., New mechanism of nonequilibrium polymer adsorption, *Science*, 255, 966, 1992.
5. Laurent, T. C., The interaction between polysaccharides and other macromolecules. 5. The solubility of proteins in the presence of dextran, *Biochem. J.*, 89, 253, 1963.
6. Laurent, T. C., The interaction between polysaccharides and other macromolecules. VI. Further studies on the solubility of proteins in dextran solutions, *Acta Chem. Scand.*, 17, 2664, 1963.
7. Procházka, K., Bednář, B., Tuzar, Z. and Kočiřík, M., Size exclusion chromatography of associating systems. I. The theoretical model, *J. Liq. Chromatogr.*, 11, 2221, 1988.
8. Flory, P. J., *Principles of Polymer Chemistry*, Cornell University Press, Ithaca, NY, 1953, 399–431.
9. Flory, P. J., *Statistical Mechanics of Chain Molecules*, Interscience, New York, 1969.
10. Scheraga, H. A., Protein structure and function, from a colloidal to a molecular view, *Carlsberg Res. Comm.*, 49, 1, 1984.
11. Bandekar, J., Amide modes and protein conformation, *Biochim Biophys. Acta*, 1120, 123, 1992.
12. Lefczynski, J. F. and Rose, G. D., Loops in globular proteins: a novel category of secondary structure, *Science*, 234, 849, 1986.

13. Vallee, B. L. and Auld, D. S., Zinc coordination, function, and structure of zinc enzymes and other proteins, *Biochemistry*, 29, 5647, 1990.

14. Swadesh, J. K., Mui, P. W., and Scheraga, H. A., Thermodynamics of the quenching of tyrosyl fluorescence by dithiothreitol, *Biochemistry*, 26, 5761, 1987.

15. Mérola, F., Rigler, R., Holmgren, A., and Brochon, J-C., Picosecond tryptophan fluoroscence of thioredoxin: evidence for discrete species in slow exchange, *Biochemistry*, 28, 3383, 1989.

16. Morgan, R. S., Tatsch, C. E., Gushard, R. H., McAdon, J. M., and Warme, P. K., Chains of alternating sulfur and π-bonded atoms in eight small proteins, *Int. J. Peptide Protein Res.*, 11, 209, 1978.

17. Lebl, M., Sugg, E. E., and Hruby, V. J., Proton N.M.R. spectroscopic evidence for sulfur-aromatic interactions in peptide, *Int. J. Peptide Protein Res.*, 29, 40, 1987.

18. Tanford, C., *Physical Chemistry of Macromolecules*, John Wiley & Sons, New York, 1961, 225–227.

19. Stenlund, B., Polyelectrolyte effects in gel chromatography, *Adv. Chromatogr.*, 14, 37, 1976.

20. Lindstrom, T., De Ruvo, A., and Soremark, C., Determination of Donnan equilibria by gel filtration, *J. Polym. Sci. Polym. Chem. Ed.*, 15, 2029, 1977.

21. Porsch, B. and Sundelöf, L.- O., Size-exclusion chromatography and dynamic light-scattering of dextrans in water: explanation of ion-exclusion behaviour, *J. Chromatogr.*, 669, 21, 1994.

22. García, R., Porcar, I., Campos, A., Soria, V. and Figueruelo, J. E., Solution properties of polyelectrolytes. X. Influence of ionic strength on the electrostatic secondary effects in aqueous size-exclusion chromatography, *J. Chromatogr. A*, 662, 61, 1994.

23. Dawkins, J. V., Gel permeation chromatography, in *Techniques in Liquid Chromatography*, Simpson, C. F., Ed., John Wiley, Chichester, 1982, chap. 12.

24. Porath, J. and Flodin, P., Gel filtration: a method for desalting and group separation, *Nature*, 183, 1657, 1959.

25. Hjertén, S. and Mosbach, R., Molecular sieve chromatography of proteins on columns of cross-linked polyacrylamide, *Anal. Biochem.*, 3, 109, 1962.

26. Arshady, R., Beaded polymer supports and gels. I. Manufacturing techniques, *J. Chromatogr.*, 586, 181, 1991.

27. Hagel, L. and Andersson, T. Characteristics of a new agarose medium for high-performance gel filtration chromatography, *J. Chromatogr.*, 285, 295, 1984.

28. Herman, D. P., Field, L. R., and Abbott, S., The size-exclusion chromatographic behavior of synthetic water-soluble polymers on diol bonded phase supports, *J. Chromatogr. Sci.*, 19, 470, 1981.

29. Pfannkoch, E., Lu, K. C., Regnier, F. E., and Barth, H. G., Characterization of some commercial high-performance size-exclusion chromatography columns for water-soluble polymers, *J. Chromatogr. Sci.*, 18, 430, 1980.

30. Majors, R. E., New chromatography columns and accessories at the 1992 Pittsburgh Conference, part II, *LC-GC*, 10, 188, 1992.

31. Majors, R. E., Recent developments in HPLC packings and columns, *J. Chromatogr. Sci.*, 18, 488, 1980.

32. Kato, Y. and Hashimoto, T., Effect of type of salt on protein-support interactions in high-performance gel filtration on TSK-GEL SW type column, *J. HRC&CC*, 6, 324, 1983.

33. Hashimoto, T., Sasaki, H., Aiura, M., and Kato, Y., High-speed aqueous gel-permeation chromatography, *J. Polym. Sci. Polym. Phys. Ed.,* 16, 1789, 1978.
34. Kulin, L.-I., Flodin, P., Ellingsen, T., and Ugelstad, J., Monosized polymer particles in size exclusion chromatography. I. Toluene as solvent, *J. Chromatogr.,* 514, 1, 1990.
35. Ehwald, R., Heese, P., and Klein, U., Determination of size limits of membrane separation in vesicle chromatography by fractionation of a polydisperse dextran, *J. Chromatogr.,* 542, 239, 1991.
36. McGreavy, C., Andrade, Jr., J. S., and Rajagopal, K., Size exclusion chromatography in pore networks, *Chromatographia,* 30, 639, 1990.
37. Moore, J. C., Gel permeation chromatography. I. A new method for molecular weight distribution of high polymers, *J. Polym. Sci. A2,* 835, 1964.
38. Casassa, E. F., Theoretical models for peak migration in gel permeation chromatography, *J. Phys. Chem.,* 75, 3929, 1971.
39. Casassa, E. F. and Tagami, Y. An equilibrium theory for exclusion chromatography of branched and linear polymer chains, *Macromolecules,* 2, 14, 1969.
40. Potschka, M., Universal calibration of gel permeation chromatography and determination of molecular shape in solution, *Anal. Biochem.,* 162, 47, 1987.
41. Giddings, J. C., Kucera, E., Russell, C. P., and Myers, M. N., Statistical theory for the equilibrium distribution of rigid molecules in inert porous networks. Exclusion chromatography, *J. Phys. Chem.,* 72, 4397, 1968.
42. Dawkins, J. V. and Yeadon, G., Macromolecular separations by liquid exclusion chromatography, *Farad. Symp. Chem. Soc.,* 15, 127, 1980.
43. Karger, B. L., Snyder, L. R., and Horváth, Cs., *An Introduction to Separation Science,* John Wiley & Sons, New York, 1973, 444–461.
44. Ellingsen, T., Aune, O., Ugelstad, J., and Hagen, S., Monosized stationary phases for chromatography, *J. Chromatogr.,* 535, 147, 1990.
45. Neddermeyer, P. A. and Rogers, L. B., Gel filtration behavior of inorganic salts, *Anal. Chem.,* 40, 755, 1968.
46. Barker, P. E., Hatt, B. W., and Holding, S. R., Dissolution of silaceous chromatographic packings in various aqueous solvents, *J. Chromatogr.,* 206, 27, 1981.
47. Regnier, F. E. and Noel, R., Glyceropropylsilane bonded phases in the steric exclusion chromatography of biological macromolecules, *J. Chromatogr. Sci.,* 14, 316, 1976.
48. Barth, H., A practical approach to steric exclusion chromatography of water-soluble polymers, *J. Chromatogr. Sci.,* 18, 409, 1980.
49. Wu, C. S., Senak, L., and Malawer, E. G., Size exclusion chromatography of poly(methyl vinyl ether-co-maleic anhydride) (PMVEMA). I. The chromatographic method, *J. Liq.Chromatogr.,* 12, 2901, 1989.
50. Wu, C. S., Senak, L., and Malawer, E. G., Size exclusion chromatography of poly(methyl vinyl ether-co-maleic anhydride) (PMVEMA). II. Absolute molecular weights and molecular weight distributions by low angle laser light-scattering and universal calibration, *J. Liq.Chromatogr.,* 12, 2919, 1989.
51. Mori, S., Elution behavior of oligomers on a polyvinyl alccohol gel column with chloroform, methanol, and their mixtures, *J. Liq. Chromatogr.,* 11, 1205, 1988.
52. Sasagawa, T. and Teller, D. C., Prediction of peptide retention times in reversed-phase HPLC, in *CRC Handbook of HPLC for the Separation of Amino Acids, Peptides, and Proteins,* Vol. II, Hancock, W. S., Ed., CRC Press, Boca Raton, FL, 1984, 53–65.

53. Swadesh, J. K., Tryptic fingerprinting on a polystyrene-divinyl benzene reversed phase column, *J. Chromatogr.*, 512, 315, 1990.
54. Dubin, P. L. and Principi, J. M., Hydrophobic parameter for aqueous size exclusion chromatography gels, *Anal. Chem.*, 61, 780, 1989.
55. Belew, M., Porath, J., Fohlman, J., and Janson, J.-C., Adsorption behavior on Sephacryl™ S-200 Superfine, *J. Chromatogr.*, 147, 205, 1978.
56. McHugh, A. J., Hydrodynamic chromatography, in *Size Exclusion Chromatography*, Hunt, B. J. and Holding, S. R., Eds., Blackie, Glasgow, 1989, chap. 10.
57. Guttman, C. M. and DiMarzio, E. A., Separation by flow. II. Application to gel permeation chromatography, *Macromolecules*, 3, 681, 1970.
58. Stegeman, G., Kraak, J. C., and Poppe, H., Hydrodynamic and size-exclusion chromatography of polymers on porous particles, *J. Chromatogr.*, 550, 721, 1991.
59. Afeyan, N. B., Gordon, N. F., Mazsaroff, I., Varady, L., Fulton, S. P, Yang, Y. B., and Regnier, F. E., Flow-through particles for the high-performance liquid chromatographic separation of biomolecules: perfusion chromatography, *J. Chromatogr.*, 519, 1 1990.
60. Potschka, M., Mechanism of size-eclusion chromatography. I. Role of convection and obstructed diffusion in size-exclusion chromatography, *J. Chromatogr.*, 648, 41, 1993.
61. Nakano, A. and Minoura, Y., Degradation of aqueous poly(acrylic acid) and its sodium salt solutions by high-speed stirring, *J. Appl. Polym. Sci.*, 22, 2207, 1978.
62. Chabreček, P., Šoltés, L., Kállay, Z., and Novák, I., Gel permeation chromatographic characterization of sodium hyaluronate and its fractions prepared by ultrasonic degradation, *Chromatographia*, 30, 201, 1990.
63. Wu, S. -L., Benedek, K., and Karger, B. L., Thermal behavior of proteins in high-performance hydro-phobic-interaction chromatography. On-line spectroscopic and chromatographic characterization, *J. Chromatogr.*, 359, 3, 1986.
64. Lu, X. M., Benedek, K., and Karger, B. L., Conformational effects in the high-performance liquid chromatography of proteins. Further studies of the reversed-phase chromatographic behavior of ribonuclease, *J. Chromatogr.*, 359, 19, 1986.
65. Goheen, S. C., The influence of pH and acetonitrile on the high performance size exclusion profile of proteins, *J. Liq. Chromatogr.*, 11, 1221, 1988.
66. Neddermeyer, P. A. and Rogers, L. B., Column efficiency and electrolyte effects of inorganic salts in aqueous gel chromatography, *Anal. Chem.*, 41, 94, 1969.
67. Kato, Y., Kido, S., Watanabe, H., Yamamoto, M., and Hashimoto, T., High resolution and high speed GPC [gel permeation chromatography] on oligomers and plasticizer, *J. Appl. Polym. Sci.*, 19, 629, 1975.
68. Dallas, G. and Abbott, S. D., New approaches to the analysis of low-molecular-weight polymers, in *Liquid Chromatographic Analysis of Food and Beverages*, Vol. 2, Charalambous, G., Ed., Academic Press, New York, 1979.
69. Swadesh, J. K., Stewart, Jr., C. W., and Uden, P. C., Comparison of liquid chromatographic methods for analysis of homologous *n*-alkyl esters of biphenyl-4,4'-dicarboxylic acid, *Analyst*, 118, 1123, 1993.
70. Kirkland, J. J. and Antle, P. E., High-performance size-exclusion chromatography of small molecules with columns of porous silica microspheres, *J. Chromatogr. Sci.*, 15, 137, 1977.

71. Benson, J. R. and Woo, D. J., Polymeric columns for liquid chromatography, *J. Chromatogr. Sci.*, 22, 386, 1984.
72. Russell, D. J., Calibration of high performance size exclusion chromatography for small epoxy molecules, *J. Liquid Chromatogr.*, 11, 383, 1988.
73. Dong, M. W. and DiCesare, J. L., Analysis of priority pollutants by very high-speed LC, *J. Chromatogr. Sci.*, 20, 517, 1982.
74. Ghijs, M., DeWaele, C., and Sandra, P., Experiments with size exclusion material in microchromatography. Part 1. Microcolumn SEC, *J. HRC*, 13, 651, 1990.
75. Gügel, A. and Müllen, K., Separation of C_{60} and C_{70} on polystyrene gel with toluene as mobile phase, *J. Chromatogr.*, 628, 23, 1993.
76. Tanford, C., *Physical Chemistry of Macromolecules*, John Wiley & Sons, New York, 1961, 145–147.
77. Flory, P. J., *Principles of Polymer Chemistry*, Cornell University Press, Ithaca, NY, 1953, 273–316.
78. Melrose, J. and Ghosh, P., Determination of the average molecular size of glycosaminoglycans by fast protein liquid chromatography, *J. Chromatogr.*, 637, 91, 1993.
79. Yau, W. W., Kirkland, J. J., and Bly, D. D., *Modern Size Exclusion Chromatography*, John Wiley, New York, 1979, chap. 9.
80. Dubin, P. L., Kaplan, J. I., Tian, B-S., and Mehta, M., Size-exclusion chromatography dimension for rod-like macromolecules, *J. Chromatogr.*, 515, 37, 1990.
81. Grubisic, Z., Rempp, P., and Benoit, H., A universal calibration for gel permeation chromatography, *J. Polym. Sci. B, Polym Lett.*, 5, 753, 1967.
82. Styring, M. G., Armonas, J. E., and Hamielec, A. E., An experimental evaluation of a new commercial viscometric detector for size-exclusion chromatography (SEC) using linear and branched polymers, *J. Liq. Chromatogr.*, 10, 783, 1987.
83. Barth, H. G., Hyphenated polymer separation techniques: present and future role, in *Chromatographic Characterization of Polymers, Hyphenated and Multidimensional Techniques*, Provder, T., Barth, H. G., and Urban, M. W., Eds., American Chemical Society, Washington, D.C., 1995, chap. 1.
84. Ouano, A. C., Gel-permeation chromatography. VII. Molecular weight detection of GPC effluents, *J. Polym. Sci. A1*, 10, 2169, 1972.
85. Yau, W. W. and Rementer, S. W., Polymer characterization by SEC-viscometry: molecular weight (MW), size (Rg) and intrinsic viscosity (IV) distribution, *J. Liq. Chromatogr.*, 13, 627, 1990.
86. Wang, P. J. and Glasbrenner, B. S., Evaluation of a commercial differential viscometer as a GPC detector and its application to polymer characterization, *J. Liq. Chromatogr.*, 11, 3321, 1988.
87. Lesec, J. and Volet, G., Data treatment in aqueous GPC with on-line viscometer and light-scattering detectors, *J. Liq. Chromatogr.*, 13, 831, 1990.
88. Balke, S. T., Cheung, P., Lew, R., and Mourey, T. H., Quantitative size exclusion chromatography: assessing new developments, *J. Liq. Chromatogr.*, 13, 2929, 1990.
89. Mourey, T. H., Miller, S. M., and Balke, S. T., Size exclusion chromatography calibration assessment utilizing coupled molecular weight detectors, *J. Liq. Chromatogr.*, 13, 435, 1990.
90. Haney, M. A., The differential viscometer. II. On-line viscosity detector for size-exclusion chromatography, *J. Appl. Polym. Sci.*, 30, 3037, 1985.

91. Lecacheux, D., Lesec, J., Quivoron, C., Prechner, R., Panaras, R., and Benoit, H., High-temperature coupling of high-speed GPC with continued viscometry. II. Ethylene-vinyl acetate copolymers, *J. Appl. Polym. Sci.*, 29, 1569, 1984.

92. Jackson, C., Computer simulation study of multi-detector size-exclusion chromatography of branched molecular mass distributions, *J. Chromatogr. A*, 662, 1, 1994.

93. Jackson, C. and Yau, W. W., Computer simulation study of multidetector size-exclusion chromatography. Flory-Schulz molecular weight distribution, in *Chromatographic Characterization of Polymers, Hyphenated and Multidimensional Techniques*, Provder, T., Barth, H. G., and Urban, M. W., Eds., American Chemical Society, Washington, D.C., 1995, chap. 6.

94. Meehan, E. and O'Donohue, S., Characterization of block copolymers using size exclusion chromatography with multiple detectors, in *Chromatographic Characterization of Polymers, Hyphenated and Multidimensional Techniques*, Provder, T., Barth, H. G., and Urban, M. W., Eds., American Chemical Society, Washington, D.C., 1995, chap. 18.

95. Mourey, T. H. and Bryan, T. G., Size-exclusion chromatography using mixed-bed columns with dimethylformamide at near-ambient conditions: comparison of μStyragel HT Linear and PL Gel mixed-bed columns, *J. Liquid Chromatogr.*, 14, 719, 1991.

96. Cheung, P., Balke, S. T., and Mourey, T. H., Data interpretation for coupled molecular-weight-sensitive detectors in SEC: interdetector transport time, *J. Liq. Chromatogr.*, 15, 39, 1992.

97. Slagowski, E. L., Fetters, L. J., and McIntyre, D., Upper molecular weight limit for the characterization of polystyrene in gel permeation chromatography, *Macromolecules*, 7, 391, 1974.

98. Wang, P. J. and Glasbrenner, B. S., Overloading and degradation study in GPC using low angle laser light-scattering detector (LALLS), *J. Liq. Chromatogr.*, 10, 3047, 1987.

99. Samay, G. and Kubin, M., Peak distortion in gel permeation chromatography at high sample concentration, *J. Appl. Polym. Sci.*, 23, 1879, 1979.

100. Czok, M., Katti, A. M., and Guiochon, G., Effect of sample viscosity in high-performance size-exclusion chromatography and its control, *J. Chromatogr.*, 550, 705, 1991.

101. Marot, G. and Lesec, J., Size exclusion chromatography of polyamides, *J. Liq. Chromatogr.*, 11, 3305, 1988.

102. Donaldson, G. R., Hlavinka, M., Bullin, J. A., Glover, C. J., and Davidson, R. R., The use of toluene as a carrier solvent for gel permeation chromatography analyses of asphalt, *J. Liq. Chromatogr.*, 11, 749, 1988.

103. Dawkins, J. V., Forrest, M. J., and Shepherd, M. J., Separation of vinyl chloride oligomers by recycle high performance size exclusion chromatography, *J. Liq. Chromatogr.*, 13, 3001, 1990.

104. Christensen, G. and Fink-Jensen, P., Gel permeation chromatography of alkyd resins, *J. Chromatogr. Sci.*, 12, 59, 1974.

105. Fuller, E. N., Porter, G. T., and Roof, L. B., Size-exclusion chromatography moves to the plant for on-line polymer analysis, *J. Chromatogr. Sci.*, 20, 120, 1982.

106. Trathnigg, B., Determination of chemical composition of polymers by SEC with coupled density and RI detection. I. Polethylene glycol and polypropylene glycol, *J. Liq. Chromatogr.*, 13, 1731, 1990.

107. Wang, P. J. and Rivard, R. J., Characterization of nylons by gel permeation chromatography and low angle laser light-scattering in 2, 2, 2-trifluoroethanol, *J. Liq. Chromatogr.*, 10, 3059, 1987.

108. Rand, W. G. and Mukherji, A. K., Size-exclusion chromatography of poly(2-vinylpyridine) and polystyrene in pyridine and *n*-methylpyrrolidine, *J. Chromatogr. Sci.*, 20, 182, 1982.

109. Billiani, J. and Lederer, K., Polypropylene characterization by high temperature SEC coupled with LALLS, *J. Liq. Chromatogr.*, 13, 3013, 1990.

110. Barlow, A., Wild, L., and Ranganath, R., Gel permeation chromatography of polyethylene. I. Calibration, *J. Appl. Polym. Sci.*, 21, 3319, 1977.

111. Ying, Q. and Ye, M., A new eluting solvent for gel permeation chromatography of isotactic polypropylene, *Macromol. Chem. Rapid Comm.*, 6, 105, 1985.

112. Evans, J. A., Gel permeation chromatography: a guide to data interpretation, *Polym. Eng. Sci.*, 13, 401, 1973.

113. Devaux, J., Delimoy, D., Daoust, D., Legras, R., and Mercier, F. J., On the molecular weight determination of a poly(aryl-ether-ether-ketone) (PEEK®), *Polymer*, 26, 1994, 1985.

114. Tokuda, T., Mori, M., and Yamada, T., Size-exclusion chromatography on polyhydroxymethacrylate gel with dimethylformamide as eluent, *J. Chromatogr. A*, 722, 123, 1996.

115. Fuchs, O. and Suhr, H. H., Solvents and non-solvents for polymers, in *Polymer Encyclopedia*, 2nd ed., Brandrup, J. and Immergut, E. H, Eds. John Wiley & Sons, New York, 1975, iv, 241–265.

116. Chang, N. and Klibanov, A. M., Protein chromatography in neat organic solvents, *Biotech. Bioeng.*, 39, 575, 1992.

117. Philo, J. S., Aoki, K. H., Arakawa, T., Narhi, L. O., and Wen, J., Dimerization of the extracellular domain of the erythropoeitin (EPO) receptor by EPO: one high-affinity and one low-affinity interaction, *Biochemistry*, 35, 1681, 1996.

118. Patapoff, T. W., Mrsny, R. J., and Lee, W. A., The application of size exclusion chromatography and computer simulation to study the thermodynamic and kinetic parameters for short-lived dissociable protein aggregates, *Anal. Biochem.*, 212, 71, 1993.

119. Nave, R., Weber, K., and Potschka, M., Universal calibration of size-exclusion chromatography for proteins in guanidinium hydrochloride including the high-molecular-mass proteins titin and nebulin, *J. Chromatogr. A*, 654, 229, 1993.

120. Ovalle, R., Nonideal size exclusion chromatography with zwitterions causes a pI-dependent elution of proteins from glycerol propyl silane-modified silica, *Anal. Biochem.*, 229, 1, 1995.

121. Itou, T., Teramoto, A., Matsuo, T., and Suga, H., Isotope effect on the order-disorder transition in aqueous schizophyllan, *Carbohydr. Res.*, 160, 243, 1987.

122. Deslandes, Y., Marchessault, R. H., and Sarko, A., Triple helical structure of $(1 \rightarrow 3)$-β-D-glucan, *Macomolecules*, 13, 1466, 1980.

123. Yanaki, T., Norisuye, T., and Fujita, H., Triple helix of schizophyllum commune polysaccharide in dilute solution. 3. Hydrodynamic properties in water, *Macromolecules*, 13, 1462, 1980.

124. Kashiwagi, Y., Norisuye, T., and Fujita, H., Triple helix of schizophyllum commune polysaccharide in dilute solution. 4. Light scattering and viscosity, *Macromolecules*, 14, 1220, 1981.

125. Hirao, T., Sato, T., Teramoto, A., Matsuo, T., and Suga, H., Solvent effects on the co-operative order-disorder transition of aqueous solutions of schizophyllan, a triple helical polysaccharide, *Biopolymers,* 29, 1867, 1990.

126. Miklautz, H., Riemann, J., and Vidic, H. J., The molecular weight distribution of heparin determined with a HPLC-LALLS technique, *J. Liq. Chromatogr.,* 9, 2073, 1986.

127. Komatsu, H., Yoshii, K., Ishimitsu, S., Okada, S., and Takahata, T., Molecular mass determination of low-molecular mass heparins. Application of wide collection angle measurements of light-scattering using a high-performance gel permeation chromatographic system equipped with a low-angle lasr light-scattering photometer, *J. Chromatogr.,* 644, 17, 1993.

128. Hasegawa, M., Isogai, A., and Onabe, F., Size-exclusion chromatography of cellulose and chitin using lithium chloride-N,N-dimethylacetamide as a mobile phase, *J. Chromatogr.,* 635, 334, 1993.

129. de Ruiter, G. A., van der Lugt, A. W, Voragen, A. G. J., Rombouts, F. M., and Notermans, S. H. W., High-performance size-exclusion chromatography and ELISA detection of extracellular polysaccharides from Mucorales, *Carbohydr. Res.,* 215, 47, 1991.

130. Papazian, L. A., A pH effect in the HPSEC separation of polyacrylamide-based copolymers, *J. Liq. Chromatogr.,* 13, 3389, 1990.

131. Mori, S., Secondary effects in aqueous size exclusion chromatography of sodium poly(styrene sulfonate) compounds, *Anal. Chem.,* 61, 530, 1989.

132. Nagy, D. J. and Terwilliger, D. A., Size exclusion chromatography/differential viscometry of cationic polymers, *J. Liq. Chromatogr.,* 12, 1431, 1989.

133. Mori, S., Copolymer analysis, in *Size Exclusion Chromatography,* Hunt, B. J. and Holding, S. R., Eds., Blackie, Glasgow, 1989, chap. 5.

134. Mori, S., Separation and detection of styrene-alkyl methacrylate and ethyl methacrylate-butyl methacrylate copolymers by liquid adsorption chromatography using a dichloroethane mobile phase and a UV detector, *J. Chromatogr.,* 541, 375, 1991.

135. Glockner, G. and van den Berg, J. H. M., Separation of copolymers according to composition with special emphasis on the effect of block structure, *J. Chromatogr.,* 550, 629, 1991.

136. Mori, S., Liquid chromatographic analysis of styrene-methacrylate and styrene-acrylate copolymer, *J. Liq. Chromatogr.,* 13, 3039, 1990.

137. Tanford, C., *Physical Chemistry of Macromolecules,* John Wiley & Sons, New York, 1961, 391.

138. Tanford, C., *Physical Chemistry of Macromolecules,* John Wiley & Sons, New York, 1961, 336–344.

139. Flory, P. J., Molecular weights and intrinsic viscosities of polyisobutylene, *J. Am. Chem. Soc.,* 65, 372, 1943.

140. Krigbaum, W. R. and Flory, P. J., Statistical mechanics of dilute polymer solutions. IV. Variation of the osmotic second coefficient with molecular weight, *J. Am. Chem. Soc.,* 75, 1775, 1953.

141. Tsitsilianis, C., Mitsiani, G., and Dondos, A., An indirect GPC calibration method for the low-molecular weight region, *J. Polymer Sci. B,* 27, 763, 1989.

142. Cantor, C. R. and Schimmel, P. R., *Biophysical Chemistry. Part II. Techniques for the Study of Biological Structure and Function,* W. H. Freeman, San Francisco, 1980, 643–659.

143. Scheraga, H. A. and Mandelkern, L., Consideration of the hydrodynamic properties of proteins, *J. Am. Chem. Soc.*, 75, 179, 1953.

144. Billiani, J., Amtmann, I., Mayr, T., and Lederer, K., Calibration of separation and instrumental peak broadening in SEC coupled with light-scattering of simple polymers, *J. Liq. Chromatogr.*, 13, 2973, 1990.

145. Billiani, J., Rois, G., and Lederer, K., A new procedure for simultaneous calibration of separation and axial dispersion in SEC, *Chromatographia*, 26, 372, 1988.

146. Sagar, A. D., Sofia, S. J., and Merrill, E. W., Estimation of inter-detector lag in multi-detection gel permeation chromatography, *J. Chromatogr.*, 635, 132, 1993.

147. Pigeon, M. G. and Rudin, A., Correction for interdetector volume in size exclusion chromatography (SEC), *J. Appl. Polym. Sci.*, 57, 287, 1995.

148. Lederer, K., Imrich-Schwarz, G., and Dunky, M., Simultaneous calibration of separation and axial dispersion in size exclusion chromatography coupled with light-scattering, *J. Appl. Polym. Sci.*, 32, 4751, 1986.

149. Gugliotta, L. M., Vega, J. R., and Meira, G. R., Instrumental broadening correction in size exclusion chromatography. Comparison of several deconvolution techniques, *J. Liq. Chromatogr.*, 13,1671, 1990.

150. Tung, L. H., Method of calculating molecular-weight-distribution functions from gel permeation chromatogram, *J. Appl. Polym. Sci.*, 10, 375, 1966.

151. Tung, L. H., Correction of instrument spreading in gel-permeation chromatography, *J. Appl. Polym. Sci.*, 13, 775, 1969.

152. Yau, W. W., Characterizing skewed chromatographic band broadening, *Anal. Chem.*, 49, 395, 1977.

153. Lederer, K., Beytollahi-Amtmann, I., and Billiani, J., Determination and correction of peak broadening in size exclusion chromatography of controlled rheology polypropylene, *J. Appl. Polym. Sci.*, 54, 47, 1994.

154. Bly, D. D., Yau, W. W., and Stoklosa, H. J., Siphon counter techniques for minimizing molecular weight errors caused by flow-rate variation in high performance liquid exclusion chromatography, *Anal. Chem.*, 48, 1256, 1976.

155. Simonsick, Jr., W. J. and Prokai, L., Size-exclusion chromatography with electrospray mass spectrometric detection, in *Chromatographic Characterization of Polymers, Hyphenated and Multidimensional Techniques*, Provder, T., Barth, H. G. and Urban, M. W., Eds., American Chemical Society, Washington, D.C., 1995, chap. 4.

156. Yau, W. W., Kirkland, J. J., and Bly, D. D., *Modern Size-Exclusion Liquid Chromatography*, John Wiley & Sons, New York, 1979, 148–151.

157. Charlesworth, J. M., Evaporative analyzer as a mass detector for liquid chromatography, *Anal. Chem.*, 50, 1414, 1978.

158. Macrae, R. and Dick, J., Analysis of carbohydrates using the mass detector, *J. Chromatogr.*, 210, 138, 1981.

159. Macrae, R., Trugo, L. C., and Dick, J., The mass detector: a new detection system for carbohydrate and lipid analyses, *Chromatographia*, 15, 476, 1982.

160. Righezza, M. and Guiochon, G., Effect of the wavelength of the laser beam on the response of an evaporative light-scattering detector, *J. Liq. Chromatogr.*, 11, 2709, 1988.

161. Schultz, R. and Engelhardt, H., The application of an evaporative light-scattering detector in polymer analysis, *Chromatographia*, 29, 517, 1990.

162. Lafosse, M., Dreux, M., Morin-Allory, L., and Colin, J. M., Some applications of a commercial light-scattering detector for liquid chromatography, *J. HRC&CC*, 8, 39, 1985.

163. Mourey, T. H. and Oppenheimer, L. H., Principles of operation of an evaporative light-scattering detector for liquid chromatography, *Anal. Chem.*, 56, 2427, 1984.

164. Stockwell, P. B. and King, B. W., A light-scattering detector for liquid chromatography, *Am. Lab.*, 23(12), 19, 1991.

165. Bruns, A., Waldhoff, H., and Winkle, W., Application of HPLC with evaporative light-scattering detection in fat and carbohydrate chemistry, *Chromatographia*, 27, 340, 1989.

166. Coulombe, S., Comparison of detectors for size exclusion chromatography of heavy oil related samples, *J. Chromatogr. Sci.*, 26, 1, 1988.

168. Colin, H., Jaulmes, A., Guiochon, G., Corno, J., and Simon, J., Construction and performance of an improved differential refractometer detector for liquid chromatography, *J. Chromatogr. Sci.*, 17, 485, 1979.

168. Dark, W. A., UV and dRI detectors in liquid chromatography: the workhorse detector, *J. Chromatogr. Sci.*, 24, 495, 1986.

169. McNair, H. M., Equipment for HPLC- V., *J. Chromatogr. Sci.*, 20, 537, 1982.

170. Terry, S. L. and Rodriguez, F., Measurement of functional group and molecular weight interrelation by gel-permeation chromatography/infra-red, *J. Polym. Sci. 21C*, 191, 1968.

171. Mori, S., Determination of the composition and molecular-weight distribution of a poly(vinyl chloride-vinyl acetate) co-polymer by gel permeation chromatography and infrared spectroscopy, *J. Chromatogr.*, 157, 75, 1978.

172. Mori, S., Wada, A., Kaneuchi, F., Ikeda, A., Watanabe, M., and Mochizuki, K., Design of a highly sensitive infrared detector and application to a high-performance size exclusion chromatography for copolymer analysis, *J. Chromatogr.*, 246, 215, 1982.

173. Vidrine, D. W., Use of subtractive techniques in interpreting on-line FTIR spectra of HPLC column eluate, *J. Chromatogr. Sci.*, 17, 477, 1979.

174. Brown, R. S., Hausler, D. W., Taylor, L. T., and Carter, R. C., Fourier transform infrared spectrometric detection in size-exclusion chromatographic separation of polar synfuel material, *Anal. Chem.*, 53, 197, 1981.

175. Cazes, J. and Fallick, G., Application of liquid chromatography to the solution of polymer problems, *Polym. News*, 3, 295, 1977.

176. Willis, J. N. and Wheeler, L., Use of a gel permeation chromatography-Fourier transform infrared spectroscopy interface for polymer analysis, in *Chromatographic Characterization of Polymers, Hyphenated and Multidimensional Techniques*, Provder, T., Barth, H. G., and Urban, M. W., Eds., American Chemical Society, Washington, D.C., 1995, chap. 19.

177. Bruckner, C. A., Foster, M. D., Lima, III, L. R., Synovec, R. E., Berman, R. J., Renn, C. N., and Johnson, E. L., Column liquid chromatography: equipment and instrumentation, *Anal. Chem.*, 66, 1R, 1994.

178. Barth, H. G., Boyes, B. E., and Jackson, C., Size exclusion chromatography, *Anal. Chem.*, 66, 595R, 1994.

179. Kurata, M., Tsunashima, Y., Iwama, M., and Kamada, K., Viscosity-molecular weight relationships and unperturbed dimensions of linear chain molecules, in *Polymer Handbook*, 2nd ed., Brandrup, J. and Immergut, E. H., Eds, John Wiley & Sons, New York, 1975, iv, 1–60.

180. Claes, P., Fowell, S., Woollin, C., and Kenney, A., On-line molecular size detection for protein chromatography, *Am. Lab.*, 22, 58, 1990.

181. Huve, P., Verrecchia, T., Bazile, D., Vauthier, C., and Couvreur, C., Simultaneous use of size exclusion chromatography and photon correlation spectroscopy for the characterization of poly(lactic acid) nanoparticles, *J. Chromatogr. A*, 675, 129, 1994.

182. Wu, C., Simultaneous calibration of size exclusion chromatography and dynamic light-scattering for the characterization of gelatin, *Macromolecules*, 26, 5423, 1993.

183. Huglin, M. B., *Determination of Molecular Weight by Light Scattering*, Springer-Verlag, New York, 1978.

184. Katime, I. A. and Quintana, J. R., Scattering properties: light and X-rays, in *Comprehensive Polymer Science*, Vol. 1, Booth, C. and Price, C., Eds., Pergamon Press, New York, 1989, 103–132.

185. Stuting, H. H., Krull, I. S., Mhatre, R., Krzysko, S. C., and Barth, H. G., High performance liquid chromatography of biopolymers using on-line laser light-scattering technique, *LC-GC*, 7, 402, 1989.

186. Wyatt, P. J., Jackson, C., and Wyatt, G. K., Absolute GPC determinations of molecular weight and sizes from light-scattering, *Am. Lab.*, 20(5), 86, 1988.

187. Frank, R., Frank, L., and Ford, N. C., Molecular characterization using a unified refractive index-light-scattering detector, in *Chromatographic Characterization of Polymers, Hyphenated and Multidimensional Techniques*, Provder, T., Barth, H. G., and Urban, M. W., Eds., American Chemical Society, Washington, D.C., 1995, chap. 9.

188. Mourey, T. H. and Coll, H., Size exclusion chromatography with two-angle laser light-scattering (SEC-TALLS) of high-molecular-weight and branched polymers, *J. Appl. Polym. Sci.*, 56, 65, 1995.

189. Wyatt, P. J., Hicks, D. L., Jackson, C., and Wyatt, G. K., Absolute gel permeation chromatographic determination of molecular weight and sizes. Part 2. Incorporation of gel permeation chromatography-size exclusion chromatographic measurements, *Am. Lab.*, 20, 108, 1988.

190. Zimm, B. H. and Stockmeyer, W. H., The dimensions of chain molecules containing branches and rings, *J. Chem. Phys.*, 17, 1301, 1949.

191. Yu, L.P. and Rollings, J. E., Low-angle laser light-scattering-aqueous size exclusion chromatography of polysaccharides: molecular weight distribution and polymer branching determination, *J. Appl. Polym. Sci.*, 33, 1909, 1987.

192. Yu, L.P. and Rollings, J. E., Quantitative branching of linear and branched polysaccharide mixtures by size exclusion chromatography and on-line low-angle laser light-scattering detection, *J. Appl. Polym. Sci.*, 35, 1085, 1988.

193. Podzimek, S., The use of GPC coupled with a multiangle laser light-scattering photometer for the characterization of polymers. On the determination of molecular weight, size, and branching, *J. Appl. Polym. Sci.*, 54, 91, 1994.

194. Beri, R. G., Walker, J., Reese, E. T., and Rollings, J. E., Characterization of chitosans via coupled size-exclusion chromatography and multiple-angle laser light-scattering techniques, *Carb. Res.*, 238, 11, 1993.

195. Tanford, C., *Physical Chemistry of Macromolecules*, John Wiley & Sons, New York, 1961, chap. 5.

196. Jackson, C. and Barth, H. G., Concerns regarding the practice of multiple detector size-exclusion chromatography, in *Chromatographic Characterization of Polymers, Hyphenated and Multidimensional Techniques*, Provder, T., Barth, H. G., and Urban, M. W., Eds., American Chemical Society, Washington, D.C., 1995. chap. 5.

197. Huglin, M. B., Specific refractive index increments of polymers in dilute solutions, in *Polymer Handbook*, 2nd ed., Brandrup, J. and Immergut, E. H., Eds., John Wiley, New York, 1975, IV, 267–308.

198. Huglin, M. B., Specific refractive index increments, in *Light Scattering from Polymer Solutions*, Huglin, M. B., Ed., Academic Press, New York, 1972, chap. 6.

199. Strazielle, C., Light scattering in mixed solvents, in *Light Scattering from Polymer Solutions*, Huglin, M. B., Ed., Academic Press, New York 1972, 649–650.

200. Balke, S. T., Mourey, T. H., and Harrison C. A., Number average molecular weights by size exclusion chromatography, *J. Appl. Polym. Sci.*, 51, 2087, 1994.

201. Dubin, P. L. and Principi, J. M., Failure of universal calibration for size exclusion chromatography of rodlike macromolecules vs. random coils and globular proteins, *Macromolecules*, 22, 1891, 1989.

202. Flory, P. J., *Principles of Polymer Chemistry*, Cornell University Press, Ithaca, NY, 1981, 610–621.

203. Ptitsyn, D. B. and Eizner, Y. E., Characteristic viscosity of polymers in good solvents, *Zhur. Fiz. Khim.*, 32, 2464, 1958.

204. Yau, W. W., Barth, H. G., and Jackson, C., Modern SEC viscometry and new experimental possibilities for polymer characterization, *Proc. Am. Chem. Soc. Div. Polym. Mat. Sci. Eng.*, 65, 146, 1991.

205. Jackson, C., Barth, H. G., and Yau, W. W., High-precision characterization of polymer conformation by SEC with an on-line viscometry and laser light scattering, *Proc. Am. Chem. Soc. Div. Polym. Mat. Sci. Eng.*, 65, 200, 1991.

206. Pang, S. and Rudin, A., SEC assesment of long chain branching frequency in polyethylenes, *Proc. Am. Chem. Soc. Div. Polym. Mat. Sci. Eng.*, 65, 95, 1991.

207. Flory, P. J. and Fox, T. G., Treatment of intrinsic viscosities, *J. Am. Chem. Soc.*, 73, 1904, 1951.

208. Flory, P. J. and Fox, T. G., Molecular configuration and thermodynamic parameters from intrinsic viscosities, *J. Polym. Sci.*, 5, 745, 1950.

209. Casassa, E. F. and Berry, G. C., Commentary and perspective. Reflections and comments on "Molecular Configuration and Thermodynamic Parameters" from intrinsic viscosities by Paul J. Flory and Thomas G Fox, [*J. of Polym. Sci.*, 5, 745 (1950)], *J. Polym. Sci. B*, 34, 203, 1996.

210. Haney, M., personal communication, 1992.

211. Dolan, J. W., Problems in size exclusion chromatography, *LC-GC*, 8, 290, 1990.

212. Johnson, J. F., Chromatography, in *Polymers: Polymer Characterization and Analysis*, Encyclopedia Reprint Series, Kroschwitz, J. I., Ed., John Wiley & Sons, New York, 78–118.

213. Yau, W. W., Ginnard, C. R., and Kirkland, J. J., Broad-range linear calibration in high-performance size-exclusion chromatography using column packings with bimodal pores, *J. Chromatogr.*, 149, 465, 1978.

214. Snyder, L. R. and Kirkland, J. J., *Introduction to Modern Liquid Chromatography*, 2nd ed., John Wiley & Sons, New York, 1979, chap. 12.

215. Dolan, J. W. and Snyder, L. R., *Troubleshooting LC Systems: A Comprehensive Approach to Troubleshooting LC Equipment and Separations*, Humana, Clifton, NJ, 1989, 412–414.

216. Soria, V., Campos, A., Garcia, R., and Parets, M. J., Solution properties of polyelectrolytes. VI. Secondary effects in aqueous size-exclusion chromatography, *J. Liq. Chromatogr.*, 13, 1785, 1990.

217. Sandford, P. A., Pittsley, J. E., Knutson, C. A., Watson, P. R., Cadmus, M. C., and Jeanes, A., Variation in *Xanthomonas campestris* NRRL B-149: characterization of xanthan products of differing pyruvic acid content, in *Extracellular Microbial Polysaccharides,* Sandford, P. A. and Laskin, A. I., Eds., ACS Symp. Ser., No. 45, American Chemical Society, Washington, D.C., 1977, 192–210.

218. Sandford, P. A., Baird, J., and Cottrell, I. W., Xanthan gum with improved dispersibility, in *Solution Properties of Polysaccharides,* Brant, D. A., Ed., ACS Symp. Ser., No. 150, American Chemical Society, Washington, D.C., 1981, 31–42.

219. Holzwarth, G., Conformation of the extracellular polysaccharide of *Xanthomonas campestris, Biochemistry,* 15, 4333, 1976).

220. Morris, E. R., Rees, D. A., Young, G., Walkinshaw, M. D., and Darke, A., Order-disorder transition for a bacterial polysaccharide in solution. A role for polysaccharide conformation in recognition between xanthomonas pathogen and its plant host, *J. Mol. Biol.,* 110, 1, 1977.

221. Milas, M. and Rinaudo, M., Conformational investigation on the bacterial polysaccharide xanthan, *Carb. Res.,* 76, 189, 1979.

222. Norton, I. T., Goodall, D. M., Morris, E. R., and Rees, D. A. Kinetic evidence for intramolecular conformational ordering of the extracellular polysaccharide (xanthan) from *Xanthomonas campestris, J. Chem. Soc. Chem. Comm.,* 545, 1980.

223. Milas, M. and Rinaudo, M., Investigation on conformational properties of xanthan in aqueous solutions, in *Solution Properties of Polysaccharides,* Brant, D. A., Ed., ACS Symp. Ser., No. 150, American Chemical Society, Washington, D.C., 1981, 25–30.

224. Southwick, J. G., Jamieson, A. M., and Blackwell, J., Quasielastic light-scattering studies of xanthan in solution, in *Solution Properties of Polysaccharides,* Brant, D. A., Ed., ACS Symp Ser., No. 150, American Chemical Society, Washington, D.C., 1981, 1–13.

225. Rees, D. A., Polysaccharide shapes and their interactions — some recent advances, *Pure Appl. Chem.,* 53, 1, 1981.

226. Jeanes, A., Pittsley, J. E., and Senti, F. R., Polysaccharide B-1459: a new hydrocolloid polyelectrolyte produced from glucose by bacterial fermentation, *J. Appl. Polym. Sci.,* 5, 519, 1961.

227. Southwick, J. G., McDonnell, M. E., Jamieson, A. M., and Blackwell, J., Solution studies of xanthan gum employing quasielastic light-scattering, *Macromolecules,* 12, 305, 1979.

228. Paoletti, S., Cesaro, A., and Delben, F., Thermally induced conformational transition of xanthan polysaccharide, *Carb. Res.,* 123, 173, 1983.

229. Norton, I. T., Goodall, D. M., Frangou, S. A., Morris, E. R., and Rees, D. A., Mechanism and dynamics of conformational ordering in xanthan polysaccharide, *J. Mol. Biol.,* 175, 371, 1984.

230. Paradossi, G. and Brant, D. A., Light scattering study of a series of xanthan fractions in aqueous solution, *Macromolecules,* 15, 874, 1982.

231. Hacche, L. S., Washington, G. E., and Brant, D. A., Light scattering investigation of the temperature-driven conformation change in xanthan, *Macromolecules,* 20, 2179, 1987.

232. Sato, T., Norisuye, T., and Fujita, H., Double-stranded helix of xanthan in dilute solution: evidence from light-scattering, *Polymer J.,* 16, 341, 1984.

233. Sato, T., Kojima, S., Norisuye, T., and Fujita, H., Double-stranded helix of xanthan in dilute solution: further evidence, *Polymer J.*, 16, 423, 1984.

234. Sato, T., Norisuye, T., and Fujita, H., Double stranded helix of xanthan: dimensional and hydrodynamic properties in 0.1 *M* aqueous sodium chloride, *Macromolecules*, 17, 2696, 1984.

235. Holzwarth, G. and Ogletree, J., Pyruvate-free xanthan, *Carb. Res.*, 76, 277, 1979.

236. Gravanis, G., Milas, M., Rinaudo, M., and Tinland, B., Comparative behavior of the bacterial polysaccharides xanthan and succinoglycan, *Carb. Res.*, 160, 259, 1987.

237. Aqualon Company, *Aqualon™ Cellulose Gum. Sodium Carboxymethylcellulose. Physical and Chemical Properties*, Wilmington, DE, Rev. 1–88.

238. Hamacher, K. and Sahm, H., Characterization of enzymatic degradation products of carboxymethyl cellulose by gel permeation chromatograph, *Carb. Polym.*, 5, 319, 1985.

239. Barth, H. G. and Regnier, F. E., High-performance gel permeation chromatography of water-soluble cellulosics, *J. Chromatogr.*, 192, 275, 1980.

240. Lecacheux, D., Mustiere, Y., Panaras, R., and Brigand, G., Molecular weight of scleroglucan and other extracellular microbial polysaccharides by size-exclusion chromatography and low angle laser light-scattering, *Carb. Polym.*, 6, 477, 1986.

241. Milas, M., Rinaudo, M., and Tinland, B., Comparative depolymerization of xanthan gum by ultrasonic and enzymic treatments. Rheological and structural properties, *Carb. Polym.*, 6, 95, 1986.

242. Tinland, B., Mazet, J., and Rinaudo, M., Characterization of water soluble polymers by multidetection size-exclusion chromatography, *Makromol. Chem., Rapid Comm.*, 2, 69, 1988.

243. Tanford, C., *Physical Chemistry of Macromolecules*, John Wiley & Sons, New York, 1961.

244. Holzwarth, G., Molecular weight of xanthan polysaccharide, *Carb. Res.*, 66, 173, 1978.

chapter seven

High performance capillary electrophoresis: an overview

Kálmán Benedek and András Guttman

0-8493-2682-6/97/$0.00+$.50
© 1997 by CRC Press, Inc.

7.1   Introduction

Recent developments in drug discovery and drug development spurred the need for novel analytical techniques and methods. In the last decade, the biopharmaceutical industry set the pace for this demand. The nature of the industry required that novel techniques should be simple, easily applicable, and of high resolution and sensitivity. It was also required that the techniques give information about the composition, structure, purity, and stability of drug candidates. Biopharmaceuticals represent a wide variety of chemically different compounds, including small organic molecules, nucleic acids and their derivatives, and peptides and proteins.

The spectrum of new analytical techniques includes superior separation techniques and sophisticated detection methods. Most of the novel instruments are hyphenated, where the separation and detection elements are combined, allowing efficient use of materials sometimes available only in minute quantities. The hyphenated techniques also significantly increase the information content of the analysis. Recent developments in separation sciences are directed towards micro-analytical techniques, including capillary gas chromatography, microbore high performance liquid chromatography, and capillary electrophoresis.

High performance capillary electrophoresis (HPCE) is the rising star of separation sciences as proven by the number of recently published books.[1-7] The technique allows the separation of almost any type of compound, regardless of the chemical nature, size, conformation, or charge. Capillary electrophoresis broke the micro separation limit and made routine separations at the nano or atto level possible. The technical basis of the high sensitivity and efficiency of HPCE is the on-column detection accomplished by using part of the separation capillary as the detector cell. The HPCE separation modes discussed in this chapter include capillary zone electrophoresis (CZE), micellar electrokinetic chromatography (MEKC), capillary isoelectric focusing (CEF), capillary isotachophoresis (CITP) and capillary gel electrophoresis (CGE). These modern capillary electrophoretic modes will be highlighted in the following chapter, where the versatility of the technique will be emphasized and supported by selected examples from the abundant literature. However, it is also important to note that no single analytical method is capable of providing full characterization of any compound. Although HPCE is fast, reproducible, and automated, it is just one of the tools for the tasks for which the analytical chemist is paid.

7.2   Fundamentals

7.2.1   Theory

In electrophoresis the migration of charged molecules occurs under a given electric field (E). The electric field is expressed as the applied voltage (V) over the total length of the capillary (L):

$$E = V/L \tag{7.1}$$

The separation of charged compounds is based on the differences in migration velocity (v) when the electric field is applied. Migration velocity is derived by dividing the length of the capillary from injection to detection (l) by the measured migration time (t):

$$v = l/t \tag{7.2}$$

The velocity of migration over the applied electric field is called the electrophoretic mobility (μ_e):

$$\mu_e = v/E \tag{7.3}$$

Inserting Equations (7.1) and (7.2) into (7.3), one obtains:

$$\mu_e = l \, L/V \, t \tag{7.4}$$

As shown above, the electrophoretic mobility can be calculated from the known and measured experimental parameters l, L, V, and t.

Electrophoresis occurs in electrolyte solutions, where a competition of two forces, the electric force F_e and the frictional force F_f, are in equilibrium. The relationship of the two forces determines the electrophoretic mobility of the compounds:

$$\mu_e \, \mu \text{ electrical force/frictional force} \tag{7.5}$$

The electrical force is

$$F_e = qE \tag{7.6}$$

where q is the charge of the ion. The frictional force is expressed as:

$$F_f = 6\pi\eta rv \tag{7.7}$$

where η is the viscosity of the solution and r is the radius of the ion. At equilibrium $F_e = F_f$, so:

$$qE = 6\pi\eta rv \tag{7.8}$$

After solving Equation (7.8) for velocity, $v = qE/6\pi\eta r$, and substituting for the velocity obtained from Equation (7.3), we can express mobility in terms of physical parameters:

$$\mu_e = q/6\pi\eta r \tag{7.9}$$

The electrophoretic mobility of a compound is a fundamental physico-chemical parameter. Values are listed in tables of physical constants as determined at full ionic charge extrapolated to infinite dilution. An empirical correction for high ionic strength and charge number has been proposed.[8] The experimentally determined electrophoretic mobility, called "effective mobility", is usually dependent on pH and buffer composition. Equation (7.9) shows that the electrophoretic mobility is a function of the charge and size of the solute. Small or highly charged ions have high mobility. Large or weakly charged molecules have low mobilities. Since charge is a function of the pH of the electrolyte, and the viscosity of the electrolyte is a function of temperature, pH and temperature are also variables in the determination of electrophoretic mobility. The role of viscosity is seen in Equation (7.8). The electrolyte also interacts with the inside wall of the capillary, introducing the most significant component affecting capillary electrophoretic separations, the electro-osmotic flow (EOF).

7.2.2   Electro-osmotic flow

Electro-osmotic flow, a fundamental component of all type of electrophoretic separations, is the bulk flow of the electrolyte inside the capillary. Figure 1 illustrates the double-layer formation and potential differences at the capillary wall. The fused silica wall is negatively charged and the positive ions from the electrolyte solution adsorb to it. Excess cations in bulk solution migrate toward the cathode, generating a net flow from the anode to cathode. Since the flow originates at the capillary walls, the flow profile is flat and plug-like as opposed to the parabolic hydrodynamic flow profile normally observed in HPLC, as is illustrated in Figure 2. The applied electric field moves the charged solute molecules to their oppositely charged electrodes. The EOF controls the overall migration time by superposition on the solute migration. The ion migrations are consequently vectors, since not only the speed of the migrating ions is important, but also the direction. Figure 3 displays the mobility vectors for cations and anions during electrophoretic separations.

The EOF can be described by the following equation:

$$\mu_{eof} = (\varepsilon/4\pi\eta)E\zeta \tag{7.10}$$

where ε is the dielectric constant, ζ is zeta potential, E is the electric field strength, and η is the viscosity of the solvent. An understanding of the nature of the EOF and control of the parameters which affect it are critical for

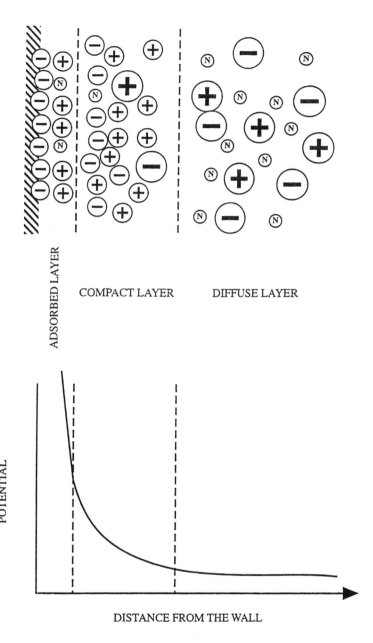

Figure 1 Model of the double layer developing at the vicinity of the silic wall. The wall is negatively charged, and the circles represent negative, positive, and neutral ions. The potential drop at the interface is also illustrated.

successful capillary electrophoresis (CE) methods development. The determining factor in the EOF control is the zeta potential of the surface. The variation of the zeta potential as a function of the pH at different concentrations of KNO_3

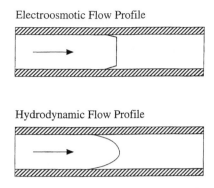

Figure 2   Flow profiles for electro-osmotic and hydrodynamic flows.

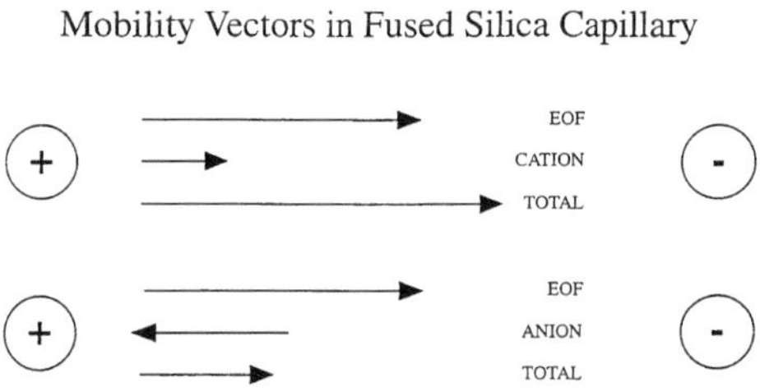

Figure 3   The mobilities of cations and anions are illustrated using vectors, as is the effect of EOF on the total mobility.

solutions is shown in Figure 4. The zeta potential below pH 6 is a strong function of the ionic strength of the electrolyte; however, above pH 6, the zeta potential is almost independent of pH but a strong function of the ionic strength.[9]

7.2.3 *Electrolytes*

Electrophoretic separations occur in electrolytes. The type, composition, pH, concentration, viscosity, and temperature of the electrolytes are all crucial parameters for separation optimization. The composition of the electrolyte determines its conductivity, buffer capacity, and ion mobility and also affects the physical nature of a fused silica surface. The general requirements for good electrolytes are listed in Table 1. Due to the complex effects of the type, concentration, and pH of the separation media buffer, conditions should be optimized for each particular separation problem.

A high electric field inside the capillary generates heat. This so-called Joule heat can destroy the efficiency of separation. An electric field corresponds to a well defined current value for a specific electrolyte system. An ohm plot can be constructed for each electrolyte system by measuring the current at increasing electric fields. For an ideal system, the ohm plot is a

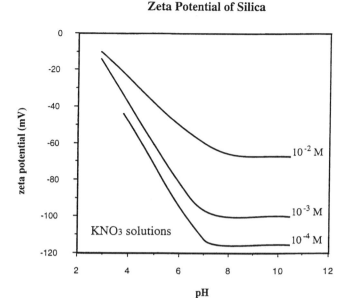

Figure 4 The variation of the zeta potential of vitreous silica as a function of pH in potassium nitrate solutions. (Based on Wiese, G. K., James, R. O., and Healy, T. W., *Trans. Farad. Soc.*, 52, 1298, 1971.)

Table 1 Characteristics of Appropriate Electrolytes for HPCE

Chemical	No interaction with the sample
	Good solubility
	High buffer capacity over wide pH range
	Low pH variation as a function of temperature
	Availability in different salt forms
	Low counterion mobility
	Mobility matching
	Good salting-in characteristics
Electrochemical	Low conductivity
Thermal	Low temperature sensitivity
Optical	Good UV transparency

linear relationship. The appearance of Joule heating is indicated by a deviation from this linearity. The construction of the ohm plot identifies the highest applicable electric current for a given electrolyte system. Figure 5 displays ohm plots of the same electrolyte when different cooling methods are applied. Proper heat dissipation can be accomplished by decreasing the internal diameter of the capillary[10,11] and/or using effective heat exchange such as forced liquid, air cooling, or Peltier blocks.[12] Each technique has its

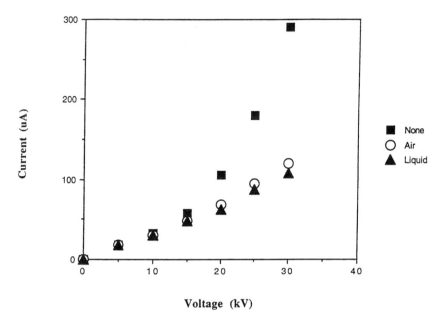

Figure 5 Ohm plots and the development of Joule heating of the same buffer system using different cooling methods.

own advantage from an engineering point of view, but active liquid cooling seems to be the most efficient way of thermostating the capillaries. A special problem regarding Joule heating arises when proteins and peptides are separated, because these solutes usually require high ionic strength buffer systems. Therefore, efficient cooling is especially important for protein separations.[12]

7.2.4   Columns

Capillary electrophoretic separations are performed in small diameter tubes, made of Teflon, polyethylene, and other materials. The most frequently used material is fused silica. Fused silica capillaries are relatively inexpensive and are available in different internal and external diameters. An important advantage of a fused silica capillary is that the inner surface can be modified easily by either chemical or physical means. The chemistry of the silica surface is well established due to the popularity of silica surfaces in gas chromatography (GC) and liquid chromatography (LC). In capillary electrophoresis, the silica surface is responsible for the EOF. Using surface modification techniques, the zeta potential and correspondingly the EOF can be varied or eliminated. Column fabrication has been done on microchips.[13]

7.2.4.1   Physical surface modifications

Adsorption is probably the simplest way to change the EOF on purpose by using appropriate additives. EOF modification by adsorption can be used on both uncoated and coated capillaries. The surface of uncoated fused silica capillaries can be changed by the adsorption of electrolytes and/or additives. The adsorption of additives can decrease, eliminate, or reverse the EOF, and the effect can be controlled by the nature and the concentration of the additives. High concentrations of salts prevent protein adsorption by competing for sites of adsorption.[14,15] It has been shown that the addition of NaCl decreases the EOF.[16] Methanol also decreases EOF, while acetonitrile increases it.[17] Cationic surfactants such as cetyl trimethyl ammonium bromide (CTAB) or tetradecyltrimethylammonium bromide (TTAB) reverse the direction of the EOF.[18-20] Nonionic surfactants such as Triton X-100, ethylene glycol, cellulose derivatives, and polyethylene glycols can also moderate or practically eliminate the EOF. Various diamines such as 1,3-diaminopropane,[21] 1,4-diaminobutane (DAB),[22,23] or 1,5-diaminopentane (DAP) have been used to prevent the adsorption of proteins.[24] The applicability of 1,3-diaminopropane as an additive over a wide pH range is illustrated in Figure 6 by the separation of basic proteins.[21] The most frequently used additives are listed in Table 2. It is important to note that because of ionic adsorption, the efficiency of the electrophoretic separation can be severely decreased by secondary adsorption such as the chelation of sample components to adsorbed Fe(III).[25] It has also been noted that phosphate ion from buffers strongly adsorbs to the silanols, influencing migration.[26,27]

7.2.4.2   Chemical surface modifications

The first surface modification for the purpose of eliminating EOF and protein adsorption was recommended by Hjérten.[28] The attachment of vinyl silanes allowed the polymerization of a variety of molecules to the surface. Most of the chemical modifications used for preparing capillaries for electrophoresis originated from the experience acquired over the years preparing GC and LC stationary phases. Chemical modification should conform to certain requirements, including the prevention of adsorption, the provision of stable and constant EOF over a wide pH range, chemical stability, ease of preparation, and reproduciblity of preparation. The effects of silanization of the inner surface of capillaries on electrophoretic separations have been extensively studied.[26,29]

7.2.5   Separation parameters

Both physical and chemical parameters can be controlled by the operator. Physical parameters include the field strength, temperature, column length and diameter, and injection mode and size. The potential can also be ramped to improve measurement of electrophoretic mobility.[30] The chemical parameters are the type and concentration of electrolyte (including buffers and additives), sample composition, and the coating of the capillary. Part of the process of optimizing a separation is to select and change these parameters to improve the speed or resolution of separations. Figure 7 illustrates the selectivity differences accomplished by varying the salt type used for the buffer preparation.

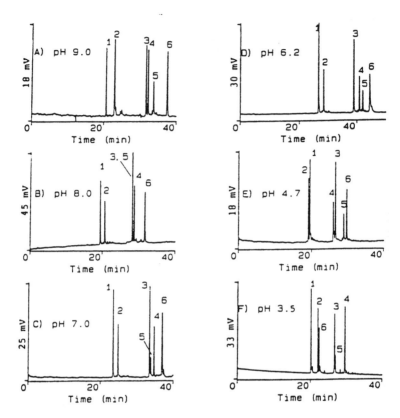

Figure 6 Separation of basic proteins on an untreated fused silica capillary with diaminopropane as buffer additive. Capillary: 75 cm (55 cm to detector) × 50 μ i.d. Buffer pHs are as noted on the figure with 30 to 60 mM DAP as an additive; 200 to 240 V/cm; peak identification: 1 = lysozyme, 2 = cytochrome, 3 = ribonuclease, 4 = α-chymotrypsin; 5 = trypsinogen, 6 = r-huIL-4. (From Bullock, J. A. and Yuan, L.-C., *J. Microcol. Sep.*, 3, 241, 1991. With permission.)

Table 2 Additives Frequently Used in HPCE
for Electro-osmotic Flow (EOF) Modification

Effect	Additives	Comments
EOF reduction	Organic solvents	Methanol, ethanol, propanol, acetonitrile
	Polymers	Methyl cellulose, PEO, polyacrylamide, PVA
	Salts	
EOF reversal	Surfactants	SDS, CTAB, Brij, Tween, quatemary amines, diaminopropane, diaminobutane, Polybrene®

Note: PEO = polyethylene oxide; PVA = polyvinyl alcohol; SDS = sodium dodecyl sulfate; CTAB = cetyl trimethyl ammonium bromide.

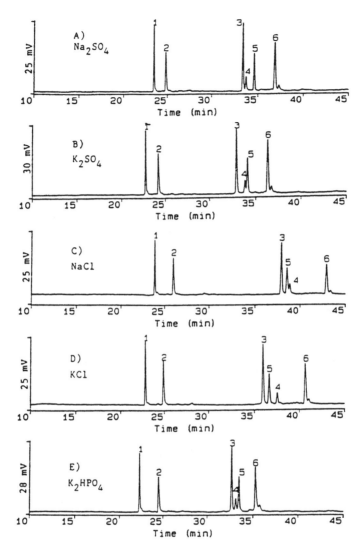

Figure 7 Effect of salt additive on selectivity. Buffer: 50 m*M* DAP, the specified salt, and titrated to pH 7 with phosphoric acid; 200 V/cm; salt concentrations (A, B, E) 40 m*M*, (C and D) 80 m*M*. For other information, see Figure 6. (From Bullock, J. A. and Yuan, L.-C., *J. Microcol. Sep.*, 3, 241, 1991. With permission.)

Sample adsorption to the silica wall is a problem in HPCE, one that is highly undesirable. As we mentioned earlier, adsorption can be minimized by proper buffer selection, additives, or chemical modification of the surface. The selection of pH is one of the simplest separation parameters to manipulate and is critical to the success of all electrophoretic separations. The pH of the media will determine the charge of the sample and the charge of the silica surface. At low values of pH, the capillary wall is protonated, the EOF is minimized, most of the sample molecules are positively charged, and electrostatic adsorption is minimized. At high values of pH, the capillary wall and most sample molecules are negatively charged, thus electrostatic repulsion eliminates or minimizes sample adsorption. The major disadvantage of doing separations at extreme values of pH is the effect on the samples, especially the effect on peptides and proteins. Aggregation, deamidation, desulfurization, and other forms of degradation may be observed at extreme values of pH. An increase in the ionic strength of the buffer is generally sufficient to prevent sample adsorption, but the EOF and Joule heat also increase at higher ionic strength.

The actual temperature of separation is determined by internal and external factors. The internal factor, as it was mentioned earlier, is the generated Joule heat. The external factor is the temperature control applied by the cooling system. A temperature increase decreases the viscosity of the electrolyte and increases the diffusion of the sample, resulting in zone broadening and a decrease in efficiency.

Special additives such as micelle forming reagents are very important in HPCE separations. They have two basic functions, namely to modify the EOF by adsorbing to the surface and also to modify the electrophoretic characteristics of the analyte. Organic solvents can be added to the buffer to change buffer polarity and viscosity.[17,31] Surfactants can form complexes with the sample molecules, just as SDS complexes to proteins in slab gel or tube electrophoresis. Surfactants may act as quasi-stationary phases in MEKC. Inclusion complexes such as cyclodextrins[32,33] and crown ethers are excellent complexing agents. Cyclodextrins are used as additives in chiral CE separations. Denaturing agents such as urea and guanidine are also popular additives. The addition of these chaotropic agents changes the structure of molecules and exposes new charged sites, which then contribute to the migration of the sample molecules.[34] A similar phenomenon occurs when complexing metal additives are used to modulate structure and selectivity.[35,36]

7.3   Instrument

As illustrated in Figure 8, the basic instrumental configuration of a CE system includes a capillary column (typically composed of fused silica), buffer reservoirs connected via platinum electrodes to a high-voltage power supply (1 to 30 kV), and a detector. Usually 20 to 150 μ i.d. and 250 to 375 μ o.d. polyimide-coated fused silica capillaries are used in CE. The polyimide coating is removed from a small section in order to provide the possibility of on-column UV, fluorescent, and radiochemical detection. Among the various on-column injection methods provided by CE technology are pressure, vacuum, gravity, or electrokinetic techniques. Displacement injection methods (pressure, vacuum) have the advantage of being quantitative; in other words, a known volume of sample can be injected. On the other hand, the use of electrokinetic injection allows sample stacking, a preconcentration procedure that is useful when very dilute samples are used. As we mentioned

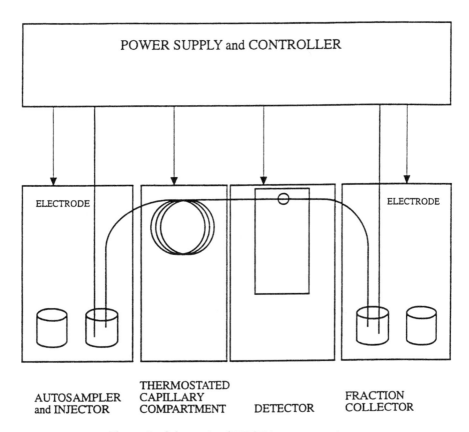

POWER SUPPLY and CONTROLLER

ELECTRODE

ELECTRODE

| AUTOSAMPLER and INJECTOR | THERMOSTATED CAPILLARY COMPARTMENT | DETECTOR | FRACTION COLLECTOR |

Figure 8   Schematic of HPCE instrumentation.

above, extra heat can develop in the column during separations due to the high current. It is important to maintain the capillary temperature constant during the separation to achieve maximum reproducibility of migration time. Some commercial instruments are capable of controlling the temperature of the capillary column by liquid or forced air cooling. Another important feature is the possibility of fraction collection and field programming for collecting nanomolar amounts of samples for subsequent microanalysis. Pulsed asymmetric field inversion has been found useful in separations of intact double-stranded DNA.[37] High-precision fraction collectors were used in the separation of restriction fragments from a plasmid digest[38] and in peptide separations.[39] As in HPLC, on-line system control and computerized data processing are available.

The detection of the migrating sample boundary in CE can be accomplished by UV, fluorescent, electrochemical, radiochemical, conductivity, and mass spectrometry (MS) means. The use of high-sensitivity detection systems is always a key issue in CE applications. The sensitivity of HPCE detectors may be at least 2 to 3 orders of magnitude better than that of HPLC detectors. Since the detection cell volume is very small, the concentration sensitivity is usually similar to that of HPLC, meaning that the relative sensitivity to a trace and major component is similar. Recently, even more sensitive detectors have been developed. A good example is the laser induced fluorescent (LIF) detector, with attomolar detection limits.[40] High sensitivity can be reached with the connection of the CE system to a specially designed electrochemical detector[41,42] or with an MS system.[43] In the latter instance, electrospray ionization is used and the connection of the capillary to the MS is done at atmospheric pressure. This on-line CE-MS combination can open new horizons in modern analytical biochemistry and pharmacology, as it unites the high efficiency and good selectivity of the CE system with extremely powerful MS detection.

Many instrumentation companies have commercialized complete or modular CE systems. The principal vendors are Beckman Instruments, Inc. (Fullerton, CA); Bio-Rad Laboratories (Hercules, CA); Dionex Co. (Sunnyvale, CA); Hewlett Packard® (Palo Alto, CA); ISCO, Inc. (Lincoln, NE); Otsuka, Ltd. (Tokyo, Japan); Perkin-Elmer™-ABI (Foster City, CA); Thermo-Separation Products (Fremont, CA); and Waters Associates (Milford, MA).

High performance capillary electrophoresis was introduced originally as an analytical tool. Now that instruments are equipped with automated fraction collection, however, capillary electrophoresis can be used for micro-preparative collection of individual peaks separated from a mixture. Using the fraction collection feature, nanomolar amounts of solute such as proteins, peptides, oligonucleotides can be collected in amounts sufficient for microsequencing. An intersample washing procedure and use of well-formed capillaries aid in the prevention of artifacts.[44]

7.4   Separation modes

7.4.1   CZE

Capillary zone electrophoresis is currently the most popular separation mode in modern high performance capillary electrophoresis. This is due to the fact that the technique is simple but very versatile. In CZE, the components of the injected sample migrate to the electrode of opposite charge due to the applied electric field. Using fused silica capillary with buffers of appropriate pH, one can cause analytes to migrate in the same direction regardless of their charge by use of electro-osmotic flow. The bulk EOF carries both the negatively and positively charged as well as the neutral molecules from the anode (+) towards the cathode (–). The negatively charged species migrate slower and the positively charged migrate faster than the EOF, while the neutral species migrate with the same velocity as the EOF. CZE has been used for the separation of inorganic anions and cations, small organic compounds, proteins and peptides, carbohydrates, and enantiomers. Studies on structural changes of peptides and proteins, such as deamidation[45] and conformational changes,[46] have also been performed. Many complex samples, such as drugs and their metabolites, hormones, and chiral molecules can be analyzed by CZE without complicated sample preparation steps.[47,48] Peptides and proteins are usually separated at low pH, when the adsorption of these compounds to the capillary wall is minimized.[26,49]

In CZE the basis of separation is differences in free solution mobility. The major separation parameters are the type, pH, and concentration of the buffer. The proper selection of buffers is not only the pre-eminent parameter to resolve species, but also has a significant role in the efficiency of separation and in the shape of the peaks evolved. Tailing and fronting of the peaks can be eliminated by matching the mobility of the analytes and the buffer.[50] As was mentioned above, the addition of surface-active agents such as SDS, CTAB, organic solvents, metals, and polymers can have a significant effect on the separation. Among other effects, polymers decrease convective band broadening. The use of appropriate additives can change selectivity, resolution, and efficiency.

7.4.2 MEKC

Micellar electrokinetic chromatography utilizes the micelle formation characteristics of detergent molecules in aqueous media. The large number of different detergents provides a variety of separation options for analysis. Anionic, cationic, and neutral detergents can be used as additives.[51] MEKC is based on the secondary equilibrium between the detergent micelles and the sample molecules. The sample components partition between the micelles and the electrolyte according to their distribution constant. Since the micelles, carrying the sample molecules, migrate according to their mass/charge ratio, differences of distribution in the micelle of the various components form the basis of separation. Consequently, the role of the micelle is similar to that of the stationary phase in liquid chromatography.[52] Sodium dodecyl sulfate (SDS) was used for the separation of molecules even at extreme values of pH where CZE could not be used.[53] The negatively charged surface of the SDS micelle can also be used for the adsorption of multivalent metal ions. The separation is similar to ion pair chromatography.[54] Vancomycin has been used as an additive to control the resolution and selectivity of the separation of nonsteroidal anti-inflammatory agents and amino acids.[55] The excellent resolution of MEKC is demonstrated by the separation of ^{16}O- and ^{18}O-containing benzoic acid molecules.[56]

7.4.3 CEF

Isoelectric focusing is one of the principal methods of protein analysis. The technique is based on the formation of a pH gradient inside the capillary. Protein molecules migrate along the pH gradient into the regions of their individual isoelectric points. The technique has remarkably high efficiency and resolving power due to the self-sharpening of peaks. The capillary version of EF (CEF) has all the advantages of on-line detection methods.[57] In CEF, the inner surface of the capillary is usually (but not necessarily[58]) coated with a nonionic, hydrophilic coating in order to minimize EOF.[59] The sample is mixed with polyions capable of forming a pH gradient, such as the ampholines.[60] This mixture is then filled into the capillary and the focusing initiated by turning on the electric field. The end of the focusing process is detected by the decrease of the electric current to almost zero. At this point all molecules in the capillary, including analytes and polyions are displaced to the pH that corresponds to their isoelectric point. Next, the contents of the capillary are moved by a pneumatic or chemical pump through the detector.[61] CEF provides a method where the separation is based almost exclusively on the isoelectric point of the sample molecules. Therefore, isozymes exhibiting very close isoelectric points can be separated.[62] An intrinsic benefit of CEF is that, due to the focusing effect, very dilute samples can be analyzed.

7.4.4   CITP

In isotachophoresis, the sample components migrate with equal velocity between the appropriately selected leading and terminator electrolytes.[63] The voltage drop in the band of a given component is constant. The voltage drop is lowest in the band of the highest mobility and highest in the band with the lowest mobility.[64] The lengths of the bands are proportional to the amount of the components.[65] Reaching equilibrium, the individual sample components are separated into distinct bands and migrate towards the electrodes with constant velocity. With isotachophoresis, either cations or anions can be determined in one run. The method is analogous to displacement chromatography.[66] Figure 9 illustrates a cationic CITP separation of proteins using three-dimensional display.[67] The concentration differences between lysozyme and ovalbumin are clearly illustrated. It is also apparent that by selecting appropriate cations as spacers we can improve the separation between the compounds of interest.

7.4.5   CGE

In the previously described electrophoretic methods, the capillary was filled with electrolytes only. Another mode of operation in capillary electrophoresis involves filling the capillary with gel or viscous polymer solutions. If desired, a column can be packed with particles and equipped with a frit.[68] This mode of analysis has been favorably used for the size determination of biologically important polymers, such as DNA, proteins, and polysaccharides. The most frequently used polymers in capillary gel electrophoresis are cross-linked or linear polyacrylamide,[69] cellulose derivatives,[70-75] agarose,[76-78] and polyethylene glycols.

Originally, polyacrylamide was used as an anticonvective additive in slab gel electrophoresis,[79] but later its molecular sieving capability was also utilized.[80] Polyacrylamide is a polymer built exclusively from monomeric units, with or without cross linking.[81] A chemically cross-linked network is

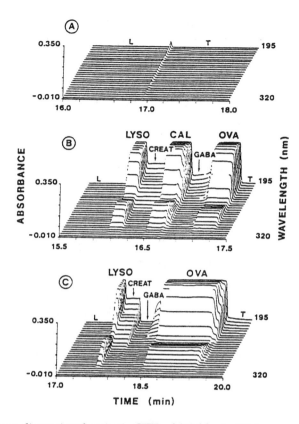

Figure 9 Three-dimensional cationic CITP of **(A)** blank; **(B)** lysozyme (LYSO), creatinine (CREAT), conalbumin (CAL), γ-amino-n-butyric acid (GABA), and ovalbumin (OVA); **(C)** OVA spiked with CREAT and GABA. Capillary: 90 cm (length to the detector, 70 cm) × 75 μ i.d.; leader: 10 mM potassium acetate and acetic acid with 0.3% HPMC, pH 4.75; terminator: 10 mM acetic acid; sample: 10 to 30 mg/ml proteins dissolved in leader without HPMC; voltage: 20 kV. (From Gebauer, P. and Thormann, W., *J. Chromatogr.*, 558, 423, 1991. With permission.)

called a "chemical gel" and a network built using noncovalent interchain interactions of linear polyacrylamide is called a "physical gel". The pore size of the chemical gel can be controlled by the degree of cross-linking. In the case of physical gels, interactions between the long polymer chains generate an organized backbone, which can be modified by changes in the experimental parameters such as concentration or temperature. The practical advantages of polyacrylamide are its charge neutrality, low UV absorption above 230 nm, and ease of preparation and handling. In CGE, cross-linked gels are prepared *in situ* in the capillary by polymerization. When physical gels are used, the low viscosity linear polymer solutions are prepared and then transferred to the capillary by pressure.[6] Both types of polyacrylamide gels can be used for the size determination of oligonucleotide, DNA, and protein molecules. The selectivity and speed of the technique provides a method for the purity assessment of oligonucleotides produced in DNA synthesizers,[82] identification of polymerase chain reaction (PCR) products,[83] and, increasingly, in DNA sequence analysis.[84] Protein separation by molecular size for purity and molecular mass determination can be accomplished, as in SDS-PAGE, by adding an appropriate amount of SDS to the separating gel.

Cellulose derivatives, such as methylcellulose, hydroxymethylcellulose,[70] hydroxyethylcellulose,[71] and hydroxypropylcellulose[72] have been used successfully for separations of large double-stranded DNA fragments by capillary electrophoresis. Rheological studies confirmed that the entanglement of cellulose polymer chains can create smaller pore-size networks than agarose gels. The use of derivatized cellulose allows the separation of double-stranded DNA molecules by size, both below and above the entanglement point.[74] It is important to note that these low viscosity polymer solutions are not only excellent molecular sieves, but they also significantly decrease the electro-osmotic flow.[85] Addition of alkali metal ions (Li^+, K^+, Na^+) also decreases the zeta potential of the capillary wall. Use of these ions as additives improves the separation of DNA molecules when cellulose derivatives are used as sieving matrices.[75] Agarose solutions in the concentration range of 0.3 to 2.6% (w:v) have been used for the separation of double-stranded DNA molecules. The agarose can be melted and transferred into the capillaries, where at lower temperature gel formation occurs without bubble formation.[76] The content of the capillary can be removed easily by warming up the capillary followed by a wash step. The UV absorbance under 230 nm of agarose gels is better than that of polyacrylamide gels. When agarose gels are used, the wall of the capillary should be coated by linear polyacrylamide to suppress or even prevent EOF.[59]

In nondenaturing gel electrophoresis, the separation is based on the size and charge of the solute. In other words, using a gel of given pore size and pH, the migration rate is controlled by the charge-to-mass ratio of the analyte. Molecules having the same charge-to-mass ratio can also be separated on nondenaturing gel columns. In this case, the principal of separation is purely the sieving effect of the gel matrix. In the case of polynucleotides, the phosphate group of each nucleotide carries a strong negative charge that is much greater than any of the charges on the bases above neutral pH. Since the charge-to-mass ratio of all polynucleotides is essentially independent of the base composition, it is nearly the same for all polynucleotides. The molecular sieving effect will cause small molecules to move faster than larger ones.

Another important application of capillary gel electrophoresis is the separation of proteins according to their molecular weight using SDS as the denaturing agent.[86] At neutral pH and in the presence of SDS (1.4 g SDS per 1 g protein) and mercaptoethanol, most multichain proteins bind SDS, disulfide linkages are broken by the reducing agent, secondary structure is lost, and the SDS-protein complex is assumed to be a random coil configuration. Proteins treated in this way have uniform shape and an almost identical charge to mass ratio, so separation is based on molecular sieving. Because of the association-dissociation phenomena of the SDS-protein complex, the gel-buffer system must contain a sufficient amount of SDS (usually 0.1%) to stabilize the SDS-protein complex. The plot of the migration time of the sample components vs. the logarithm of the molecular weight is approximately linear. Hence, if a protein of unknown molecular weight is analyzed by electrophoresis vs. known standards, the molecular weight of the unknown can be calculated to an average accuracy of about 10%. In the instance of glycoproteins and lipoproteins, the so-called Ferguson method should be used to achieve higher precision in molecular weight determination, since for these compounds the SDS-protein binding ratio is irregular.[87]

7.4.6   Affinity electrophoresis

Capillary gel columns can also be used with complexing agents to achieve special selectivities. As in affinity electrophoresis, the additive can either be covalently bound to the gel matrix or incorporated into the polymeric fiber gel matrix.[88] The latter method is always preferable since it requires no special chemistry. Under an applied electric field, the complex will migrate according to its overall charge in the gel when the complexing agent is not bound to, or entrapped in the matrix. Thus, it can slow down or speed up the separation procedure depending on the complex formation constant and the electrokinetic behavior of the complex. High resolution of DNA restriction fragments by capillary gel affinity electrophoresis has been achieved by adding ethidium bromide to the gel-buffer system. Ethidium bromide is a selective intercalating agent for double-stranded DNA molecules. Since ethidium bromide is positively charged, it causes a reduction in electrophoretic mobility of DNA when intercalating into the two strands. Affinity electrophoresis is also an excellent tool for studying complex formation and the effects of solution parameters such as pH, temperature, and concentration on complex formation. Antigen-antibody, drug-receptor, enzyme-substrate,[89] peptide-antibiotics,[90] protein-sugar,[91] protein-amino acid,[92] and other types of complex formations can be studied by HPCE.

7.5   Applications

High performance capillary electrophoresis in its current form is a new technique. Its feasibility has been proven by the analysis and separation of small ions, drugs, chiral molecules, polymers, and biopolymers.[93] We are learning more every day about the small tricks of the trade of the technique, and the efficiency and reproducibility of the methods are improving.

7.5.1   Small molecules

The initial driving force behind the development of HPCE came from the need of better and faster separations for biologically important molecules used in biotechnology research and in the biopharmaceutical industry. Traditional electrophoresis is an established method in life sciences; however, the classical pharmaceutical industry has responded very quickly and positively to the availability of a novel analytical tool. The analysis of environmental samples such as waste water has also become a major area of application for HPCE.

7.5.1.1 Ions

The separation, identification, and quantitation of small inorganic ions are always challenges and have great importance from the point of view of environmental awareness and process control. The separation of ions is traditionally accomplished by ion chromatography, an HPLC-based separation method. Most of the practical separation techniques in ion analysis use indirect UV detection. The introduction of CZE was a natural application for the analysis of small inorganic ions, and most of the established detection methods were applicable for electrophoresis. CZE methods can be used for the analysis of both anions and cations.[94-100] Inorganic ion content of waste water,[101] of ocular lenses,[102] and inorganic cations at the ppb level[103] were determined by CZE using indirect spectrophotometric determination. The power of electrophoresis is illustrated in Figure 10, with the separation of 30 different anions using indirect detection. Complex formations of ions have been used for selectivity and detection enhancement. Lanthanides were separated using hydroxyisobutyric acid as the complexing agent and creatinine as a UV-adsorbing co-ion for indirect detection.[104] Crown ethers were used as selectivity enhancers for the separation of K^+/NH_4^+ and Ba^{2+}/Sr^{2+}.[105] Chelates of a variety of ions were separated by MEKC, indicating the unique resolving power of secondary equilibrium.[106-108] The organic acid content of wines,[109] and the cation content of apple vinegar and mineral water,[50] cola beverages and fermentation broth,[105] and industrial waste water[110] were also analyzed by HPCE. Indirect fluorescence from a fluorescein-containing solution was used to detect cyanide, cyanate, thiocyanate, and nitrate.[111]

7.5.1.2 Drugs and other bioactive compounds

Samples of the traditional pharmaceutical industry represent a chemically most heterogeneous molecular pool. GC, HPLC, and ion pair chromatography are the most commonly used and established techniques for the analysis of these small organic molecules. The compounds included in this group can be charged or neutral molecules with different hydrophobicities. Their analysis consequently requires great experience from the separation scientist. During the production and storage of drugs, degradation products are generated that might analyze differently than the parent molecule, exhibiting different migration or spectrophotometric characteristics. All the required steps of analytical method validation, such as verification of reproducibility, linearity, and sensitivity, can be done with HPCE-based analytical methods.

HPCE separations utilizing the MEKC mode allow the electrophoretic separations of neutral components using detergent micelles. The advantage of using detergents is that in most cases the sample cleanup and solubilization step can be eliminated because of the presence of the detergent. Penicillins,[112,113]

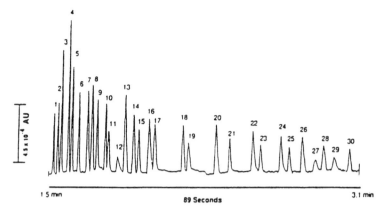

Figure 10 Capillary ion analysis of 30 anions: 1 = thiosulfate, 2 = bromide, 3 = chloride, 4 = sulfate, 5 = nitrite, 6 = nitrate, 7 = molybdate, 8 = azide, 9 = tungstate, 10 = monofluorophosphate, 11 = chlorate, 12 = citrate, 13 = fluoride, 14 = formate, 15 = phosphate, 16 = phosphite, 17 = chlorite, 18 = galactarate, 19 = carbonate, 20 = acetate, 21 = ethanesulphonate, 22 = propionate, 23 = propanesulphonate, 24 = butyrate, 25 = butanesulphonate, 26 = valerate, 27 = benzoate, 28 = D-glutamate, 29 = pentanesulphonate and 30 = D-gluconate. Experimental conditions: fused silica capillary, 60 cm (Ld 52 cm) × 50 μ i.d., voltage 30 kV, indirect UV detection at 254 nm, 5 mM chromate, 0.5 mM NICE-Pak OFM Anion-BT, adjusted to pH 8.0, with 100 mM NaOH. (From Jones, W. R. and Jandik, P., *J. Chromatogr.*, 546, 445, 1991. With permission.)

tetracyclines,[114] sulfonamides,[115-117] cephalosporins,[118,119] barbiturates,[120] antihistamines,[121] corticosteroids,[122,123] catecholamines such as dopamine,[124] and painkillers[125] are just some of the representative drugs that have been analyzed by HPCE. Herbicides from the chloro-, hydroxy-, methoxy-, and thiomethyl-s-triazine groups were separated at pH 2.5 to 8.3.[126] A mixture of 19 s-triazines, including Ameline, Simazine, Atrazine, and Prometon, could be separated at pH 2.2. Figure 11 shows electropherograms of anti-inflammatory drugs using coated and uncoated capillaries at different pH. It is important to note that the migration order changes depending upon whether the capillary is coated or not. The reversal of migration order can be beneficial from a preparative point of view.

7.5.1.3 Chiral separations

Capillary electrophoresis employing chiral selectors has been shown to be a useful analytical method to separate enantiomers. Conventionally, instrumental chiral separations have been achieved by gas chromatography and by high performance liquid chromatography.[127] In recent years, there has been considerable activity in the separation and characterization of racemic pharmaceuticals by high performance capillary electrophoresis, with particular interest paid to using this technique in modern pharmaceutical analytical laboratories.[128-130] The most frequently used chiral selectors in CE are cyclodextrins, crown ethers, chiral surfactants, bile acids, and protein-filled

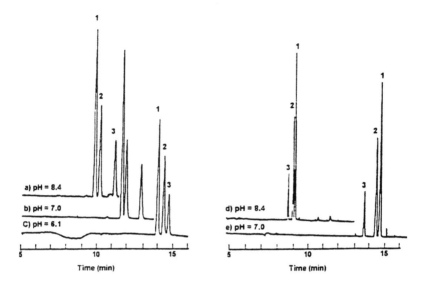

Figure 11 Separation of anti-inflammatories by CZE at various pHs in a 40-cm polyacrylamide-coated (left) and a 70-cm uncoated (right) capillary. Experimental conditions: 275 V/cm; UV = 215 nm; buffers: 20 mM borate-100 mM boric acid, pH 8.4 (46 µA); 30 mM phosphate-9 mM borate, pH 7.0 (70 µA); 80 mM MES-30 mM Tris, pH 6.1 (20 µA); peak identification: 1 = naproxen, 2 = ibuprofen, 3 = tolmetin. (From Wainwright, A., *J. Microcol Sep.*, 2, 166, 1990. With permission.)

columns. The most popular ones are native and derivatized cyclodextrins, which have been successfully employed in chiral separations using isotachophoresis, CZE, MEKC, and capillary gel electrophoresis. Cyclodextrins (CDs) are nonionic cyclic polysaccharides containing glucose units shaped like a torus or hollow truncated cone. The cavity is relatively hydrophobic while the external faces are hydrophilic, with the edge of the torus of the larger circumference containing secondary hydroxyl groups connected to the chiral carbons.[131] These secondary hydroxyl groups can be derivatized in order to increase the size of the cavity or the solubility of the cyclodextrin, e.g., by permethylation or hydroxypropylation.

A general, systematic approach can be developed for the optimization of chiral separations of weakly acidic and basic compounds in capillary electrophoresis using different natural and derivatized cyclodextrins as chiral selectors based on the Vigh's theory.[132] Vigh's theory suggests there are three types of chiral separations: nonionoselective (only the non-dissociated enantiomer complexes selectively), ionoselective (only the dissociated enantiomer complexes selectively, and duoselective (both enantiomers complex selectively). Based on his theory, in most instances there is an existing pH (high pH, low pH, and pH = solute pK) and cyclodextrin type and concentration that result in satisfactory chiral separation of the solute. Figure 12 shows the resolution surface of dansyl phenylalanine as a function

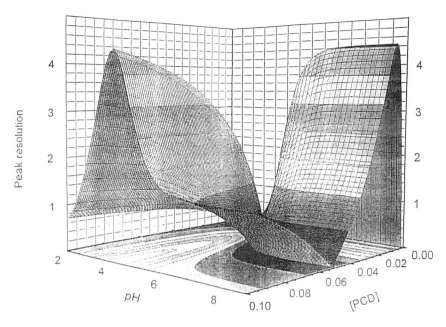

Figure 12 Resolution map of dansyl-phenylalanine as a function of pH and cyclo-dextrin concentration.

of the separation buffer pH and the CD concentration.[133] It should be empha-sized that no single capillary electrophoresis buffer, containing any chiral pseudophase, can be considered as a universal system for carrying out chiral separations. This is due to the fact that capillary electrophoresis separations, based on diastereomeric complexation between the pseudophase and the enantiomers, are very dependent upon the extent of complexation. In case of cyclodextrins as chiral selectors, factors such as the tightness of the cyclo-dextrin cavity and the ability to form hydrogen bonds between the solute and the hydroxyl groups on the rim of the cyclodextrin appear to be impor-tant in affecting chiral selectivity. Figure 13 shows an optimized separation of propranolol. As another example of selectivity manipulation, cyclodex-trins can also be used as complexing agents in polyacrylamide capillary gel electrophoresis.[134] By incorporation of cyclodextrins into the gel matrix, the complexing agent is practically immobilized, as the cyclodextrin has no charge. This is particularly useful when the pore size of the gel is smaller than the size of β-CD (<14 nm). Using the special selectivity of CDs, closely related species can be separated. The enantiomers of the calcium channel blocker veraprimil and its major metabolite norveraprimil were separated using MEKC in a bile salt-polyoxyethylene ether-mixed micellar phase.[135] Hydroxypropyl-b-cyclodextrin was used for the separation of mianserin and the diuretic chlorthalidone.[68] Vancomycin was used to control selectivity and improve resolution in the chiral separation of amino acids.[55] Figure 14 shows

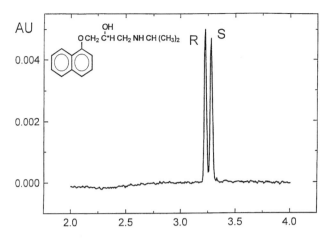

Figure 13 Separation of propranolol isomers. (From Benedek, K. and Guttman, A., *J. Chromatogr.*, 680, 375, 1994. With permission.)

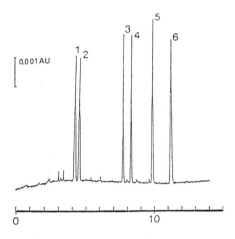

Figure 14 Chiral separation of dansyl amino acids by chiral polyacrylamide gel electrophoresis. Peak identification: 1 = dansyl-L-glutamic acid, 2 = dansyl-D-glutamic acid, 3 = dansyl-L-serine, 4 = dansyl-D-serine, 5 = dansyl-L-leucine, 6 = dansyl-D-leucine. (From Guttman, A., Paulus, A., Cohen, A. S., Grinberg, N. and Karger, B. L., *J. Chromatogr.*, 448, 41, 1988. With permission.)

the chiral separation of a mixture of dansylated amino acids based on the chiral recognition of the cyclodextrin incorporated into the polyacrylamide gel matrix.

7.5.2   *Biopolymers*

The development of the biotechnology industry presents new and novel molecules derived from nature. The utilization of these molecules as pharmaceutical products presents an analytical challenge of a magnitude greater than ever confronted. The drug candidate and closely-related molecules have to be identified and characterized and their stability established. All of these tasks can use the extra resolving power of HPCE. One of the driving forces behind the development of HPCE is actually to fulfill the needs of the biotechnology industry. Biomolecules such as amino acids, peptides and proteins, and nucleic acids and carbohydrates are key molecular classes of focus in the biotechnology industry. Often these molecules or their derivatives, metabolites, or degradation products are available only in small quantities. On-line detection, ease of sample handling, and convenient sample preparation are significant features dearly valued during the analysis of these molecules. Capillary electrophoresis has been used successfully for purity determination, for quantitation of protein and peptide content at various steps of the manufacturing process, for structural studies such as peptide mapping, for microheterogeneity determinations, for a variety of binding studies, and for following chemical biochemical reactions or conformational changes, stability studies, and diagnostics.

7.5.2.1   Separation of amino acids, peptides, and proteins

Amino acids are interesting molecules by themselves from an analytical point of view for two reasons. They are inherently enantiomeric and are the building blocks of peptides and proteins. The separation of amino acids is usually done through a derivatization process due to the fact that the absorbance in the UV is low. The most frequently used derivatization is done by fluorescent tagging. Sensitivity can reach the subfemtomole level.[136-139] Temperature control can be used to separate conformers.[140] Two conformers of Tyr-Pro-Phe-Asp-Val-Val-Gly-NH_2 and four conformers of Tyr-Pro-Phe-Gly-Tyr-Pro-Ser-NH_2 were separated at subzero temperatures by including glycerol as an antifreeze component of the buffer.

The analysis of peptides should provide three pieces of information: the size of the peptide, the amino acid composition, and the sequence of those amino acids. In traditional HPLC methods, the separation is based on the interactions between the peptide and the adsorptive surface. These interactions occur with a specific area of the peptide while the other parts of the molecule are inconsequential from the standpoint of separation. In HPCE, the separation of peptides is based on the total charge differences defined by their structure and composition. An electrophoretic method provides a separation method which can be designed to be orthogonal to RP-HPLC, with different selectivity and migration order. HPCE has became a new method for peptide mapping complementing the traditionally used HPLC methods. Figure 15 illustrates the effect of ionic strength on the separation of tryptic digest of human growth hormone.[141] The variety of peptides analyzed by HPCE is demonstrated by the separation of neuropeptides,[142] hormones,[24,143] synthetic peptides,[26] and endorphins.[144,145] A tryptic digest of erythropeitin was separated, and fractions were collected and analyzed by time-of-flight mass spectrometry.[39]

The major problem with proteins in HPCE is their tendency to adsorb to the silica surface. This can be minimized or prevented by modifying the

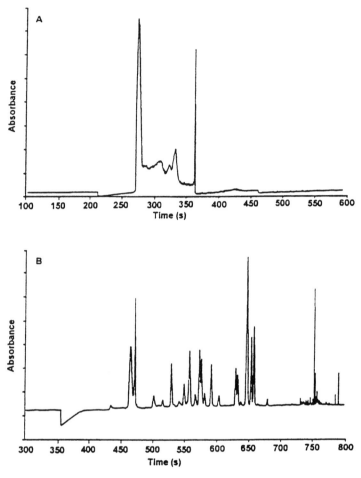

Figure 15 CZE separation of hGH tryptic digest in pH 8.1 tricine buffer: **(A)** 10 mM, **(B)** 100 mM. Capillary: 100 cm (80 cm to detector) × 50 μ i.d.; 30 kV; t = 30°C; detection: UV = 200 nm. (From Nielsen, R. G. and Rickard, E. C., *J. Chromatogr.*, 516, 99, 1990. With permission.)

zeta potential of the capillary surface by changing the buffer pH, by use of additives, or by chemical modification of the surface that can eliminate or reverse the charge of the silica. Basic proteins were separated by CZE using 1,3-diaminopropane as additive as shown in Figures 6 and 7. All of these methods are viable, and the particular protein group determines which option provides the best and most reliable solution. Working with proteins, the possibility of aggregation and/or conformational changes also has to be taken into consideration. HPCE is also a tool for measuring physico-chemical constants such as binding constants for antigen-antibody systems, drug-receptor interactions, folding-unfolding,[46] and mechanisms of denaturation. Protein separations have been performed using different modes of HPCE either with uncoated[14,146-148] or coated capillaries.[29,149-151] Examples of protein separations include the characterization of human recombinant tissue plasminogen activator glycoform,[152] human recombinant erythropoietin,[153] and human growth hormone.[154] The coupling of the proteins' abscisic acid and bovine serum albumin was optimized using HPCE.[155] Whey proteins were separated by HPCE.[156] Figure 16 shows the separation of r-EPO glycoforms by CZE, where the effect of additives such as 1,4-diaminobutane and urea on the improvement of separation is well demonstrated. Peptide mapping of proteins is an area where the unique selectivity of HPCE has been utilized for tissue plasminogen activator[157] and human growth hormone.[141,158] The addition of urea as a solubilizing agent is also applicable for HPCE separations as shown for the separation of hydrophobic membrane proteins.[159] Isoelectric focusing in a methylcellulose-coated capillary was used for the separation of the human red cell glucose transporter, a transmembrane protein.[160]

Capillary gel electrophoresis also offers an automated option to replace SDS-PAGE. Results acquired by CGE are similar to those of the traditional SDS-PAGE method, with the significant advantage that the results are also quantitative due to the direct detection capability of CGE. Different sieving matrices such as polyethylene glycols,[161-163] dextrans,[164] and polyacrylamides[69] have been suggested, and it seems that their performance is comparable. CGE of proteins not only provides quantitative data, but the method is also much faster then the traditional SDS-PAGE. Figure 17 illustrates that CGE separations easily can be accomplished within 3 minutes.[165]

7.5.2.2   Separation of DNA

For the separation of single-stranded DNAs, chemical denaturants such as urea and formamide must be used in the polyacrylamide gel to prevent the formation of DNA secondary structures.[166] In capillary gel electrophoresis of polynucleotides, 7 to 9 M urea or 35 to 70% formamide are used as denaturing agents.[167] Methyl cellulose was used as the sieving agent in the separation of apolipoprotein B gene fragments.[168] Conformational polymorphism was detectable in DNA fragments in nondenaturing buffers.[169] Asymmetric field inversion proved useful in preventing DNA agglomeration and artifactual peaks.[37] Figure 18 shows a separation of a polydeoxyadenylic acid mixture, p(dA)$_{40-60}$. This figure demonstrates the extremely high resolving power of the method, especially in the base number range of DNA primers and probes (20- to 80-mers). The average resolution between the peaks is RS > 2.0. The same kinds of gel columns are being used for DNA sequencing of up to several hundreds of bases using laser-induced fluorescence detection methods.[170]

Figure 19 shows an electropherogram of a DNA restriction fragment mixture over a broad base-pair range (72 to 1353).[171] It is important to note that all of the 11 sample components have the same charge-mass ratio. A typical semilogarithmic plot of molecular weight vs. migration time is helpful for the determination of the size and molecular weight of an unknown DNA within the range of the plot. Figure 20 compares the separations of the

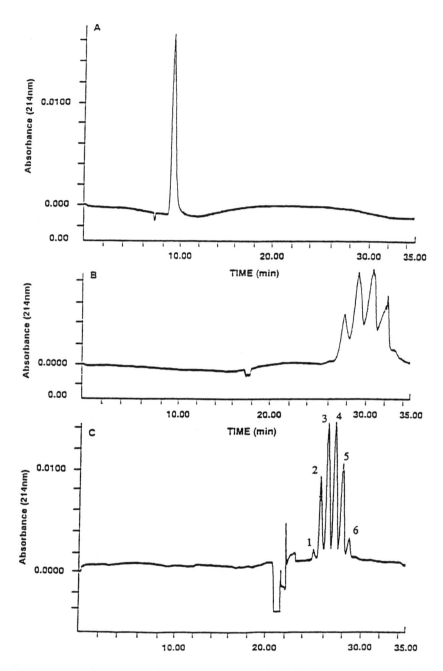

Figure 16 Capillary electrophoretic separations of r-HuEPO at 1 mg/ml in 50 cm ×
75 μ i.d. uncoated capillary at 10 kV. Buffers used were **(A)** 10 m*M* tricine/10 m*M*
NaCl, pH 6.2; **(B)** 10 m*M* tricine/10 m*M* NaCl/2.5 m*M* 1,4-diaminobutane, pH 6.2;
and **(C)** 10 m*M* tricine/10 m*M* NaCl/2.5 1,4-DAB/7 *M* urea, pH 6.2. (From Watson,
E. and Yao, F., *Anal. Biochem.*, 210, 389, 1993. With permission.)

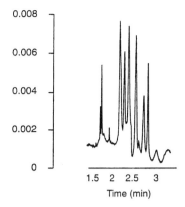

Figure 17 SDS-capillary gel electrophoresis of protein standards. Separation conditions: 27 cm × 50 μm i.d. uncoated capillary; 888 V/cm; gel; eCAP 200. (From Benedek, K. and Guttman, A., *J. Chromatogr.*, 680, 375, 1994. With permission.)

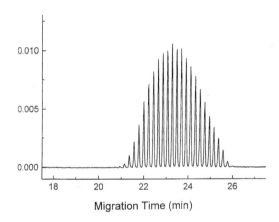

Figure 18 Separation of polydeoxyadenylic acid mixture. (Courtesy of Beckman Instruments; Fullerton, CA.)

PBR320 DNA-Hae III digest restriction fragments mixture on a replaceable polyacrylamide gel-filled capillary with and without ethidium bromide.[88] By the addition of only 1 μg/ml ethidium bromide and using the same sieving polymer concentration, a wider molecular weight range of DNA molecules can be separated, as is shown in Figure 20B.

7.5.2.3 *Carbohydrate separations*
Carbohydrates play a major role in protein bioactivity, bioavailability, and antigenicity; therefore, the understanding of the glycosylation of protein molecules is very important in the development of effective glycoprotein therapeutics.[172] In recent years, there has been considerable activity in the development of simple, rapid, and reliable separation methods for the analysis of

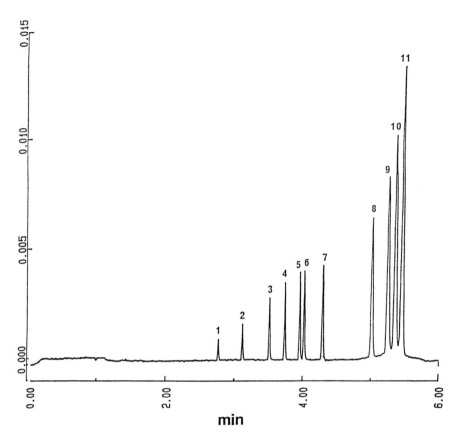

Figure 19 Capillary gel electrophoresis of DNA-Hae-digested restriction fragments of FX 174. Peak identification: 1 = 72, 2 = 118, 3 = 194, 4 = 234, 5 = 271, 6 = 281, 7 = 310, 8 = 603, 9 = 872, 10 = 1078, 11 = 1353 base pairs. (From Guttman, A., *App. Theor. Electrophoresis,* 3, 91, 1991. With permission.)

complex carbohydrates released from a variety of glycoconjugates including glycoproteins, glycolipids, proteoglycans, and glycosaminoglycans. Recently, capillary electrophoresis was introduced as a new alternative to HPLC and slab gel electrophoresis with the advantages of fast analysis time, automation, on-column injection, and detection.[173]

Because many carbohydrates are not charged and have no significant absorbance of UV light at the common wavelength range of capillary electrophoresis detection systems (200 to 600 nm), a derivatization procedure is required before analysis. This derivatization involves the stoichiometric labeling of the reducing end of the oligosaccharide with the labeling reagent.[174] The derivatization agent can be UV-active or fluorescent, as well as charged or uncharged. When it is uncharged, a secondary equilibration with borate complexation is used to achieve electrophoretic mobility differences.[175] The labeling of the oligosaccharides, linear homo-oligomers such

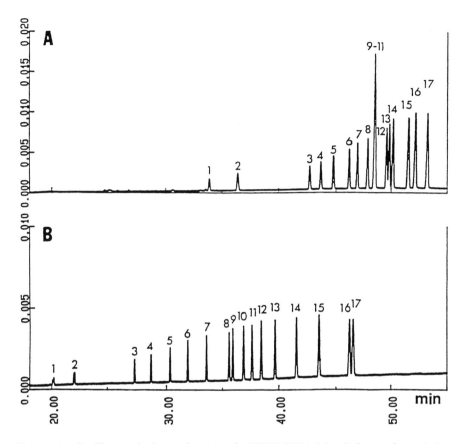

Figure 20 Capillary gel electrophoresis of pBR322 DNA-Msp I digested restriction fragments. **(A)** Without ethidium bromide, **(B)** 1 µg/ml ethidium bromide. Peak identification: (1) 26, (2) 34, (3) 67, (4) 76, (5) 90, (6) 110, (7) 123, (8) 147, (9) 157, (10) 160, (11) 160, (12) 180, (13) 190, (14) 201, (15) 217, (16) 238, (17) 242 base pairs. (From Guttman, A. and Cooke, N., *Anal. Chem.*, 63, 2038, 1991. With permission.)

as wheat starch digest or branched carbohydrates released form glycoproteins, is usually performed through reductive amination[176] with a fluorescent dye, 8-amino-1,3,6-naphthalene trisulfonate (ANTS), as a labeling tag followed by reduction using $NaCNBH_4$. The fluorescent-labeled oligosaccharides are then separated and quantified by capillary electrophoresis using the laser induced fluorescent (LIF) detection method employing a 360-nm He-Cd laser. Using very high concentration linear polyacrylamide gels, separation of oligosaccharides as high as 50 units has been achieved.[177] Complexation of heparin to a variety of peptides has been demonstrated using HPCE.[178] High-mannose oligosaccharides were derivatized with ANTS and separated in a polyethyleneoxide-filled capillary.[179] Derivatization with 2-aminoacridone has also been used.[180] Others have shown that the use of a gel or a kind of sieving matrix is not essential in the case of the electrophoretic

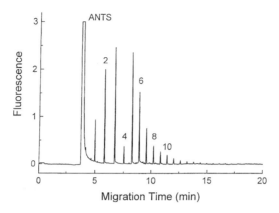

Figure 21 Separation of ANTS-labeled oligosaccharides. (From Guttman, A., Cooke, N., and Star, C. M., *Electrophoresis,* 15, 1518, 1994. With permission.)

separation of complex carbohydrates. Oligosaccharides up to 25 U were separated as shown in Figure 21 by simply using low pH buffers in CZE mode, taking advantage of the different charge-to-mass ratios of the ANTS-labeled, different chain length oligosaccharides.[175,181]

7.6   Conclusions

High performance capillary electrophoresis is an exceptionally versatile and sensitive analytical method complementary to HPLC. The versatility is seen in the availability of modes that separate by charge (CZE and isotachophoresis); by complexation with detergent (MEKC), chiral additives (chiral CE), or antibodies (affinity CE); by isolectric point (CEF); or by molecular weight (CGE). Versatility is also seen in on-line detection, which permits direct mass spectral, spectrophotometric, or electrochemical analysis. The sensitivity of electrophoretic methods extends to the attomol range, which is particularly useful for scarce samples, such as drug metabolites and degradation products. Environmental applications are also well recognized. Recent developments in the separation of erythrocytes, cells from peritoneal dialysis, and other particles from 10 nm to 50 µ promise even wider application of electrophoretic techniques.[182]

References

1. Camilleri, P., in *Capillary Electrophoresis, Theory and Practice*, CRC Press, Boca Raton, FL, 1993.
2. Grossman, P. D. and Colburn, J. C. Eds., *Capillary Electrophoresis, Theory and Practice*, Academic Press, San Diego, CA, 1992.
3. Kuhn, R. and Hofstetter-Kuhn, S., *Capillary Electrophoresis: Principles and Practice*, Springer-Verlag, Berlin, 1993.
4. Weinberger, R., *Practical Capillary Electrophoresis*, Academic Press, Boston, MA, 1993.

5. Guzman N.A., *Capillary Electrophoresis Technology*, Marcel Dekker, New York, 1993.
6. Landers, J. P., *Handbook of Capillary Electrophoresis*, CRC Press, Boca Raton, FL, 1993.
7. Li, S. F. Y., *Capillary Electrophoresis: Principles, Practice and Applications*, Elsevier, Amsterdam, 1993.
8. Friedl, W., Reijenga, J. C., and Kenndler, E., Ionic strength and charge number correction for mobilities of multivalent organic anions in capillary electrophoresis, *J. Chromatogr. A*, 709, 163, 1995.
9. Wiese, G. K., James, R. O., and Healy, T.W., Effect of particle size on colloid stability, *Trans. Farad Soc.*, 52, 302, 1971.
10. Jorgenson, J. W. and Lukacs, K. D., Zone electrophoresis in open-tubular glass capillaries, *Anal. Chem.*, 53, 1298, 1981.
11. Jorgenson, J. W. and Lukacs, K. D., Zone electrophoresis in open-tubular glass capillaries: preliminary data on performance, *J. High Res. Chromatogr., Chromatogr. Comm.*, 4, 230, 1981.
12. Nelson, R. J., Paulus, A., Cohen, A. S., Guttman, A., and Karger, B. L., Use of Peltier thermoelectric devices to control column temperature in high-performance capillary electrophoresis, *J. Chromatogr.*, 480, 111, 1989.
13. Jacobsen, S. C., Moore, A. W., and Ramsey, J. M., Fused quartz substrates for microchip electrophoresis, *Anal. Chem.*, 67, 2059, 1993.
14. Lauer, H. H. and McManigill, D., Capillary zone electrophoresis of proteins in untreated fused silica tubing, *Anal.Chem.*, 58, 166, 1986.
15. Bushey, M. M. and Jorgenson, J. W., Capillary electrophoresis of proteins in buffers containing high concentrations of zwitterionic salts, *J. Chromatogr.*, 480, 301, 1989.
16. Fujiwara, S. and Honda, S., Determination of cinnamic acid and its analogs by electrophoresis in a fused silica capillary tube, *Anal. Chem.*, 58, 1811, 1986.
17. Fujiwara, S. and Honda, S., Effect of addition of organic solvent on the separation of positional isomers in high-voltage capillary zone electrophoresis, *Anal. Chem.*, 59, 487, 1987.
18. Tsuda, T., Modification of electro-osmotic flow with cetyltrimethylammonium bromide in capillary zone electrophoresis, *J. High Resolut. Chromatogr.*, 10, 622, 1987.
19. Muijselaar, W. G. H. M., de Bruijn, C. H. M. M., and Everaerts, F. M., Capillary zone electrophoresis of proteins with a dynamic surfactant coating: influence of a voltage gradient on the separation efficiency, *J. Chromatogr.*, 605, 115, 1992.
20. Wiktorowicz, J. E. and Colburn, J. C., Separation of cationic proteins via charge reversal in capillary electrophoresis, *Electrophoresis*, 11, 769, 1990.
21. Bullock, J.A. and Yuan, L.-C., Free solution capillary electrophoresis of basic proteins in uncoated fused silica capillary tubing, *J. Microcol. Sep.*, 3, 241, 1991.
22. Landers, J. P., Oda, R. P., Madden, B. J., and Spelsberg, T. C., High-performance capillary electrophoresis of glycoproteins: the use of modifiers of electro-osmotic flow for analysis of microheterogeneity, *Anal. Biochem.*, 205, 115, 1992.
23. Nielsen, R. G., Riggin, R. M., and Rickard, E. C., Capillary zone electrophoresis of peptide fragments from trypsin digestion of biosynthetic human growth hormone, *J. Chromatogr.*, 480, 393, 1990.
24. McLaughlin, G. M., Nolan, J. A., Lindahl, J. L., Palmieri, R. H., Anderson, K. W., Morris, S. C., Morrison, J. A., and Bronzert, T. J., Pharmaceutical drug separations by HPCE: practical guidelines, *J. Liq. Chromatogr.*, 15, 961, 1992.

25. Gassner, B., Friedl, W., and Kenndler, E., Wall adsorption of small anions in capillary zone electrophoresis induced by cationic trace constituents of the buffer, *J. Chromatogr.*, 680, 25, 1994.

26. McCormick, R. M., Capillary zone electrophoretic separation of peptides and proteins using low pH buffers in modified silica capillaries, *Anal. Chem.*, 60, 2322, 1988.

27. Tran, A. D., Park, S., Lisi, P. J., Huynh, O. T., Ryall, R. R., and Lane, P. A., Separation of carbohydrate-mediated microheterogeneity of recombinant human erythropoietin by free solution capillary electrophoresis. Effects of pH, buffer type and organic additives, *J. Chromatogr.*, 542, 459, 1991.

28. Hjertén, S., High-performance electrophoresis. Elimination of electroendosmosis and solute adsorption, *J. Chromatogr.*, 347, 191, 1985.

29. Towns, J. K. and Regnier, F. E., Capillary electrophoretic separations of proteins using nonionic surfactant coatings, *Anal. Chem.*, 63, 1126, 1991.

30. Williams, B. A. and Vigh, G., Effect of the initial potential ramp on the accuracy of electrophoretic mobilities in capillary electrophoresis, *Anal. Chem.*, 67, 3079, 1995.

31. Zhu, M., Hansen, D.L., Burd, S., and Gannon, F., Factors affecting free zone electrophoresis and isoelectric focusing in capillary electrophoresis, *J. Chromatogr.*, 480, 311, 1989.

32. Terabe, S., Ozaki, H., Otsuka, K., and Ando, T., Electrokinetic chromatography with 2-O-carboxymethyl-β-cyclodextrin as a moving "stationary" phase, *J. Chromatogr.*, 332, 211, 1985.

33. Guttman, A., Novel separation scheme for capillary electrophoresis of enantiomers, *Electrophoresis*, 16, 1900, 1995.

34. Kuhn, R. and Wagner, H., Application of free flow electrophoresis to the preparative purification of basic proteins from an *E. coli* cell extract, *J. Chromatogr.*, 481, 343, 1989.

35. Mosher, R. A., The use of metal ion-supplemented buffers to enhance the resolution of peptides in capillary zone electrophoresis, *Electrophoresis*, 11, 765, 1990.

36. Fanali, S., Ossicini, L., Foret, F., and Boček, P., Resolution of optical isomers by capillary zone electrophoresis: study of enantiomeric and distereoisomeric cobalt (III) complexes with ethylenediamine and amino acid ligands, *J. Microcol. Sep.*, 1, 190, 1989.

37. Heller, C., Pakleza, C., and Viovy, J. L., DNA Separation with field inversion capillary electrophoresis, *Electrophoresis*, 16, 1423, 1995.

38. Müller, O., Foret, F., and Karger, B. L., Design of a high-precision fraction collector for capillary electrophoresis, *Anal. Chem.*, 67, 2974, 1995.

39. Boss, H. J., Rohde, M. F., and Rush, R. S., Multiple sequential fraction collection of peptides and glycopeptides by high-performance capillary electrophoresis, *Anal. Biochem.*, 230, 123, 1995.

40. Wu, S. and Dovichi, N. J., Capillary zone electrophoresis separation and laser-induced fluorescence detection of zeptomole quantities of fluorescein thiohydantoin derivatives of amino acids, *Talanta*, 39, 173, 1992.

41. Mikkers, F. E. P., Everaerts, F. M., and Verheggen, Th. P. E. M., High-performance zone electrophoresis, *J. Chromatogr.*, 169, 11, 1979.

42. Wallingford, R. A. and Ewing, A. G., Capillary zone electrophoresis with electrochemical detection, *Anal. Chem.*, 59, 1762, 1987.

43. Moseley, M. A., Deterding, L. J., Tomer, K. B., and Jorgenson, J. W. Coupling of capillary zone electrophoresis and capillary liquid chromatography with coaxial continuous-flow fast atom bombardment tandem sector mass spectrometry, *J. Chromatogr.*, 480, 197, 1989.

44. Guttman, A. and Schwartz, H. E., Artifacts related to sample introduction in capillary gel electrophoresis affecting separation performance and quantitation, *Anal. Chem.*, 34, 2279, 1995.

45. Nielsen, R. G., Sittampalam, G. S., and and Rickard, E. C., Capillary zone electrophoresis of insulin and growth hormone, *Anal. Biochem.*, 177, 20, 1989.

46. Rush, R. S., Cohen, A. S., and Karger, B. L., Influence of column temperature on the electrophoretic behavior of myoglobin and α-lactalbumin in high-performance capillary electrophoresis, *Anal. Chem.*, 63, 1346, 1991.

47. Altria, K. D. and Simpson, C. F., Analysis of some pharmaceuticals by high voltage capillary zone electrophoresis, *Pharm. Biomed. Anal.*, 6, 801, 1988.

48. Advis, J. P., Hernandez, L., and Guzman, N. A., Analysis of brain neuropeptides by capillary electrophoresis: determination of luteinizing hormone-releasing hormone from ovine hypothalamus, *Peptide Res.*, 2, 389, 1989.

49. Grossman, P. D., Wilson, K. J., Petrie, G., and Lauer, H. H., Effect of buffer pH and peptide composition on the selectivity of peptide separations by capillary zone electrophoresis, *Anal. Biochem.*, 173, 265, 1988.

50. Beck, W. and Engelhardt, H., Capillary electrophoresis of organic and inorganic cations with indirect UV detection, *Chromatographia*, 33, 313, 1992.

51. Terabe, S., Otsuka, K., and Ando, T., Electrokinetic chromatography with micellar solution and open-tubular capillary, *Anal. Chem.*, 57, 834, 1985.

52. Burton, D. E., Sepaniak, M. J., and Maskarinec, M. P., Evaluation of the use of various surfactants in micellar electrokinetic capillary chromatography, *J. Chromatogr. Sci.*, 25, 514, 1987.

53. Terabe, S., Otsuka, K., and Ando, T., Band broadening in electrokinetic chromatography with micellar solutions and open-tubular capillaries, *Anal. Chem.*, 61, 251, 1989.

54. Cohen, A. S., Paulus, A., and Karger, B. L. High-performance capillary electrophoresis using open tubes and gels, *Chromatographia*, 24, 15, 1987.

55. Rundlett, K. L. and Armstrong, D. W., Effect of micelles and mixed micelles on efficiency and selectivity of antibiotic-based capillary electrophoretic separations, *Anal. Chem.*, 67, 2088, 1995.

56. Terabe, S., Yashima, T., Tanaka, N., and Araki, M., Separation of oxygen isotopic benzoic acids by capillary zone electrophoresis based on isotope effects on the dissociation of the carboxyl group, *Anal. Chem.*, 60, 1673, 1988.

57. Hjertén, S., Isoelectric focusing in capillaries, in *Capillary Electrophoresis*, Grossman, P. D. and Colburn, J. C. Eds., Academic Press, San Diego, 1992.

58. Mazzeo, J. R. and Krull, I. S., Examination of variables affecting the performance of isoelectric focusing in uncoated capillaries, *J. Microcol. Sep.*, 4, 29, 1992.

59. Hjertén, S., Elenbring, K., Kilár, F., Liao, J.-L, Chen, A. J. C., Siebert, C. J., and Zhu, M.-D., Carrier-free zone electrophoresis, displacement electrophoresis and isoelectric focusing in a high-performance electrophoresis apparatus, *J. Chromatogr.*, 403, 47, 1987.

60. Righetti, P. G., *Isoelectric Focusing: Theory, Methodology and Applications*, Elsevier, New York, 1983.

61. Hjertén, S. and Zhu, D. M., The electrophoretic counterpart of narrow-bore high performance liquid chromatography, *J. Chromatogr.*, 327, 517, 1985.

62. Kilár, F. and Hjertén,S., Fast and high resolution analysis of human serum transferrin by high performance isoelectric focusing in capillaries, *Electrophoresis*, 10, 23, 1989.

63. Boček, P., Deml, M., Gebauer, P., and Dolnik, V., *Analytical Isotachophoresis*, VCH Publishers, New York, 1988.

64. Jorgenson, J. W., Electrophoresis, *Anal. Chem.*, 58, 743A, 1986.

65. Everaerts, F. M. and Verheggen, T. P. E. M., *New Directions in Electrophoretic Methods*, Jorgenson, J. W. and Phillips, M., Eds., American Chemical Society, Washington, D.C., 1987, 199.

66. Horváth, Cs., Nahum, A., and Frenz, J. H., High-performance displacement chromatography, *J. Chromatogr.*, 218, 365, 1981.

67. Gebauer, P. and Thormann, W., Isotachophoresis of proteins in uncoated open-tubular fused-silica capillaries with a simple approach for column conditioning, *J. Chromatogr.*, 558, 423, 1991.

68. Lelièvre, F., Yan, C., Zare, R. N., and Gareil, P., Capillary electrochromatography: operating characteristics and enantiomeric separations, *J. Chromatography A*, 723, 145, 1996.

69. Heiger, D. N., Cohen, A. S., Aharon, S., and Karger, B. L., Separation of DNA restriction fragments by high performance capillary electrophoresis with low and zero cross-linked polyacrylamide using continuous and pulsed electric fields, *J. Chromatogr.*, 516, 33, 1990.

70. Strege, M. and Lagu, A., Separation of DNA restriction fragments by capillary electrophoresis using coated fused silica capillaries, *Anal. Chem.*, 63, 1233, 1991.

71. Grossmann, P. D. and Soane, D. S., Experimental and theoretical studies of DNA separations by capillaryelectrophoresis in entangled polymer solutions, *Biopolymers*, 31, 1221, 1991.

72. Schwartz, H. E., Ulfelder, K., Sunzeri, F. J., Busch, M. P., and Brownlee, R. G., Analysis of DNA restriction fragments and polymerase chain reaction products towards detection of the AIDS (HIV-1) virus in blood, *J. Chromatogr.*, 559, 267, 1991.

73. Chrambach, A. and Aldroubi, A., Relative efficiency of molecular sieving in solutions of four polymers, *Electrophoresis*, 14, 18, 1993.

74. Barron, A. E., Soane, D. S., Blanch, H. W., Capillary electrophoresis of DNA in uncross-linked polymer solutions, *J. Chromatogr. A*, 652, 3, 1993.

75. Nathakarnkitkool, S., Oefner, P. J., Bartsch, G., Chin, M. A., and Bonn, G. K., High-resolution capillary electrophoretic analysis of DNA in free solution, *Electrophoresis*, 13, 18, 1992.

76. Boček, P. and Chrambach, A., Capillary electrophoresis of DNA in agarose solutions at 40°C, *Electrophoresis*, 12, 1059, 1991.

77. Boček, P. and Chrambach, A., Capillary electrophoresis in agarose solutions: extension of size separations to DNA of 12 kb in length, *Electrophoresis*, 13, 31, 1992.

78. Boček, P. and Chrambach, A., Electrophoretic size separations in liquidified agarose of polystyrene particles and circular DNA, *Electrophoresis*, 12, 620, 1991.

79. Raymond, S. and Weintraub, L., Acrylamide as a supporting medium for zone electrophoresis, *Science*, 130, 711, 1959.

80. Hjertén, S., Free zone electrophoresis, *Chromatogr. Rev.*, 9, 122, 1962
81. Bekturov, E. A. and Bakanova, Z. Kh., *Synthetic Water-Soluble Polymers in Solution*, Huthig, Heidelberg, 1981.
82. Karger, B. L., High performance capillary electrophoresis, *Nature*, 339, 641, 1989.
83. Lu, W., Han, D. S., Yuan, J., and Andrieu, J. M., Multi-target PCR analysis by capillary electrophoresis and laser-induced fluorescence, *Nature*, 368, 269, 1994.
84. Smith, R. D., Olivares, J. A., Nguyen, N. T., and Udseth, H. R., Capillary zone electrophoresis-mass spectrometry using an electrospray ionization interface, *Anal. Chem.*, 60, 436, 1988.
85. Ulfelder, K., Anderson, K., and Schwartz, H. E., Analysis of PCR products and DNA restriction fragments to detect AIDS (HIV-1) virus in blood, poster presentation, HPCE '91, San Diego, CA, 1991.
86. Guttman, A., Nolan, J. A., and Cooke, N., Capillary sodium dodecyl sulfate gel electrophoresis of proteins, N., *J. Chromatogr.*, 632, 171, 1993.
87. Guttman, A., Shieh, P., Lindahl, J., and Cooke, N., Capillary sodium dodecyl sulfate gel electrophoresis of proteins. II. On the Ferguson method of polyethylene oxide gels, *J. Chromatogr.*, 676, 227, 1994.
88. Guttman, A. and Cooke, N., Capillary gel affinity electrophoresis of DNA fragments, *Anal. Chem.*, 63, 2038, 1991.
89. Chu, Y.-H., Avila, L. Z., Biebuyck, H. A., and Whitesides, G. M., Use of affinity capillary electrophoresis to measure binding constants of ligands to proteins, *J. Med. Chem.*, 35, 2915, 1992.
90. Chu, Y.-H. and Whitesides, G. M., Affinity capillary electrophoresis can simultaneously measure binding constants of multiple peptides to vancomycin, *J. Org. Chem.*, 57, 3524, 1992.
91. Honda, S., Taga, A., Suzuki, K., Suzuki, S., and Kakehi, K., Determination of the association constant of monovalent mode protein-sugar interaction by capillary zone electrophoresis, *J. Chromatogr.*, 597, 377, 1992.
92. Kilár, F. and Fanali, S., Separation of tryptophan-derivative enantiomers with iron-free human serum transferrin by capillary zone electrophoresis, *Electrophoresis*, 16, 1510, 1995.
93. Engelhardt, H., Beck, W., Kohr, J., and Schmitt, T., Capillary electrophoresis: methods and scope, *Angewandte Chemie*, 32, 629, 1993.
94. Jandik, P. and Jones, W. R., Optimization of detection sensitivity in the capillary electrophoresis of inorganic anion, *J. Chromatogr.*, 546, 431, 1991.
95. Bondoux, G., Jandik, P., and Jones, W. R., New approach to the analysis of low levels of anions in water, *J. Chromatogr.*, 602, 79, 1992.
96. Jones, W. R. and Jandik, P., Controlled changes of selectivity in the separation of ions by capillary electrophoresis, *J. Chromatogr.*, 546, 445, 1991.
97. Romano, J. P. and Krol, J., Capillary ion electrophoresis, an environmental method for the determination of anions in water, *J. Chromatogr.*, 640, 403, 1993.
98. Jackson, P. E. and Haddad, P. R., Optimization of injection technique in capillary ion electrophoresis for the determination of trace level anions in environmental samples, *J. Chromatogr.*, 640, 481, 1993.
99. Weston, A., Brown, P. R., Jandik, P., Heckenberg, A. L., and Jones, W. R., Factors affecting the separation of inorganic metal cations by capillary electrophoresis, *J. Chromatogr.*, 593, 289, 1992.

100. Weston, A., Brown, P. R., Jandik, P., Heckenberg, A. L., and Jones, W. R., Optimization of detection sensitivity in the analysis of inorganic cations by capillary ion electrophoresis using indirect photometric detection, *J. Chromatogr.*, 608, 395, 1992.

101. Oehrle, S. A., Blanchard, R. D., Stumpf, C. L., and Wulfeck, D. L., Environmental monitoring of wastewater using capillary ion electrophoresis, *J. Chromatogr. A*, 680, 645, 1994.

102. Shi, H., Zhang, R., Chandrasekher, G., and Ma, Y., Simultaneous detection and quantitation of sodium, potassium, calcium and magnesium in ocular lenses by high-performance capillary electrophoresis with indirect photometric detection, *J. Chromatogr. A*, 680, 653 1994.

103. Barger, W. R., Mowery, R. L., and Wyatt, J. R., Separation and indirect detection by capillary zone electrophoresis of ppb (w/w) levels of aluminum ions in solutions of multiple cations, *J. Chromatogr. A*, 680, 659, 1994.

104. Foret, F., Fanali, S., Nardi, A., and Boček, P., Capillary zone electrophoresis of rare earth metals with indirect UV absorbance detection, *Electrophoresis*, 11, 780, 1990.

105. Bächmann, K., Boden, J., and Haumann, I., Indirect fluorimetric detection of alkali and alkaline earth metal ions in capillary zone electrophoresis with cerium (III) as carrier electrolyte, *J. Chromatogr.*, 626, 259, 1992.

106. Saitoh, T., Hoshino, H., and Yotsuyanagi, T., Micellar electrokinetic capillary chromatography of porphinato chelates as a spectrophotometric approach to subfemtomole detection of metal chelates, *Anal. Sci.*, 7, 495, 1991.

107. Saitoh, K., Micellar electrokinetic chromatography, *Kaguka (Kyoto)*, 45, 884, 1990.

108. Saitoh, K., Kiyohara, C., and Suzuki, N., Mobilities of metal β-diketonato complexes in micellar electrokinetic chromatography, *J. High Resolut. Chromatogr.*, 14, 245, 1991.

109. Kenney, B. F., Determination of organic acids in food samples by capillary electrophoresis, *J. Chromatogr.*, 546, 423, 1991.

110. Kuhn, R., Erni, F., Bereuter, T., and Häusler, J., Chiral recognition and enantiomeric resolution based on host-guest complexation with crown ethers in capillary zone electrophoresis, *Anal. Chem.*, 64, 2815, 1992.

111. Martí, V., Aguilar, M., and Yeung, E. S., Indirect fluorescence detection of free cyanide and related compounds by capillary electrophoresis, *J. Chromatogr. A*, 709, 367, 1995.

112. Swartz, M. E., Method development and selectivity control for small molecule pharmaceutical separations by capillary electrophoresis, *J. Liq. Chromatogr.*, 14, 923, 1991.

113. Nishi, H. and Terabe, S., Application of micellar electrokinetic chromatography to pharmaceutical analysis, *Electrophoresis*, 11, 691, 1990.

114. Zhang, C.-X., Sun, Z.-P., Ling, D.-K., and Zhang, Y.-J., Separation of tetracycline and its degradation products by capillary zone electrophoresis, *J. Chromatogr.*, 627, 281, 1992.

115. Johansson, I. M., Pavelka, R., and Henion, J. D., Determination of small drug molecules by capillary electrophoresis-atmospheric pressure ionization mass spectrometry, *J. Chromatogr.*, 559, 515, 1991.

116. Ng, C. L., Lee, H.K., and Li, S. F. Y., Systematic optimization of capillary electrophoretic separation of sulphonamides, *J. Chromatogr.*, 598, 133, 1992.

117. Ackermans, M. T., Beckers, J. L., Everaerts, F. M., Hoogland, H., and Tomassen, M. J. H., Determination of sulphonamides in pork meat extracts by capillary zone electrophoresis, *J. Chromatogr.*, 596, 101, 1992.

118. Miyashita, Y., Terabe, S., and Nishi, H., Separation of antibiotics and corticosteroids by MECC, Appl. Brief DS-766, Beckman Instruments, Fullerton, CA, 1990.

119. Wainwright, A., Capillary electrophoresis applied to the analysis of pharmaceutical compounds, *J. Microcol. Sep.*, 2, 166, 1990.

120. Thormann, W., Meier, P., Marcolli, C., and Binder, F., Analysis of barbiturates in human serum and urine by high-performance capillary electrophoresis-micellar electrokinetic capillary chromatography with on-column multi-wavelength detection, *J. Chromatogr.*, 545, 445, 1991.

121. Ong, C. P., Ng, C. L., Lee, H. K., and Li, S. F. Y., Determination of antihistamines in pharmaceuticals by capillary electrophoresis, *J. Chromatogr.*, 588, 335, 1991.

122. Terabe, S., Ishihama, Y., Nishi, H., Fukuyama, T., and Otsuka, K., Effect of urea addition in micellar electrokinetic chromatography, *J. Chromatogr.*, 545, 359, 1991.

123. Nishi, H. and Matsuo, M., Separation of corticosteroids and aromatic hydrocarbons by cyclodextrin-modified micellar electrokinetic chromatography, *J. Liq. Chromatogr.*, 14, 973, 1991.

124. Olefirowicz, T. W. and Ewing, A. G., Capillary electrophoresis in 2 and 5 μm diameter capillaries: application to cytoplasmic analysis, *Anal. Chem.*, 62, 1872, 1990.

125. Fazio, S., Vivilecchia, R., Lesueur, L., and Sheridan, J., Capillary zone electrophoresis: some promising pharmaceutical applications, *Am. Biotechnol. Lab.*, 8, 10, 1990.

126. Schmitt, Ph., Garrison, A. W., Freitag, D., and Kettrup, A., Separation of s-triazine herbicides and their metabolites by capillary zone electrophoresis as a function of pH, *J. Chromatogr. A*, 723, 169, 1996.

127. Ahuja, S., *Chiral Separations by Liquid Chromatography*, American Chemical Society, Washington, D.C, 1991.

128. Vespalec, R. and Boček, P., Chiral separations by capillary electrophoresis: present state of the art, *Electrophoresis*, 15, 755, 1994.

129. Novotny, M., Soini, M., and Stefansson, M., Chiral separation through capillary electromigration methods, *Anal. Chem.*, 66, 646A, 1994.

130. Bereuter, T. L., Enantioseparation by capillary electrophoresis, *LC-GC*, 12, 10, 1994.

131. Szejtli, J., *Cyclodextrins and their Inclusion Complexes*, Academic Press, Budapest, 1982.

132. Rawjee, Y. Y and Vigh, Gy., A peak resolution model for the capillary electrophoretic separation of the enantiomers of weak acids with hydroxypropyl β-cyclodextrin-containing background electrolytes, *Anal. Chem.*, 66, 619, 1994.

133. Rawjee, Y. Y and Vigh, Gy., Efficiency optimization in capillary electrophoretic chiral separations using dynamic mobility matching, *Anal. Chem.*, 66, 3777, 1994.

134. Guttman, A., Paulus, A., Cohen, A. S., Grinberg, N., and Karger, B. L., Use of complexing agents for selective separation in high-performance capillary electrophoresis. Chiral resolution via cyclodextrins incorporated within polyacrylamide gel columns, *J. Chromatogr.*, 448, 41, 1988.

135. Clothier, Jr., J. G. and Tomellini, S. A., Chiral separation of veraprimil and related componds using micellar electrokinetic capillary chromatography with mixed micells of bile salt and polyoxyethylene ethers, *J. Chromatogr. A,* 723, 179, 1996.
136. Kuhr, W. G. and Yeung, E. S., Indirect fluorescence detection of native amino acids in capillary zone electrophoresis, *Anal. Chem.,* 60, 1832, 1988.
137. Ong, C. P., Ng, C. L., Lee, H. K., and Li, S. F. Y., Separation of Dns-amino acids and vitamins by micellar electrokinetic chromatography, *J. Chromatogr.,* 559, 537, 1991.
138. Cheng, Y.-F. and Dovichi, N. J., Subattomole amino acid analysis by capillary zone electrophoresis and laser-induced fluorescence, *Science,* 242, 562, 1988.
139. Liu, J., Hsieh, Y.-Z., Wiesler, D., Novotny, M., Design of 3-(4-carboxybenzoyl)-2-quinolinecarboxaldehyde as a reagent for ultrasensitive determination of primary amines by capillary electrophoresis using laser fluorescence detection, *Anal. Chem.,* 63, 408, 1991.
140. Ma, S., Kálmán, F., Kálmán, A., Thunecke, F., and Horváth, Cs., Capillary zone electrophoresis at subzero temperatures. I. Separation of the *cis* and *trans* conformers of small peptides, *J. Chromatogr. A,* 716, 167, 1995.
141. Nielsen, R. G. and Rickard, E. C., Method optimization in capillary zone electrophoretic analysis of hGH tryptic digest fragments, *J. Chromatogr.,* 516, 99 1990.
142. Satow, T., Machida, A., Funakushi, K., and Palmieri, R., Effects of the sample matrix on the separation of peptides by high performance capillary electrophoresis, *HRC&CC,* 14, 276, 1991.
143. van de Goor, T. A. A. M., Janssen, P. S. L., van Nispen, J. W., van Zeeland, M. J. M., and Everaerts, F. M., Capillary electrophoresis of peptides: analysis of adrenocorticotropic hormone-related fragments, *J. Chromatogr.,* 545, 379, 1991.
144. Grossman, P. D., Colburn, J. C., Lauer, H. H., Nielsen, R. G., Riggin, R. M., Sittampalam, G. S., and Rickard, E. C., Application of free-solution capillary electrophoresis to the analytical scale separation of proteins and peptides, *Anal. Chem.,* 61, 1186, 1989.
145. Grossman, P. D., Colburn, J. C., and Lauer, H. H., A semiempirical model for the electrophoretic mobilities of peptides in free-solution capillary electrophoresis, *Anal. Biochem.,* 179, 28, 1989.
146. Lee, K.-J. and Heo, G. S., Free solution capillary electrophoresis of proteins using untreated fused-silica capillaries, *J. Chromatogr.,* 559, 317, 1991.
147. Emmer, Å., Jansson, M., and Roeraade, J., Improved capillary zone electrophoretic separation of basic proteins, using a fluorosurfactant buffer additive, *J. Chromatogr.,* 547, 544, 1991.
148. Green, J. S. and Jorgenson, J. W., Minimizing adsorption of proteins on fused silica in capillary zone electrophoresis by the addition of alkali metal salts to the buffers, *J. Chromatogr.,* 478, 63, 1989.
149. Towns, J. K., Bao, J., and Regnier, F. E., Synthesis and evaluation of epoxy polymer coatings for the analysis of proteins by capillary zone electrophoresis, *J. Chromatogr.,* 599, 227, 1992.
150. Bruin, G. J. M., Chang, J. P., Kuhlman, R. H., Zegers, K., Kraak, J. C., and Poppe, H., Capillary zone electrophoretic separations of proteins in polyethylene glycol-modified capillaries, *J. Chromatogr.,* 471, 429, 1989.
151. Swedberg, S. A., Characterization of protein behavior in high-performance capillary electrophoresis using a novel capillary system, *Anal. Biochem.,* 185, 51, 1990.

152. Yim, K. W., Fractionation of the human recombinant tissue plasminogen activator (rtPA) glycoforms by high-performance capillary zone electrophoresis and capillary isoelectric focusing, *J. Chromatogr.*, 559, 401, 1991.

153. Watson, E. and Yao, F., Capillary electrophoretic separation of human recombinant erythropoietin (r-HuEPO) glycoforms, *Anal. Biochem.*, 210, 389, 1993.

154. Frenz, J., Wu, S.-L., and Hancock, W. S., Characterization of human growth hormone by capillary electrophoresis, *J. Chromatogr.*, 480, 379, 1988.

155. Pédron, J., Maldiney, R., Brault, M., and Miginiac, E., Monitoring of hapten-protein coupling reactions by capillary electrophoresis: improvement of abscisic acid-bovine serum albumin coupling and determination of molar coupling ratios, *J. Chromatogr. A*, 723, 381, 1996.

156. Kinghorn, N. M., Paterson, G. R., and Otter, D. E., Quantification of the major bovine whey proteins using capillary electrophoresis, *J. Chromatogr. A*, 723, 371, 1996.

157. Palmieri, R., Peptide mapping by capillary electrophoresis, Appl. Brief DS-774, Beckman Instruments, Fullerton, CA, 1990.

158. Rickard, E. C., Strohl, M. M., and Nielsen, R. G., Correlation of electrophoretic mobilities from capillary electrophoresis with physicochemical properties of proteins and peptides, *Anal. Biochem.*, 197, 197, 1991.

159. Josic, D., Zeilinger, K., Reutter, W., Bottcher, A., and Schmitz, G., High-performance capillary electrophoresis of hydrophobic membrane proteins, *J. Chromatogr.*, 516, 89, 1990.

160. Englund, A.-K., Lundahl, P., Elenbring, K., Ericson, C., and Hjérten, S., Capillary and rotating-tube isoelectric focusing of a transmebrane protein, the human red cell glucose transporter, *J. Chromatogr. A*, 711, 217, 1995.

161. Bode, H. J., SDS-polyethyleneglycol electrophoresis: a possible alternative to SDS-polyacrylamide gel electrophoresis, *FEBS Lett.*, 65, 56, 1976.

162. Benedek, K. and Thiede, S., High performance capillary electrophoresis of proteins using sodium dodecyl sulfate-poly(ethylene oxide), *J. Chromatogr.*, 676, 209, 1994.

163. Guttman, A. and Nolan, J., Comparison of the separation of proteins by sodium dodecyl sulfate-slab gel electrophoresis and capillary sodium dodecyl sulfate-gel electrophoresis, *Anal. Biochem.*, 221, 285, 1994.

164. Ganzler, K., Greve, K. S., Cohen, A. S., Karger, B. L., Guttman, A., and Cooke, N. C., High-performance capillary electrophoresis of SDS-protein complexes using UV-transparent polymer networks, *Anal. Chem.*, 64, 2665, 1992.

165. Benedek, K. and Guttman, A., Ultra-fast high-performance capillary sodium dodecyly sulfate gel electrophoresis of proteins, *J. Chromatogr.*, 680, 375, 1994.

166. Andrews, A. T., *Electrophoresis*, Claredon Press, Oxford, 1986.

167. Baba, Y., Tomisaki, R., Sumita, C., Morimoto, I., Sugita, S., Tsuhako, M., Miki, T., and Ogihara, T., Rapid typing of variable number of tandem repeat locus in the human apolipoprotein B gene for DNA diagnosis of heart disease by polymerase chain reaction and capillary electrophoresis, *Electrophoresis*, 16, 1437, 1995.

168. Arakawa, H., Nakashiro, S., Maeda, M., and Tsuji, A., Analysis of single-strand DNA conformation polymorphism by capillary electrophoresis, *J. Chromatogr. A*, 722, 359, 1996.

169. Guttman, A., Separation of DNA by capillary gel electrophoresis, in *Handbook of Capillary Electrophoresis*, Landers, J. P., Ed., CRC Press, Boca Raton, FL, 1994, chap. 6.

170. Pentoney, Jr., S. L., Konrad, K. D., and Kaye, W., A single-fluor approach to DNA sequence determination using high performance capillary electrophoresis, *Electrophoresis*, 13, 467, 1992.
171. Guttman, A., *Appl. Theor. Electrophoresis*, 3, 91, 1991.
172. Varki, A., Biological roles of oligosaccharides: all of the theories are correct, *Glycobiology*, 2, 97, 1993.
173. Karger, B. L., Cohen, A. S., and Guttman, A., High-performance capillary electrophoresis in the biological sciences, *J. Chromatogr.*, 492, 585, 1989.
174. Jackson, P., The use of polyacrylamide-gel electrophoresis for the high-resolution separation of reducing saccharides labeled with the fluorophore 8-aminonaphthalene-1,3,6-trisulfonic acid. Detection of picomolar quantities by an imaging system based on a cooled charge-coupled device, *Biochem. J.*, 270, 705, 1990.
175. Suzuki, S., Kakehi, K., and Honda, S., Two-dimensional mapping of N-glycosidically linked asialo-oligosaccharides from glycoproteins as reductively pyridylaminated derivatives using dual separation modes of high-performance capillary electrophoresis, *Anal. Biochem.*, 205, 227, 1992.
176. Chiesa, C. and Horváth, Cs., Capillary zone electrophoresis of malto-oligosaccharides derivatized with 8-aminonaphthalene-1,3,6-trisulfonic acid, *J. Chromatogr.*, 645, 337, 1993.
177. Liu, J., Shirota, O., Wiesler, D., and Novotny, M., Ultrasensitive fluorometric detection of carbohydrates as derivatives in mixtures separated by capillary electrophoresis, *Proc. Nat. Acad. Sci.*, 88, 2302, 1991.
178. Heegard, N. H. H. and Robey, F. A., Use of capillary zone electrophoresis to evaluate the binding of anionic carbohydrates to synthetic peptides derived from human serum amyloid P component, *Anal. Chem.*, 64, 2479, 1992.
179. Guttman, A. and Pritchett, T., Capillary gel electrophoresis separation of high-mannose type oligosaccharides derivatized by 1-aminopyrene-3,6,8-trisulfonic acid, *Electrophoresis*, 16, 1906, 1995.
180. Camilleri, P., Harland, G. B., and Okafo, G., High resolution and rapid analysis of branched oligosaccharides by capillary electrophoresis, *Anal. Biochem.*, 230, 115, 1995.
181. Guttman, A., Cooke, N., and Star, C. M., Capillary electrophoresis separation of oligosaccharides. I. Effect of operational variables, *Electrophoresis*, 15, 1518, 1994.
182. Grümmer, G., Knippel, E., Budde, A., Brockman, H., and Treichler, J., An electrophoretic instrumentation for the multi-parameter analysis of cells and particles, *Instr. Sci. Technol.*, 23, 265, 1995.

Index